Dynamics Reported

Expositions in Dynamical Systems

Dynamical Systems are a rapidly developing field with a strong impact on applications. Dynamics Reported is a series of books of a new type. Its principal goal is to make available current topics, new ideas and techniques. Each volume contains about four or five articles of up to 60 pages. Great emphasis is put on an excellent presentation, well suited for advanced courses, seminars etc. such that the material becomes accessible to beginning graduate students. To explain the core of a new method contributions will treat *examples* rather than general theories, they will describe *typical results* rather than the most sophisticated ones. Theorems are accompanied by *carefully written proofs*. The presentation is as *self-contained* as possible.

Authors will receive 5 copies of the volume containing their contributions. These will be split among multiple authors.

Authors are encouraged to prepare their manuscripts in Plain TEX or LATEX. Detailed information and macro packages are available via the Managing Editors.

Manuscripts and correspondence should be addressed to the Managing Editors:

C.K.R.T. Jones
Division of Applied Mathematics
Brown University
Providence, Rhode Island 02912
USA
e-Mail: ckrtj@cfm.brown.edu

U. Kirchgraber
Mathematics
Swiss Federal Institute
of Technology (ETH)
CH-8092 Zürich, Switzerland
e-Mail: kirchgra@math.ethz.ch

H.O. Walther
Mathematics
Ludwig-Maximilians University
D–80333 Munich
Federal Republic of Germany
e-Mail: Hans-Otto.Walther
@mathematik.
uni-muenchen.dbp.de

C. K. R. T. Jones U. Kirchgraber H. O. Walther

(Managing Editors)

Dynamics Reported

Expositions in Dynamical Systems

New Series: Volume 4

With Contributions of
A.M. Blokh, A. Celletti, L. Chierchia,
C. Liverani, M.P. Wojtkowski, T. Wanner

 Springer-Verlag
Berlin Heidelberg New York
London Paris Tokyo
Hong Kong Barcelona
Budapest

ISBN-13: 978-3-642-64748-2   Springer-Verlag Berlin Heidelberg New York
ISBN-13: 978-0-387-64748-2   Springer Verlag New York Berlin Heidelberg

Library of Congress Cataloging-in-Publication Data
Dynamics reported: expositions in dynamical systems/C.K.R.T. Jones, U. Kirchgraber, H.O. Walther,
managing editors: with contributions of R. Fournier ... [et al.]. p. cm.
ISBN-13: 978-3-642-64748-2   e-ISBN-13: 978-3-642-61215-2
DOI: 10.1007/978-3-642-61215-2
I. Differentiable dynamical systems. I. Kirchgraber, Urs, 1945– . II. Walther, Hans-Otto.
III. Bielawsi, R. QA614.8D96 1991 003′.85–dc20 91-23213 CIP

Softcover reprint of the hardcover 1st edition 1995

Typesetting with TEX: Lewis & Leins, Berlin.
Produktion: PRODUserv Springer Produktions-Gesellschaft, Berlin
SPIN 10100470   41/3020-5 4 3 2 1 0 –

Preface

DYNAMICS REPORTED reports on recent developments in dynamical systems.

Dynamical systems of course originated from ordinary differential equations. Today, dynamical systems cover a much larger area, including dynamical processes described by functional and integral equations, by partial and stochastic differential equations, etc. Dynamical systems have involved remarkably in recent years. A wealth of new phenomena, new ideas and new techniques are proving to be of considerable interest to scientists in rather different fields. It is not surprising that thousands of publications on the theory itself and on its various applications are appearing

DYNAMICS REPORTED presents carefully written articles on major subjects in dynamical systems and their applications, addressed not only to specialists but also to a broader range of readers including graduate students. Topics are advanced, while detailed exposition of ideas, restriction to *typical* results – rather than the *most general* ones – and, last but not least, lucid proofs help to gain the utmost degree of clarity.

It is hoped, that *DYNAMICS REPORTED* will be useful for those entering the field and will stimulate an exchange of ideas among those working in dynamical systems

Summer 1991
Christopher K.R.T Jones
Urs Kirchgraber
Hans-Otto Walther

Managing Editors

Table of Contents

The "Spectral" Decomposition for One-Dimensional Maps

Alexander M. Blokh

A Constructive Theory of Lagrangian Tori and Computer-assisted Applications
A. Celletti, L. Chierchia

Ergodicity in Hamiltonian Systems
C. Liverani, M.P. Wojtkowski

Linearization of Random Dynamical Systems

Thomas Wanner

The "Spectral" Decomposition for One-Dimensional Maps

Alexander M. Blokh

Department of Mathematics, University of Alabama at Birmingham, UAB Station, Birmingham, AL 35294, USA

Abstract. We construct the "spectral" decomposition of the sets $\overline{Per\, f}$, $\omega(f) = \cup\omega(x)$ and $\Omega(f)$ for a continuous map $f : [0, 1] \to [0, 1]$. Several corollaries are obtained; the main ones describe the generic properties of f-invariant measures, the structure of the set $\Omega(f) \setminus \overline{Per\, f}$ and the generic limit behavior of an orbit for maps without wandering intervals. The "spectral" decomposition for piecewise-monotone maps is deduced from the Decomposition Theorem. Finally we explain how to extend the results of the present paper for a continuous map of a one-dimensional branched manifold into itself.

1. Introduction and Main Results

1.0. Preliminaries

Let $T : X \to X$ be a continuous map of a compact space into itself (in what follows we consider continuous maps only). For $x \in X$ the set $orb\, x \equiv \{T^i x : i \geq 0\}$ is called *the orbit of x* or *the x-orbit*. The set $\omega(x)$ of all limit points of the x-orbit is called *the ω-limit set of x* or *the limit set of x*. Topological dynamics studies the properties of limit sets. Let us define some objects playing an important role here. A point $x \in X$ is called *non-wandering* if for any open $U \ni x$ there exists $n > 0$ such that $T^n U \cap U \neq \emptyset$. The set $\Omega(T)$ of all non-wandering points is called *the non-wandering set*; clearly, $\Omega(T)$ is closed.

Let us give an important example. A point $p \in X$ is called *periodic* if $T^n p = p$ for some positive integer n. Such an n is called *a period of p* and the set $orb\, p = \bigcup_{i \geq 0} T^i p$ is called *a cycle*. The set of all periodic points of T is denoted by $Per\, T$. Clearly, *periodic points are non-wandering*.

We denote the set $\bigcup_{x \in X} \omega(x)$ by $\omega(T)$. The following assertion explains the role of the set $\Omega(T)$.

Assertion 1.1. *For any open set $U \supset \Omega(T)$ and a point $x \in X$ there exists N such that $T^n x \in U$ for all $n > N$, and so $\omega(T) \subset \Omega(T)$.*

Sometimes it is important to know where a point $x \in X$ spends not all the time but almost all the time. The following definition is useful for considering this problem: a point $x \in X$ is called *recurrent* if $x \in \omega(x)$. The set of all recurrent points is denoted by $R(T)$. The set $\overline{R(T)} \equiv C(T)$ is called *the center of T* (here $\overline{Z}$ is *the closure of the set Z*).

Assertion 1.2 (see, e.g., [Ma]). *For any open $U \supset C(T)$ and $x \in X$ the following property holds:* $\lim_{n \to \infty} card\{i \le n : T^i x \in U\} \cdot n^{-1} = 1.$

Let us summarize the connection between the sets $Per\, T, R(T), C(T), \omega(T)$ and $\Omega(T)$ as follows:

$$Per\, T \subset R(T) \subset \omega(T) \subset \Omega(T) \tag{1.1}$$

$$\overline{Per\, T} \subset \overline{R(T)} = C(T) \subset \omega(T) \subset \Omega(T). \tag{1.2}$$

It is useful to split the sets $\Omega(T)$ and $\omega(T)$ into components such that for any $x \in X$ the set $\omega(x)$ belongs to one of them. The remarkable example of such a splitting is the famous Smale spectral decomposition theorem [S] (see also [B4]). The aim of this paper is to show that *in the one-dimensional case for any continuous map there exists a decomposition which is in a sense analogous to that of Smale.*

1.1. Historical Remarks

We start with the history of the subject. From now on fix an arbitrary continuous map $f : [0, 1] \to [0, 1]$. Speaking of maximality, minimality and ordering among sets we mean that sets are ordered by inclusion. The following definitions are due to A.N.Sharkovskii [Sh3–6]. Let $\omega(x)$ be an infinite limit set maximal among all limit sets. The set $\omega(x)$ is called *a set of genus 1* if it contains no cycles; otherwise it is called *a set of genus 2*. A cycle maximal among limit sets is called *a set of genus 0* (see [B14]; periodic attractors and isolated periodic repellers are the most important and well-known examples of sets of genus 0).

In [Sh3–6] A.N.Sharkovskii has in fact constructed the decomposition of the set $\omega(f)$ into sets of genus 0,1 and 2. He studied mostly properties of the partially ordered family of limit sets belonging to a maximal limit set. Furthermore, he obtained a number of fundamental results on properties of the sets $\Omega(f), \omega(f), C(f)$ and $\overline{Per\, f}$. Here we formulate some of Sharkovskii's results which we need.

Theorem Sh1 [Sh2]. $C(f) \equiv \overline{R(f)} = \overline{Per\, f} = \Omega(f|\Omega(f)).$

Theorem Sh2 [Sh5]. *A point x belongs to $\omega(f)$ if and only if at least one of the following properties holds:*
1) for any $\varepsilon > 0$ there exists $n > 0$ such that $(x - \varepsilon, x) \cap f^n(x - \varepsilon, x) \ne \emptyset$;
2) for any $\varepsilon > 0$ there exists $n > 0$ such that $(x + \varepsilon, x) \cap f^n(x + \varepsilon, x) \ne \emptyset$;
3) $x \in Per\, f$.
In particular, $\omega(f)$ is closed and so $\overline{Per\, f} \subset \omega(f)$.

The main idea of the proofs here is to consider a special kind of recurrence which may occur for maps of the interval and also to use the following

Property C. *If $I = [a, b]$ is an interval and either $fI \supset I$ or $fI \subset I$ or points a and b move under the first iteration of f in different directions then there is $y \in I$ such that $fy = y$.*

We illustrate this approach considering Theorem Sh1. Indeed, let U be an interval complementary to $\overline{Per\, f}$. Then by Property C for any n either $f^n x > x \,(\forall x \in U)$ or $f^n x < x \,(\forall x \in U)$. Suppose that for some n and $x \in U$ we have $f^n x > x$, $f^n x \in U$. Then $f^n(f^n x) > f^n x > x$, i.e. $f^{2n} x > x$; moreover, if $f^{kn} x > x$ then $f^{(k+1)n} x = f^{kn}(f^n x) > f^n x > x$ which proves that $f^{in} x \geq f^n x > x$ for all i. Now suppose that there exists $y \in U$ and m such that $f^m y \in U$ and $f^m y < y$. Then by the same arguments $f^{jm} y < y$ for any j. This implies that $f^{mn} x > x$ and $f^{mn} y < y$; so by Property C there is a periodic point in the interval $(x, y) \subset U$, which is a contradiction.

Now the definition of a non-wandering point implies that if $z \in U$ is non-wandering then it never enters U again since otherwise it returns to U to the right of itself and at the same time by definition of a non-wandering point there exist points in U which are sufficiently close to the place of the first returning of z into U and are mapped into a small neighborhood of z by the corresponding iteration of f which is impossible by what we have just shown. Clearly this implies that there are no recurrent points of f inside U and, moreover, $\Omega(f|\Omega(f)) = \overline{Per\, f}$.

One of the most well-known and surprising results about one-dimensional dynamics is, perhaps, the famous Sharkovskii theorem. To state it let us consider the set of all positive integers with the following *Sharkovskii ordering*:

$$3 \prec 5 \prec 7 \prec \ldots \prec 2 \cdot 3 \prec 2 \cdot 5 \prec 2 \cdot 7 \ldots \prec 2^3 \prec 2^2 \prec 2 \prec 1 \qquad (*)$$

Theorem Sh3 [Sh1]. *Let $m \prec n$ and f have a cycle of minimal period m. Then f has a cycle of minimal period n.*

We say that m is *$\prec$-stronger than n* if $m \prec n$ and $m \neq n$. We say that f *is a map of type m* [Bl7] if the $\prec$-strongest period of cycles of f is m; in other words, m is the largest period which appears in terms of the Sharkovskii ordering. Such a period does not exist if the periods of cycles of f are exactly $1, 2, 2^2, 2^3, \ldots$; then say that f is *of type 2^∞* [Bl1].

For piecewise-monotone continuous maps the splittings of the sets $\Omega(f)$, $\omega(f)$ and $C(f)$ in fact analogous to that of Sharkovskii were constructed later by using the different techniques (see the articles of Jonker and Rand [JR1–JR2], Nitecki [N2] and the books of Preston [P1–P2]). For piecewise-monotone maps with a finite number of discontinuities the construction of the splitting is due to Hofbauer [H2–H3].

1.2. A Short Description of the Approach Presented

The approach in this paper is different from the one of Sharkovskii and the "piecewise-monotone" approach; it is based on the author's articles [Bl1–Bl12]. First we need more definitions. Let $T : X \to X$ and $F : Y \to Y$ be maps of compact spaces. If there exists a *surjective* map $\phi : X \to Y$ such that $\phi \circ T = F \circ \phi$ then it is said that ϕ *semiconjugates* T to F and ϕ is called *a semiconjugation* between T and F; if ϕ is a homeomorphism then it is said that ϕ *conjugates* T to F and ϕ is called *a conjugation* between T and F.

Roughly speaking, our approach to one-dimensional maps is the following: we propose models for different kinds of limit sets, study the properties of these models, extend the properties to the limit sets and the map itself, and also obtain some corollaries. In the rest of this Section we formulate the main results of the present paper. The proofs will be given in Sections 2–11. At the end of this Section we apply the results to piecewise-monotone maps and explain how to extend the decomposition to continuous maps of one-dimensional branched manifolds.

An interval I is called *periodic (of period k)* or *k-periodic* if $J, \ldots, f^{k-1}J$ are pairwise disjoint and $f^k J = J$ (if it is known only that $f^k J \subset J$ then J is called *a weakly periodic interval*). The set $\bigcup_{i=0}^{k-1} f^i J \equiv orb\, J$ is called *a cycle of intervals* if J is periodic and *a weak cycle of intervals* if J is weakly periodic (here k is a period of J).

Let us explain briefly how we will classify limit sets. Fix an infinite set $\omega(x)$ and consider the family $\mathcal{A}$ of all cycles of intervals $orb\, I$ such that $\omega(x) \subset orb\, I$. There are two possibilities.

1) *Periods of sets $orb\, I \in \mathcal{A}$ are not bounded.* Then there exist ordered cycles of intervals containing $\omega(x)$ with periods tending to infinity. This allows us to semiconjugate $f|\omega(x)$ to a transitive translation in a compact group and implies many properties of $f|\omega(x)$. The set $\omega(x)$ corresponds to a Sharkovskii's set of genus 1.

2) *Periods of sets $orb\, I \in \mathcal{A}$ are bounded.* Then there exists a minimal cycle of intervals $orb\, J \in \mathcal{A}$. It is easy to see that all points $y \in \omega(x)$ have the following property: if U is a neighborhood of y in $orb\, J$ then $\overline{orb\, U} = orb\, J$ (otherwise $\overline{orb\, U}$ generates a cycle of intervals $orb\, K$ such that $\omega(x) \subset orb\, K \subset orb\, J$, $orb\, K \neq orb\, J$ which is a contradiction). The idea is to consider all the points $z \in orb\, J$ with this property. They form a set B which is another example of a maximal limit set. The set B is a Sharkovskii's set of genus 2.

In the following Subsections 1.3–1.10 we are going to formulate and discuss the main results of Sections 3–10 correspondingly.

1.3. Solenoidal Sets

Let us proceed more precisely. Let $T : X \to X$ be a map of a compact metric space (X, d) into itself. The map T is said to be *transitive* if there exists an x such that $\omega(x) = X$, to be *minimal* if for any $x \in X$ we have $\omega(x) = X$, to be *topologically mixing* (or simply *mixing*) if for any open U, V there exists an N such that $T^n U \cap V \neq \emptyset$ for any $n > N$.

We will also need the definition of the topological entropy; the notion was introduced in [AKMcA] but we give the definition following Bowen[B1]. A set $E \subset X$ is said to be (n, ε)-*separated* if for any two distinct points $x, y \in E$ there exists $k, 0 \leq k < n$ such that $d(T^k x, T^k y) > \varepsilon$. By $S_n(\varepsilon)$ we denote the largest cardinality of an (n, ε)-separated subset of X. Let $S(\varepsilon) \equiv \limsup n^{-1} \cdot \ln S_n(\varepsilon)$. Then the limit $h(T) = \lim_{\varepsilon \to 0} S(\varepsilon)$ exists and is called *the topological entropy* of T (see [B1]). Now let us turn back to interval maps.

Let $I_0 \supset I_1 \supset \ldots$ be periodic intervals with periods $m_0, m_1, \ldots$. Obviously m_{i+1} is a multiple of m_i for all i. If $m_i \to \infty$ then the intervals $\{I_j\}_{j=0}^{\infty}$ are said to be *generating* and any invariant closed set $S \subset Q = \bigcap_{j \geq 0} orb\, I_j$ is called *a solenoidal set*; if Q is nowhere dense then we call Q *a solenoid*. In the sequel we use the following notation:

$\bigcap_{j \geq 0} orb\, I_j \equiv Q(\{I_j\}_{j=0}^{\infty}) \equiv Q;$

$Q \cap \overline{Per\, f} \equiv S_p(Q) \equiv S_p;$

$Q \cap \omega(f) \equiv S_\omega(Q) \equiv S_\omega;$

$Q \cap \Omega(f) \equiv S_\Omega(Q) \equiv S_\Omega$.

Observe that $S_p \subset S_\omega \subset S_\Omega$ and all these sets are invariant and closed (for S_ω it follows from Theorem Sh2).

One can use a transitive translation in an Abelian zero-dimensional infinite group as a model for the map on a solenoidal set. Namely, let $D = \{n_i\}_{i=0}^\infty$ be a sequence of integers, with n_{i+1} a multiple of n_i for all i, and $n_i \to \infty$. Let us consider a group $H(D)$, defined by $H(D) \equiv \{(r_0, r_1, \ldots) : r_{i+1} \equiv r_i \pmod{m_i}(\forall i)\}$ where r_i is an element of a group of residues modulo m_i for any i. The group operation is defined in a trivial way; now denote by τ the (minimal) translation in $H(D)$ by the element $(1, 1, \ldots)$.

Theorem 3.1[Bl4,Bl7]. *Let $\{I_j\}_{j=0}^\infty$ be generating intervals with periods $\{m_i\}_{i=0}^\infty = D$, $Q = \bigcap_{j\geq0} \operatorname{orb} I_j$. Then there exists a continuous map $\phi : Q \to H(D)$ with the following properties:*

1) ϕ semiconjugates $f|Q$ to τ (i.e. $\tau \circ \phi = \phi \circ f$ and ϕ is surjective);

2) there exists a unique set $S \subset S_p$ such that $\omega(x) = S$ for any $x \in Q$ and, moreover, S is the set of all limit points of S_Ω and $f|S$ is minimal;

3) if $\omega(z) \cap Q \neq \emptyset$ then $S \subset \omega(z) \subset S_\omega$;

4) for any $\mathbf{r} \in H(D)$ the set $J = \phi^{-1}(\mathbf{r})$ is a connected component of Q and:

   a) if $J = \{a\}$ then $a \in S$;

   b) if $J = [a, b], a \neq b$ then $\emptyset \neq S \cap J \subset S_\Omega \cap J \subset \{a, b\}$;

5) $S_\Omega \setminus S$ is at most countable and consists of isolated points;

6) $h(f|Q) = 0$.

It should be noted that the best known example of a solenoid is the Feigenbaum attractor ([CE],[F]) for which generating intervals have periods $\{2^i\}_{i=0}^\infty$. If for a solenoid or a solenoidal set generating intervals have periods $\{2^i\}_{i=0}^\infty$ then we call it 2-*adic*.

1.4. Basic Sets

Let us turn to another type of maximal infinite limit set. Let $\{J_i\}_{i=1}^l$ be an ordered collection of disjoint intervals (one can imagine these intervals lying on the real line in such a way that $J_1 < J_2 < \ldots < J_l$, $J_i \cap J_r = \emptyset$ for $i \neq r$); set $K = \bigcup_{i=1}^l J_i$. A continuous map $\psi : K \to K$ which permutes the intervals $\{J_i\}_{i=1}^l$ cyclically is called *non-strictly periodic* (or *l-periodic*). Note that this term concerns a map, not an interval; we speak of non-strictly periodic maps to distinguish them from *periodic maps* which are traditionally those with all points periodic. An example of a non-strictly periodic map is a map of the interval restricted to a weak cycle of intervals.

Now let $\psi : K \to K$ and $\psi' : K' \to K'$ be non-strictly l-periodic maps (so that K and K' are unions of l intervals). Let $\phi : K \to K'$ be a monotone semiconjugation between ψ and ψ' and $F \subset K$ be a ψ-invariant closed set such that $\phi(F) = K'$, for any $x \in K'$ we have $\operatorname{int} \phi^{-1}(x) \cap F = \emptyset$ and so $\phi^{-1}(x) \cap F \subset \partial\phi^{-1}(x)$, $1 \leq \operatorname{card}\{\phi^{-1}(x) \cap F\} \leq 2$. Then we say that ϕ *almost conjugates* $\psi|F$ to ψ' or ϕ is an *almost conjugation* between $\psi|F$ and ψ'. Remark that here $\operatorname{int} Z$ is *an interior* of a set Z and ∂Z is *a boundary* of a set Z.

Finally let I be an n-periodic interval, $\operatorname{orb} I = M$. Consider a set $\{x \in M :$ for any relative neighborhood U of x in M we have $\overline{\operatorname{orb} U} = M\}$; it is easy to see that this is a closed invariant set. It is called *a basic set* and denoted by $B(M, f)$ provided it is infinite. Now we can formulate

Theorem 4.1[Bl4,Bl7]. *Let I be an n-periodic interval, $M = \text{orb}\, I$ and $B = B(M, f)$ be a basic set. Then there exist a transitive non-strictly n-periodic map $g : M' \to M'$ and a monotone map $\phi : M \to M'$ such that ϕ almost conjugates $f|B$ to g. Furthermore, B has the following properties:*
a) B is a perfect set;
b) $f|B$ is transitive;
c) if $\omega(z) \supset B$ then $\omega(z) = B$ (i.e. B is a maximal limit set);
d) $h(f|B) \geq \ln 2 \cdot (2n)^{-1}$;
e) $B \subset \overline{\text{Per}\, f}$;
f) there exist an interval $J \subset I$, an integer $k = n$ or $k = 2n$ and a set $\widetilde{B} = \overline{\text{int}\, J \cap B}$ such that $f^k J = J$, $f^k \widetilde{B} = \widetilde{B}$, $f^i \widetilde{B} \cap f^j \widetilde{B}$ contains no more than 1 point $(0 \leq i < j < k)$, $\bigcup_{i=0}^{k-1} f^i \widetilde{B} = B$ and $f^k|\widetilde{B}$ is almost conjugate to a mixing interval map (one can assume that if $k = n$ then $I = J$).

So we use transitive non-strictly periodic maps as models for the map f on basic sets. Note that Theorems 3.1 and 4.1 allow us to establish the connection between sets of genus 1 and solenoidal sets, and between sets of genus 2 and basic sets (see Assertion 4.2 in Section 4). Moreover, Theorems 3.1 and 4.1 easily imply that sets of genus 0 and limit solenoidal sets may be characterized as those $\omega(x)$ for which the inclusion $\omega(y) \supset \omega(x)$ implies that $h(f|\omega(y)) = 0$ for any y (see Assertion 4.3 in Section 4).

Now we can construct the "spectral" decomposition for the sets $\overline{\text{Per}\, f}$ and $\omega(f)$. However, to extend the decomposition to the set $\Omega(f)$ we need the following definition. Let $B = B(\text{orb}\, I, f)$ be a basic set and A be the set of all endpoints x of the intervals of $\text{orb}\, I$ with the following properties:
1) $x \in \Omega(f)$;
2) there exists an integer n such that $f^n x \in B$ and if m is the least such integer then $x, fx, \ldots, f^{m-1}x \notin \text{int}(\text{orb}\, I)$.

We denote the set $B \cup A$ by $B'(\text{orb}\, I, f)$ and call it *an Ω-basic set*.

Let us consider an example of a nontrivial Ω-basic set (cf. [Sh2], [Y]). Construct a map $f : [0, 1] \to [0, 1]$ in the following way: fix 6 points $c_0 = 0 < c_1 < \ldots < c_5 = 1$, define $f|\bigcup_{i=0}^{5} c_i$ and then extend f on each interval $[c_i, c_{i+1}]$ as a linear function. Namely:
1) $c_0 = 0, f c_0 = 2/3$;
2) $c_1 = 1/3, f c_1 = 1$;
3) $c_2 = 1/2, f c_2 = 5/6$;
4) $c_3 = 2/3, f c_3 = 1/6$;
5) $c_4 = 5/6, f c_4 = 5/6$;
6) $c_5 = 1, f c_5 = 1$.

It is easy to see that the interval $I = [1/6, 1]$ is f-invariant and the point $5/6$ is fixed. Let us show that there exists a basic set $B = B(\text{orb}\, I, f)$ and $1/2, 5/6 \in B$. For the moment we know nothing about the cardinality of the set $\{x \in \text{orb}\, I : \text{for any relative neighborhood } U \text{ of } x \text{ in } \text{orb}\, I \text{ we have } \overline{\text{orb}\, U} = \text{orb}\, I\}$, so let us denote this set by L; by the definition of a basic set we need to prove that L is infinite. Indeed, any left semi-neighborhood of $5/6$ covers the whole interval I after some iterations of f, so $5/6 \in L$.

On the other hand $f(1/2) = 5/6$ and the f-image of any right semi-neighborhood of $1/2$ covers some left semi-neighborhood of $5/6$. So $1/2 \in L$ as well. But it is easy to see that there are infinitely many points $z \in (1/2, 5/6)$ such that $f^n z = 1/2$ for some n and the f^n-image of any neighborhood of z covers some right semi-neighborhood of $1/2$ which implies that $z \in L$; so L is infinite and by the definition $L = B(\mathrm{orb}\, I, f) = B$ is a basic set.

Furthermore, the map f coincides with the identity on $[5/6, 1]$, at the same time $f[1/6, 1/2] = [5/6, 1] = f[5/6, 1]$ and f-image of any right semi-neighborhood of $1/6$ is some right semi-neighborhood of $5/6$. So by the definition we see that there are no points of B in $[1/6, 1/2)$; in particular, $1/6 \notin B$. Moreover, it is easy to see that there are no periodic points of f in $[0, 1/2)$. Indeed, there are no periodic points in $[1/6, 1/2]$ because $f[1/6, 1/2] = [5/6, 1] = f[5/6, 1]$. On the other hand there are no periodic points in $[0, 1/6)$ because $f[0, 1/6) \subset [1/6, 1] = f[1/6, 1]$. So $\mathrm{Per}\, f \cap [0, 1/2] = \emptyset$.

Now let us show that $1/6 \in B' \setminus B$ where $B' = B'(\mathrm{orb}\, I, f)$. Indeed, we have already seen that $f(1/6) = 5/6 \in B$. So by the properties of the point $5/6$ established above we see that for any open $U \ni 1/6$ there exists m such that $f^m U \supset I = [1/6, 1]$. It proves that $1/6 \in \Omega(f)$. Now the definition implies that $1/6 \in B' \setminus B$. It remains to note that by the definition $B' \setminus B$ consists only of some endpoints of intervals from $\mathrm{orb}\, I$, i.e. in our case of some of the points $1/6, 1$. Clearly, $1 \notin B'$ and so $\{1/6\} = B' \setminus B$.

1.5. The Decomposition and Main Corollaries

Now we can formulate the Decomposition Theorem. Let us denote by X_f the union of all limit sets of genus 0.

Theorem 5.4 (Decomposition Theorem)[Bl4,Bl7]. *Let $f : [0, 1] \to [0, 1]$ be a continuous map. Then there exist an at most countable family of pairs of basic and Ω-basic sets $\{B_i \subset B_i'\}$ and a family of collections of solenoidal sets $\{S^{(\alpha)} \subset S_p^{(\alpha)} \subset S_\omega^{(\alpha)} \subset S_\Omega^{(\alpha)} \subset Q^{(\alpha)}\}_{\alpha \in A}$ with the following properties:*

1) $\Omega(f) = X_f \cup (\bigcup_\alpha S_\Omega^{(\alpha)}) \cup (\bigcup_i B_i')$;

2) $\omega(f) = X_f \cup (\bigcup_\alpha S_\omega^{(\alpha)}) \cup (\bigcup_i B_i)$;

3) $\overline{\mathrm{Per}\, f} = X_f \cup (\bigcup_\alpha S_p^{(\alpha)}) \cup (\bigcup_i B_i)$;

4) the set $S_\Omega^{(\alpha)} \setminus S^{(\alpha)}$ is at most countable set of isolated points, the set $\{\alpha : \mathrm{int}\, Q^{(\alpha)} \neq \emptyset\}$ is at most countable and $S^{(\alpha)} = Q^{(\alpha)}$ for all other $\alpha \in A$;

5) intersections in this decomposition are possible only between different basic or Ω-basic sets, each three of them have an empty intersection and the intersection of two basic or two Ω-basic sets is finite.

Note that in statement 5) of Decomposition Theorem we do not take into account intersections between a basic set and an Ω-basic set with the same subscript and also between different solenoidal sets with the same superscript.

The Decomposition Theorem in the full formulation is somewhat cumbersome but the idea is fairly clear and may be expressed in the following rather naive version of the Decomposition Theorem:

*for any continuous map $f : [0, 1] \to [0, 1]$ the non-wandering set $\Omega(f)$ and related sets
(like $\omega(f)$ and $\overline{Per\, f}$) are unions of the set X_f, solenoidal sets and basic sets.*

The main corollaries of this picture of dynamics are connected with the following
problems.
1) What is the generic limit behavior of orbits for maps without wandering intervals (we
   call an interval I *wandering* if $f^n I \cap f^m I = \emptyset$ for $n > m \geq 0$ and I does not tend to
   a cycle)(Section 6)?
2) What is the related structure of the sets $\Omega(f)$, $\omega(f)$ and $\overline{Per\, f}$(Section 7)?
3) How does the dynamics of a map depend on its set of periods of cycles (Section 9)?
4) What are the generic properties of invariant measures (Section 10)?

Note that in order to study the generic properties of invariant measures we establish in
Section 8 some important properties of transitive and mixing interval maps. In Section
11 we also investigate the connection between the results of the present paper and some
recent results of Block and Coven [BC] and Xiong Jincheng [X].

In the following Subsections 1.6-1.10 we outline the way we are going to obtain the
aforementioned corollaries of the Decomposition Theorem in Sections 6–10 of the paper
correspondingly. In Subsections 1.11-1.12 we describe the decomposition for piecewise-
monotone interval maps and in Subsection 1.13 we discuss further generalizations.

1.6. The Limit Behavior and Generic Limit Sets for Maps Without Wandering Intervals

In this Subsection we describe the results of Section 6. We start with the reformulation of
the Decomposition Theorem for maps without wandering intervals. Namely, Theorem 3.1
implies that if a map f does not have wandering intervals then in the notation from the
Decomposition Theorem for any $\alpha \in A$ we have $\{S^{(\alpha)} = S_p^{(\alpha)} = S_\omega^{(\alpha)} = S_\Omega^{(\alpha)} = Q^{(\alpha)}\}_{\alpha \in A}$;
in other words all solenoidal sets are in fact solenoids (recall that solenoids are nowhere
dense intersections of cycles of generating intervals which in particular implies that the
map on a solenoid is topologically conjugate with the translation in the corresponding
group). This makes the formulation of the Decomposition Theorem easier, so let us
restate it in this case.

Decomposition Theorem for interval maps without wandering intervals. *Let $f :
[0, 1] \to [0, 1]$ be a continuous map without wandering intervals. Then there exist an
at most countable family of pairs of basic and Ω-basic sets $\{B_i \subset B_i'\}$ and a family of
solenoids $\{Q^{(\alpha)}\}_{\alpha \in A}$ with the following properties:*
1) $\Omega(f) = X_f \cup (\bigcup_\alpha Q^{(\alpha)}) \cup (\bigcup_i B_i')$;
2) $\omega(f) = X_f \cup (\bigcup_\alpha Q^{(\alpha)}) \cup (\bigcup_i B_i)$;
3) $\overline{Per\, f} = X_f \cup (\bigcup_\alpha Q^{(\alpha)}) \cup (\bigcup_i B_i)$;
*4) intersections in this decomposition are possible only between different basic or Ω-
   basic sets, each three of them have an empty intersection and the intersection of two
   basic or two Ω-basic sets is finite.*

A set A which is a countable intersection of open subsets of a compact metric space X
is said to be a G_δ-set. A set G containing a dense G_δ-set is said to be *residual*. A property

which holds for a residual subset of a compact metric space is said to be *topologically generic*. One of the corollaries of the aforementioned version of the Decomposition Theorem is the description of generic limit sets for maps without wandering intervals.

First let us explain why the concept of wandering interval appears naturally while studying the problem in question. Indeed, consider a pm-map g without *flat spots* (i.e. intervals I such that fI is a point). Take any point $x \in [0, 1]$ with an infinite orbit not tending to a cycle. Then instead of points z from the set $\bigcup_{i,j}^{\infty} g^{-i}(g^j x)$ we can "paste in" intervals $I(z)$ in such a way that a new map will have a wandering interval $I(x)$ and that $orb_f I(x)$ will have essentially the same structure as $orb_g x$. Therefore to consider the problem in question one should forbid the existence of wandering intervals. This remark makes the following Theorem 6.2 quite natural.

Theorem 6.2(cf.[B11],[B18]). *Let $f : [0, 1] \to [0, 1]$ be a continuous map without wandering intervals. Then there exists a residual subset $G \subset [0, 1]$ such that for any $x \in G$ one of the following possibilities holds:*
1) $\omega(x)$ is a cycle;
2) $\omega(x)$ is a solenoid;
3) $\omega(x) = orb\, I$ is a cycle of intervals.

Remark. Note, that possibility 3) of Theorem 6.2 will be essentially specified in Section 10 where we show that in fact generic points x for which $\omega(x) = orb\, I$ is a cycle of intervals may be chosen in such a way that the set of all limit measures of time averages of iterates of δ-measure δ_x coincides with the set of all invariant measures of $f|orb\, I$ (precise definitions will be given in Subsection 1.10).

1.7. Topological Properties of Sets $\overline{Per\, f}$, $\omega(f)$ and $\Omega(f)$

In this Subsection we summarize the results of Section 7. The following Theorem 7.6 which is the main theorem of Section 7 describes the structure of the set $\Omega(f) \setminus \overline{Per\, f}$.

Theorem 7.6. *Let $U = (a, b)$ be an interval complementary to $\overline{Per\, f}$. Then up to the orientation one of the following four possibilities holds.*
1) $\Omega(f) \cap U = \emptyset$.
2) $\Omega(f) \cap U = \{x_1 < x_2 < \ldots < x_n\}$ is a finite set, $card(orb\, x_1) < \infty, \ldots,$ $card(orb\, x_{n-1}) < \infty$, $(\bigcup_{i=1}^{n-1} x_i) \cap \omega(f) = \emptyset$ and there exist periodic intervals $J_i = [x_i, y_i]$ such that $x_i \in B'(orb\, J_i, f)$ for $1 \le i \le n - 1$ and $J_i \supset J_{i+1}$ for $1 \le i \le n - 2$. Moreover, for x_n there exist two possibilities: a) x_n belongs to a solenoidal set; b) x_n belongs to an Ω-basic set $B'(orb\, J_n, f)$ where $J_n = [x_n, y_n] \subset J_{n-1}$.
3) $\Omega(f) \cap U = (\bigcup_{i=1}^{\infty} x_i) \cap x$, $x_1 < x_2 < \ldots$, $x = \lim x_i$, and there exist generating intervals $J_i = [x_i, y_i]$ such that:
a) $x_i \in B'(orb\, J_i, f)$, $card(orb\, x_i) < \infty\ (\forall i)$ and $(\bigcup_{i=1}^{\infty} x_i) \cap \omega(f) = \emptyset$;
b) $x \in S_\omega(\{orb\, J_i\}_{i=1}^{\infty}) = \omega(f) \cap (\bigcap_{i=1}^{\infty} orb\, J_i)$.
4) $\Omega(f) \cap U = \bigcup_{i=1}^{\infty} x_i$, $x_1 < x_2 < \ldots$, $\lim x_i = b$, $card(orb\, x_i) < \infty\ (\forall i)$, $(\bigcup_{i=1}^{\infty} x_i) \cap \omega(f) = \emptyset$ and there exist periodic intervals $J_i = [x_i, y_i]$ such that $x_i \in B'(orb\, J_i, f), J_i \supset J_{i+1}\ (\forall i)$ and $\bigcap_{i=1}^{\infty} J_i = b$. Moreover, either periods of

J_i tend to infinity, $\{J_i\}$ are generating intervals and b belongs to the corresponding solenoidal set, or periods of J_i do not tend to infinity and b is a periodic point.

In any case $card\{\omega(f) \cap U\} \leq 1$.

Remark. Several results close to Theorem 7.6 are obtained also in Chapters 4 and 5 of [BCo]. Our approach is different from that of [BCo] since we apply the Decomposition Theorem and related technique. It allows to get some additional facts (in particular, it allows to describe the cases when limit sets of points from $\Omega(f) \cap U$ are solenoidal and to investigate the case when the set $\Omega(f) \cap U$ is infinite).

We also prove in Section 7 that $\omega(f) = \bigcap_{n \geq 0} f^n \Omega(f)$ ([Bl1],[Bl7]; see [BCo] for an alternative proof). Finally, we extend for any continuous interval map a result from [Yo] where it is proved that if f is a pm-map and $x \in \Omega(f) \setminus \overline{Per\,f}$ then there exists $n > 0$ and a turning point c such that $f^n c = x$ (the similar result was obtained in the recent paper [Li], see Theorem 2 there).

1.8. Properties of Transitive and Mixing Maps

Theorem 4.1 implies that properties of a map on basic sets are closely related to properties of transitive and mixing interval maps. We investigate these properties in Section 8 and give here a summary of the corresponding results.

The following lemma shows the connection between transitive and mixing maps of the interval.

Lemma 8.3[Bl7]. Let $f : [0, 1] \to [0, 1]$ be a transitive map. Then one of the following possibilities holds:
1) the map f is mixing and, moreover, for any $\varepsilon > 0$ and any non-degenerate interval U there exists m such that $f^n U \supset [\varepsilon, 1 - \varepsilon]$ for any $n > m$;
2) the map f is not mixing and there exists a fixed point $a \in (0, 1)$ such that $f[0, a] = [a, 1]$, $f[a, 1] = [0, a]$, $f^2|[0, a]$ and $f^2|[a, 1]$ are mixing.

In any case $\overline{Per\,f} = [0, 1]$.

It turns out that mixing interval maps have quite strong expanding properties: any open interval under iterations of a mixing map $f : [0, 1] \to [0, 1]$ eventually covers any compact subset of $(0, 1)$. More precisely, let $A(f) \equiv A$ be the set of those endpoints of $[0, 1]$ whith no preimages interior to $[0, 1]$.

Lemma 8.5[Bl7]. If $f : [0, 1] \to [0, 1]$ is mixing then there are the following possibilities for A:
1) $A = \emptyset$;
2) $A = \{0\}$, $f(0) = 0$;
3) $A = \{1\}$, $f(1) = 1$;
4) $A = \{0, 1\}$, $f(0) = 0$, $f(1) = 1$;
5) $A = \{0, 1\}$, $f(0) = 1$, $f(1) = 0$.

Moreover, if I is a closed interval, $I \cap A = \emptyset$, then for any open U there exists n such that $f^m U \supset I$ for $m > n$ (in particular, if $A = \emptyset$ then for any open U there exists n such that $f^n U = [0, 1]$).

Remark. Results closely related to those of Lemmas 8.3 and 8.5 were also obtained in [BM1, BM2].

In fact Lemma 8.5 is one of the tools in the proof of Theorem 8.7 where we show that mixing interval maps have the specification property. It is well-known ([Si1-Si2], [DGS]) that this implies a lot of generic properties of invariant measures of a map, and we will rely on them in the further study of interval dynamics.

Let us give the exact definition. Let $T : X \to X$ be a map of a compact infinite metric space (X, d) into itself. A dynamical system (X, T) is said to have *the specification property* or simply *specification* [B2] if for any $\varepsilon > 0$ there exists an integer $M = M(\varepsilon)$ such that for any $k > 1$, for any k points $x_1, x_2, \ldots, x_k \in X$, for any integers $a_1 \leq b_1 < a_2 \leq b_2 < \ldots < a_k \leq b_k$ with $a_i - b_{i-1} \geq M$, $2 \leq i \leq k$ and for any integer p with $p \geq M + b_k - a_1$ there exists a point $x \in X$ with $T^p x = x$ such that $d(T^n x, T^n x_i) \leq \varepsilon$ for $a_i \leq n \leq b_i, 1 \leq i \leq k$.

Theorem 8.7[Bl4,Bl7]. *If $f : [0, 1] \to [0, 1]$ is mixing then f has the specification property.*

Remark. In Section 8 we in fact introduce a slightly stronger version of the specification property (i-specification property) related to the properties of interval maps and prove that mixing maps of the interval have i-specification.

1.9. Corollaries Concerning Periods of Cycles for Interval Maps

Here we formulate two results concerning periods of cycles for interval maps which are proved in Section 9. We explain also how the famous Misiurewicz theorem on maps with zero entropy is connected with our results.

Well-known properties of the topological entropy and Theorem Sh1 imply that $h(f) = h(f|\overline{Per\ f})$. However, it is possible to get a set D such that $h(f) = h(f|D)$ using essentially fewer periodic points of f. Indeed, let A be some set of positive integers, $K_f(A) \equiv \{y \in Per\ f : \text{minimal period of } y \text{ belongs to } A\}$.

Theorem 9.1[Bl4,Bl7]. *The following two properties of A are equivalent:*
1) $h(f) = h(f|\overline{K_f(A)})$ for any f;
2) for any k there exists $n \in A$ which is a multiple of k.

In Theorem 9.4 we study how the sets $\Omega(f), \Omega(f^2), \ldots$ vary for maps with a fixed set of periods of cycles. In [CN] this problem was investigated for an arbitrary continuous map of the interval and it was proved that $\Omega(f) = \Omega(f^n)$ for any odd n and any continuous interval map. The following theorem is related to the results of [CN].

Theorem 9.4[Bl4,Bl8]. *Let $n \geq 0, k \geq 1$ be fixed and f have no cycles of minimal period $2^n(2k + 1)$. Then the following statements hold:*
1) if $B = B(orb\ I, f)$ is a basic set and I has a period m then $2^n(2k + 1) \prec m \prec 2^{n-1}$;

2) $\Omega(f) = \Omega(f^{2^n})$;

3) if f is of type $2^m, 0 \leq m \leq \infty$, then $\Omega(f) = \Omega(f^l)$ $(\forall l)$.

Remark. Another proof of statements 2) and 3) of Theorem 9.4 is given in Chapter 4 of [BCo]. The statement 3) was also proved in [N3] and [Zh].

Note that we use here the Sharkovskii ordering "$\prec$" (see (*) in the beginning of this Section) in the strict sense (i.e. $m \prec n$ implies $m \neq n$). The assertion close to statement 1) of Theorem 9.4 was proved in [N2], Theorem 1.10.

Now we explain the connection between the famous Misiurewicz theorem on maps with zero entropy and our results. Bowen and Franks [BF] have proved that if f is a map of type $m, m \neq 2^n (0 \leq n \leq \infty)$ then $h(f) > 0$. The converse was proved first for pm-maps [MiS] and then for arbitrary continuous maps of the interval into itself [Mi2]. Let us show how to deduce the converse assertion from our results.

If $T : X \to X$ is a map of a compact metric space X with specification then there exists N such that for any $n > N$ there exists a periodic point $y \in X$ of a minimal period n. Let f be of type $2^n, n \leq \infty$. Suppose that f has a basic set $B = B(orb\, I, f)$. By Theorem 4.1 properties of the map restricted on B are close to properties of the corresponding mixing interval map. Furthermore, by Theorem 8.7 this mixing interval map has specification, and so by the property of maps with specification mentioned above one can find integers l, k such that for any $m > l$ there exists an f-cycle of minimal period km , which contradicts the fact that f is of type 2^n. Thus f has no basic sets and $h(f) = 0$ by well-known properties of the topological entropy and the fact that the entropy of f on a solenoidal set is 0 (see Theorem 3.1). Note that now the Decomposition Theorem implies that infinite limit sets of a map with zero entropy are 2-adic solenoidal sets (another proof of this assertion follows from Misiurewicz's papers [Mi2–Mi3]).

1.10. Invariant Measures for Interval Maps

We describe here the results from Section 10. To investigate the properties of invariant measures it is natural to consider the restriction of f to a component of the decomposition. We start by study $f|B$ for a basic set B. By Theorems 4.1 and 8.7 we may apply the results of [Si1-Si2],[DGS] where a lot of generic properties of maps with specification are established. To formulate the theorem which summarizes the results from [Si1-Si2], [DGS] we need some definitions.

Let $T : X \to X$ be a map of a compact metric space (X, d) into itself. By $M(X)$ we denote the set of all Borel normalized measures on X (i.e. $\mu(X) = 1$ for any $\mu \in M(X)$). *The weak topology* on the set $M(X)$ is defined by taking the sets

$$V_\mu(f_1, \ldots, f_k; \varepsilon_1, \ldots, \varepsilon_k) = \{\nu \in M(X) : |\int f_j\, d\mu - \int f_j\, d\nu| < \varepsilon_j, j = 1, \ldots, k\}$$

$$(**)$$

as a basis of open neighborhood for $\mu \in M(X)$ with $\varepsilon_j > 0$ and $f_j \in C(X)$ where $C(X)$ is a space of all continuous functions defined on X. The map T transports every measure $\mu \in M(X)$ into another measure $T_*\mu \in M(X)$.

If $\mu = T_*\mu$ then μ is said to be *invariant*. The set of all T-invariant measures $\mu \in M(X)$ with the weak topology is denoted by $M_T(X) \equiv M_T$. A measure $\mu \in M(X)$ is said to be *non-atomic* if $\mu(x) = 0$ for any point $x \in X$. *The support of μ is the* minimal closed set $S \equiv supp\,\mu$ such that $\mu(S) = 1$. A measure $\mu \in M_T$ whose $supp\,\mu$ coincides with one closed periodic orbit is said to be *a CO-measure* ([DGS], Section 21); if $a \in Per\,T$ then the corresponding CO-measure is denoted by $\nu(a)$. The set of all CO-measures which are concentrated on cycles with minimal period p is denoted by $P_T(p)$.

For any $x \in X$ let $\delta_x \in M(X)$ be the corresponding δ-measure (i.e. $\delta_x(x) = 1$). Let $V_T(x)$ be the set of limit measures of time averages $N^{-1} \cdot \sum_{j=0}^{N-1} T_*^j \delta_x$; it is well-known ([DGS], Section 3) that $V_T(x)$ is a non-empty closed and connected subset of M_T. A point $x \in X$ is said to have *maximal oscillation* if $V_T(x) = M_T$. If $V_T(x) = \{\mu\}$ then a point x is said to be *generic* for μ.

A measure $\mu \in M_T$ is said to be *strongly mixing* if $\lim_{n \to \infty} \mu(A \cap T^{-n}B) = \mu(A) \cdot \mu(B)$ for all measurable sets A, B. A measure μ is said to be *ergodic* if there is no set B such that $TB = B = T^{-1}B$, $0 < \mu(B) < 1$.

We summarize some of the results from [Si1-Si2] and [DGS] in the following

Theorem DGS[Si1-Si2],[DGS]. *Let $T : X \to X$ be a continuous map of a compact metric space X into itself with the specification property. Then the following statements are true.*

For any positive integer l the set $\bigcup_{p \geq l} P_T(p)$ is dense in M_T.

2) The set of ergodic non-atomic invariant measures μ with $supp\,\mu = X$ is residual in M_T.

3) The set of all invariant measures which are not strongly mixing is a residual subset of M_T.

4) Let $V \subset M_T$ be a non-empty closed connected set. Then the set of all points x such that $V_T(x) = V$ is dense in X (in particular, every measure $\mu \in M_T$ has generic points).

5) The set of points with maximal oscillation is residual in X.

Let us return to interval maps. In fact an interval map on one of its basic set does not necessarily have the specification property. However, applying Theorem DGS and some of the preceding results (Theorem 4.1, Lemma 8.3 and Theorem 8.7) we prove in Theorem 10.3 that the restriction of an interval map to its basic set has all the properties 1)-5) stated in Theorem DGS. The fact that statement 5) of Theorem DGS holds for mixing maps (since mixing maps have specification by Theorem 8.7) allows us to specify the third possible type of generic behavior of an orbit for maps without wandering intervals (as was explained after the formulation of Theorem 6.2).

Namely, in fact the following theorem holds.

Theorem 6.2' (cf.[Bl1],[Bl8]). *Let $f : [0, 1] \to [0, 1]$ be a continuous map without wandering intervals. Then there exists a residual subset $G \subset [0, 1]$ such that for any $x \in G$ one of the following possibilities holds:*

1) $\omega(x)$ is a cycle;

2) $\omega(x)$ is a solenoid;

3) $\omega(x) = orb\,I$ is a cycle of intervals and $V_f(x) = M_{f|orb\,I}$.

Furthermore, the statement 5) of Theorem 10.3 leads to the consideration of typical behavior of orbits showing the difference between topological and metric typical behavior even for the "simplest" (and still very rich in the dynamical sense) one-dimensional dynamical smooth systems – namely for unimodal maps with so-called *negative Schwarzian* (see the definition later in Subsection 1.12). Indeed, for a unimodal map f with negative Schwarzian it follows from [BL2] (see also [Ke] and [GJ]) that there is a unique set A such that $\omega(x) = A$ for a.e. x.

Moreover, the similarity in typical (in metric sense) behavior of orbits may by also confirmed by the fact that interval maps in question are ergodic with respect to Lebesgue measure (see [BL2]). Let us show that this implies the following conclusion: there is a closed set $V \subset M_f$ such that for a.e. x we have $V_f(x) = V$ where f is a unimodal negative Schwarzian map.

Indeed, let us choose a countable basis of neighborhoods in M_f and denote it by $\{W_i\}$; denote also $M_f \setminus W_i$ by F_i for any i. Then by the ergodicity of f we see that for any i the measure $m_i \equiv \lambda\{x : V_f(x) \subset F_i\}$ is either 0 or 1. Let $\mathcal{M} = \{i : m_i = 1\}$, $V = \bigcap_{i \in \mathcal{M}} F_i$. Then by the definition for a.e. x we have $V_f(x) \subset V$.

At the same time it is easy to see by the construction that the set $\mathcal{L} = \{i : i \notin \mathcal{M}\}$ coincides with the set of all i such that $W_i \cap V \neq \emptyset$. Moreover, for any $l \in \mathcal{L}$ we have by the construction that $\lambda\{x : V_f(x) \subset F_l\} = 0$ and thus $\lambda\{x : V_f(x) \cap W_l \neq \emptyset\} = 1$. So for a.e. x and for any $l \in \mathcal{L}$ we have $V_f(x) \cap W_l \neq \emptyset$ which easily implies $V_f(x) \supset V$ for a.e. x. Together with the conclusion from the preceding paragraph it implies that in fact $V_f(x) = V$ for a.e. x. This fact in turn easily implies that for a map in question there is a unique closed subset A' of the interval such that for a.e. x the corresponding minimal center of attraction (i.e. the minimal closed set containing supports of all measures from $V_f(x)$) is A' (another proof of this statement is given in [Ke]).

It shows that for maps in question typical in a metric sense behavior of orbits is to some extent similar: a.e. point has the same limit set and the same minimal center of attraction. One could say that "the culmination" of such a similarity is achieved on the family of unimodal negative Schwarzian mixing maps having measure absolutely continuous with respect to Lebesgue measure (a.c.i.m.); in this case as it follows from the ergodicity of a map f the only measure in typical $V_f(x)$ is the aforementioned a.c.i.m. However, Theorem 10.3.5) implies that the set of all points with maximal oscillation is residual here, so we can conclude that *the residual set of points with maximal oscillation for a negative Schwarzian mixing map with a.c.i.m. has Lebesgue measure zero* (clearly this is true in fact for any mixing interval map with ergodic a.c.i.m. μ such that $supp\,\mu = [0, 1]$). This leads to the following

Problem 1. *Does there exist a unimodal mixing map with negative Schwarzian such that the residual set of all points with maximal oscillation has full Lebesgue measure?*

Let us return to the consideration of the results of Section 10. We prove there the following Theorem 10.4 and Corollary 10.5.

Theorem 10.4. *Let μ be an invariant measure. Then the following properties of μ are equivalent;*
1) there exists $x \in [0, 1]$ such that $supp\,\mu \subset \omega(x)$;
2) the measure μ has a generic point;
3) the measure μ can be approximated by CO-measures.

Remark. In fact one can deduce Theorem 10.4 for a non-atomic invariant measure directly from Theorem DGS and the above mentioned Theorem 4.1, Lemma 8.3, Theorem 8.7; this version of Theorem 10.4 was obtained in [Bl4],[Bl7]. Note that even this preliminary version implies the following

Corollary 10.5[Bl4],[Bl7]. *CO-measures are dense in all ergodic invariant measures of f.*

In what follows we need the definition of a piecewise-monotone continuous map. A continuous map $f : [0, 1] \to [0, 1]$ is said to be *piecewise-monotone (a pm-map)* if there exist $n \geq 0$ and points $0 = c_0 < c_1 < \ldots < c_{n+1} = 1$ such that for any $0 \leq k \leq n$ f is monotone on $[c_k, c_{k+1}]$ and $f[c_k, c_{k+1}]$ is not a point (so $[c_k, c_{k+1}]$ is not a flat spot); by monotone we always mean non-strictly monotone. Each c_k, $1 \leq k \leq n$ is called *a turning point of f*; we denote the set $\{c_k\}_{k=1}^n$ by $C(f)$. Each interval $[c_k, c_{k+1}]$ is called *a lap.*

We need also the definition of the measure-theoretic entropy. Let $T : X \to X$ be a map of a compact metric space (X, d) into itself. *A partition* is a finite family $\alpha = \{A_i\}_{i=1}^N$ of disjoint measurable sets such that $\bigcup_{i=1}^N A_i = X$. If $\alpha = \{A_i\}_{i=1}^N$ and $\beta = \{B_j\}_{j=1}^M$ then $\alpha \vee \beta$ is the partition $\{A_i \cap B_j\}_{i=1, j=1}^{i=N, j=M}$. In the similar way $\vee_{k=1}^n \alpha_k$ is defined where each α_k is a partition. Furthermore, if $\alpha = \{A_i\}_{i=1}^N$ is a partition then $T^{-k}\alpha \equiv \{T^{-k}A_i\}_{i=1}^N$, $(\alpha)^t \equiv \vee_{k=0}^t T^{-k}\alpha$.

Let $\mu \in M_T$ be a T-invariant measure, α be a partition. The quantity $H_\mu(\alpha) = \sum_{A \in \alpha} \mu(\alpha) \cdot \ln \mu(\alpha)$ is called *the entropy of the partition* α. Then there exists $\lim_{N \to \infty} N^{-1} \cdot H_\mu((\alpha)^{N-1}) \equiv h_\mu(\alpha, T)$ which is called *the entropy of α with respect to T*. The quantity $h_\mu(T) = sup\{h_\mu(\alpha, T) : \alpha$ is a partition with $H_\mu(\alpha) < \infty\}$ is called *the measure-theoretic entropy of T (with respect to μ)*[K]. The following Theorem DG1 and Corollary DG2 play an important role in the theory of dynamical systems.

Theorem DG1[Di],[Go]. $h(T) = \sup_{\mu \in M_T} h_\mu(T) = sup\{h_\mu(T) : \mu \in M_T$ is ergodic $\}$.

Corollary DG2. $h(T) = \sup_{x \in X} h(T|\omega(x)) = \sup_{x \in R(T)} h(T|\omega(x)) = h(T|\overline{R(T)}) = h(T|C(T))$.

Now let us return to one-dimensional maps. In his recent paper [H4] F.Hofbauer has proved statements 1)–3) of Theorem 10.3 for pm-maps. He used a technique which seems to be essentially piecewise-monotone. Moreover, he has proved that for a pm-map f the set of all f-invariant measures μ such that $h_\mu(f) = 0$ is a residual subset of $M_{f|B}$. This result can be deduced also from Theorem 10.3 and the following theorem of Misiurewicz and Szlenk.

Theorem MiS[MiS]. *If $f : [0, 1] \to [0, 1]$ is a pm-map then the entropy function h as a map from M_f to the set of real numbers defined by $h(\mu) = h_\mu(f)$ is upper-semicontinuous.*

Indeed, by Theorem 10.3.1) and Theorem MiS the set $h^{-1}(0) \cap M_{f|B}$ is a dense G_δ-subset of $M_{f|B}$. However, the corresponding problem for an arbitrary continuous map of the interval (not necessarily a pm-map) has not been solved yet since there is no result similar to Theorem MiS and concerning arbitrary continuous interval maps. By Theorem 4.1 and Lemma 8.3 it is sufficient to consider a mixing map of the interval so the natural question is whether entropy h as a map from M_f to the set of real numbers is upper-semicontinuous provided that $f : [0, 1] \to [0, 1]$ is mixing.

Suppose that the answer is affirmative. Then by Theorem DG1 for any mixing f there exists a measure μ of maximal entropy. However, [GZ] contains an example of a mixing map without any such measure. So we get to the following

Problem 2. *Let $g : [0, 1] \to [0, 1]$ be a mixing map.*
1) Do measures $\mu \in M_g$ with zero entropy form a residual subset of M_g?
2) What are the conditions on g that would imply the upper-semicontinuity of the entropy
   h as a function from M_g to the set of real numbers or at least the existence of a measure
   of maximal entropy for g?

This problem is connected with the following problem which is due to E. M. Coven; to formulate it we need the definition of *an entropy-minimal* map. Namely a continuous map $f : X \to X$ of a compact space X into itself is called *entropy-minimal* if the only invariant closed set $K \subset X$ such that $h(f|K) = h(f)$ is X itself (see [CS]). It is shown in [CS] that entropy-minimal maps are necessarily transitive and at the same time if a transitive map has a unique measure of maximal entropy with support coinciding with the entire compact space then this map is entropy-minimal. So by the results of Hofbauer ([H1]) we see that transitive pm-maps of the interval into itself are entropy-minimal. On the other hand in [CS] the example of a mixing interval map which is not entropy-minimal is given (so by Theorem 8.7 this is an example of a map with specification but without entropy-minimality and hence without measure of maximal entropy).

The problem is to describe the class of transitive interval maps which is wider than all transitive pm-maps but still all maps from this class are entropy-minimal. The following conjectures are connected with the problem; the idea here is to find out what properties of pm-maps being inherited by arbitrary interval maps would imply the corresponding consequences (the existence of a measure of maximal entropy or at least the fact that a map in question is entropy-minimal).

Conjecture(cf. Lemma 8.3). *Let $g : [0, 1] \to [0, 1]$ be a mixing map and suppose for any open U there exists a positive integer n such that $g^n U = [0, 1]$. Then:*
1) the map g is entropy-minimal;
2) the entropy h as a function from M_g to the set of real numbers is upper-semicontinuous
   and g possesses a measure of maximal entropy.

1.11. The Decomposition for Piecewise-Monotone Maps

For pm-maps we can make our results more precise. In fact we are going to illustrate how our technique works applying it to pm-maps. It should be mentioned that the version of the "spectral" decomposition we present here for pm-maps is quite close to that of Nitecki (see [N2]); however we deduce the Decomposition Theorem for pm-maps as an easy consequence of the Decomposition Theorem for arbitrary continuous maps and the properties of basic and solenoidal sets. The properties of pm-maps will not be discussed in Sections 2–11.

Let f be a pm-map with the set of turning points $C(f) = \{c_i\}_{i=1}^m$.

Lemma PM1. *Let* $A = \overline{\bigcup_{c \in C(f)} orb\, c}$, $t' = \inf A$, $t'' = \sup A$. *Then* $f[t', t''] \subset [t', t'']$.

Proof. Left to the reader. □

Lemma PM2. *Let* $\{I_j\}_{j=1}^\infty$ *be generating intervals with periods* $\{m_j\}_{j=1}^\infty$. *Then in the notation of Theorem 3.1* $S = S_p = S_\omega$ *(i.e. for any* $x \in Q = \cap orb\, I_j$ *we have* $\omega(x) = S = Q \cap \overline{Per\, f} = S_p = Q \cap \omega(f) = S_\omega$).

Proof. By Theorem 3.1 $S \subset S_p = Q \cap \overline{Per\, f} \subset S_\omega = Q \cap \omega(f)$. Suppose that $S_\omega \setminus S \neq \emptyset$ and $x \in S_\omega \setminus S$. Then we can make the following assumptions.

1) Replacing x if necessary by an appropriate preimage of x we can assume that $x \notin \{orb\, c : c \in C(f)\}$. Indeed, the fact that $x \in S_\omega$ implies that there exists y such that $x \in \omega(y)$. But $f|\omega(y)$ is surjective, so for any $n > 0$ there exists $x_{-n} \in \omega(y)$ such that $f^n x_{-n} = x$. Moreover, $x \notin S$, so $x_{-n} \notin S$ too.

Now suppose that there are some points $c \in C(f)$ such that for some $m = m(c)$ we have $f^m c = x$. Clearly $x \notin Per\, f$; thus the number $m(c)$ is well defined. Take the maximum M of the numbers $m(c)$ over all $c \in C(f)$ which are preimages of x under iterations of f and then replace x by x_{M+1}. Obviously $x_{M+1} \notin \{orb\, c : c \in C(f)\}$.

2) We can assume x to be an endpoint of non-degenerate component $[x, y]$, $x < y$ of Q.

3) We can assume i to be so large that $orb\, I_i \cap C(f) = Q \cap C(f) \equiv C'$, where $I_i = [x_i, y_i] \ni x$. Indeed, if $c \in C(f) \setminus Q$ then there exists $j = j(c)$ such that $c \notin orb\, I_j$. So if N is the maximum of such $j(c)$ taken over all $c \in C(f) \setminus Q$ then for any $i > N$ we have $orb\, I_i \cap C(f) = Q \cap C(f)$.

4) By Theorem 3.1 $\omega(c') = S$ for any $c' \in C'$. So $x \notin \omega(c')$ for any $c' \in C'$. Now the first assumption implies that $x \notin \{\bigcup orb\, c' : c' \in C'\} = D \supset S$. Thus we can assume i to be so large that $[x_i, x + \varepsilon] \cap D = \emptyset$ for some $\varepsilon > 0$.

Let $A = I_i \cap D$, $t' = \inf A$, $t'' = \sup A$. Note that then by the assumption 4) $x \notin [t', t'']$. The fact that $x \in S_\omega$ implies that there exists a point z such that $x \in \omega(z)$; by Theorem 3.1 $S \subset \omega(z)$. At the same time $S \subset D$ and so $S \cap I_i \subset D \cap I_i \subset [t', t'']$. But by Lemma PM1 the interval $[t', t'']$ is f^{m_i}-invariant and so by the properties of solenoidal sets the fact that $S \subset \omega(z)$ implies that $\omega(z) \cap I_i \subset [t', t'']$. Thus we see that $x \in \omega(z) \cap I_i \subset [t', t'']$ which contradicts the assumption 4). □

Note that together Lemma PM2 and the Decomposition Theorem imply Theorem N[N1].

Lemma PM3. *Let* $g : [0, 1] \to [0, 1]$ *be a continuous map,* I *be an* n-*periodic interval,* $B = B(orb\, I, g)$ *be a basic set,* $J \subset I$ *be another* n-*periodic interval. Furthermore, let* $I = L \cup J \cup R$ *where* L *and* R *are the components of* $I \setminus J$. *Then at least one of the functions* $g^n|L$, $g^n|R$ *is not monotone. In particular, if* g *is a pm-map then* $C(g) \cap orb\, I \supset C(g) \cap orb\, J$ *and* $C(g) \cap orb\, I \neq C(g) \cap orb\, J$.

Remark. It is easy to give an example of intervals $I \supset J$ such that $fI = I$, $fJ = J$ and a basic set $B = B(orb\, I, g)$ exists. Indeed, consider a mixing pm-map f with a fixed point a and then "glue in" intervals instead of a and all preimages of a under iterations of f. It is quite easy to see that this may be done in such a way that we will get a new map g with the required property; J will be an interval which replaces a itself.

Proof. We may assume $n = 1$ and J to be an interval complementary to B. Suppose that $g|L$ and $g|R$ are monotone. By Theorem 4.1 there exists a transitive map $\psi : [0, 1] \to [0, 1]$ which is a monotone factor of the map g; in other words g is semiconjugate to ψ by a monotone map ϕ. By the definition of a basic set $\phi(J) = a$ is a point a. The monotonicity of ϕ and the fact that $g|L$ and $g|R$ are monotone imply that $\psi|[0, a]$ and $\psi|[a, 1]$ are monotone; moreover, $\psi(a) = a$. Clearly, it contradicts the transitivity of ψ. $\square$

We need the following definition: if $B = B(orb\, I, f)$ is a basic set then the period of I is called *the period of B* and is denoted by $p(B)$. To investigate the decomposition for a pm-map let us introduce the following ordering in the family of all basic sets of $f : B(orb\, I_1, f) = B_1 \succ B(orb\, I_2, f) = B_2$ if and only if $orb\, I_1 \supset orb\, I_2$. The definition is correct for a continuous map of the interval. So it is possible to analyze the structure of the decomposition via $\succ$-ordering in the continuous case. We do not follow this way to avoid unnecessary complexity and consider only pm-maps.

For any set $D \subset C(f)$ consider the family $G(D)$ of all basic sets $B(orb\, I, f)$ such that $D = orb\, I \cap C(f)$. Let us investigate the properties of the family $G(D)$ with the $\succ$-ordering. Fix a subset $D \subset C(f)$ and suppose that $B_1 = B(orb\, I_1, f) \in G(D), B_2 = B(orb\, I_2, f) \in G(D)$. Then either $orb\, I_1 \supset orb\, I_2$ or $orb\, I_1 \subset orb\, I_2$. Indeed, otherwise let $J = I_1 \cap I_2 \neq \emptyset$ and let for instance $p(B_1) \leq p(B_2)$. It is easy to see that the period of J is equal to $p(B_2)$. Now Lemma PM3 implies $C(f) \cap orb\, I_2 \neq C(f) \cap orb\, J$ which is a contradiction.

Thus we may assume $orb\, I_1 \supset orb\, I_2$; by Lemma PM3 it implies that $p(B_1) < p(B_2)$. So if $D \subset C(f)$ and $G(D)$ is infinite then $G(D) = \{B_1 \succ B_2 \succ \ldots\}$. Moreover, assume that $B_i = B(orb\, I_i, f)$; then $Q(D) \equiv \cap orb\, I_i$ is a solenoidal set, $D \subset Q(D)$, for any $z \in D$ we have by Lemma PM2 that $\omega(z) = S(D) \equiv Q(D) \cap \overline{Per\, f}$ and the corresponding group is $H(\{p(B_i)\}_{i=1}^{\infty})$. Let us show that there is no basic set $B(orb\, J, f) = B \notin G(D)$ such that $orb\, J \subset orb\, I_1$.

Indeed, let $B = B(orb\, J, f)$ be such a basic set. Let $E = orb\, J \cap C(f)$; then $\emptyset \neq E \subset D, E \neq D$. At the same time it is easy to see that $Q(D) \subset orb\, J$. Indeed, let $z \in E \subset D$. Then by what we have shown in the previous paragraph $\omega(z) = S(D) \subset Q(D)$ and at the same time $\omega(z) \subset orb\, J$ as well; in other words, $\omega(z) \subset Q(D) \cap orb\, J$, which implies $Q(D) \subset orb\, J$ and contradicts the fact that $E \neq D$.

Note that if $G(D) = \{B_1 \succ B_2 \succ \ldots\}$ then the well-known methods of one-dimensional symbolic dynamics easily yield that $f|B_i$ is semiconjugate by a map ϕ to a one-sided shift $\sigma : M \to M$ and $1 \leq card\, \phi^{-1}(\xi) \leq 2$ for any $\xi \in M$. Indeed, let $B_i = B(orb\, I_i, f), B_{i+1} = B(orb\, I_{i+1}, f), orb\, I_{i+1} \subset orb\, I_i$ and $\mathcal{R}$ be a collection of components of the set $orb\, I_i \setminus orb\, I_{i+1}$. Then for each interval $J \in \mathcal{R}$ a map $f|J$ is monotone and for some finite subset $\mathcal{F} = \mathcal{F}(J)$ of $\mathcal{R}$ we have $fJ \supset J'$ if $J' \in \mathcal{F}$ and $fJ \cap J' = \emptyset$ if $J' \notin \mathcal{F}$. Construct an oriented graph X with vertices which are elements of $\mathcal{R}$ and oriented edges connecting $J \in \mathcal{R}$ with $J' \in \mathcal{R}$ if and only if $fJ \supset J'$. This graph generates a one-sided shift $\sigma : M \to M$ in the corresponding topological Markov chain. Let $K = \{x : f^n x \in orb\, I_i \setminus orb\, I_{i+1}\}$. Then $f|K$ is monotonically semiconjugate to

$\sigma : M \to M$ (*monotonically* means here that a preimage of any point is an interval, probably degenerate) and B coincides with ∂K; in other words, to get a set B from K we need to exclude from K interiors of all non-degenerate intervals which are components of K.

Let us return to the properties of the family of all basic sets. If $D \subset C(f)$ then $G(D)$ is either infinite or finite. Let $\{D^i\}_{i=1}^k$ be all D^i such that $G(D)$ is infinite and $\{\widetilde{B}_r\}_{r=1}^R$ be all basic sets belonging to finite sets $G(D)$. The family of all possible sets $D \subset C(f)$ is finite so $R < \infty$, $k < \infty$ and basic sets from $\{G(D^i)\}_{i=1}^k$ together with the collection $\{\widetilde{B}_r\}_{r=1}^R$ form the family of all basic sets. Note that $D^i \cap D^j = \emptyset (i \neq j)$. Indeed, otherwise $\emptyset \neq D^i \cap D^j \subset Q(D^i) \cap Q(D^j)$; by the Decomposition Theorem this is only possible if $Q(D^i) = Q(D^j)$. But $D^r = C(f) \cap Q(D^r)$ $(r = i, j)$ and thus $D^i = D^j$ which is a contradiction. Moreover, if $E \subset C(f)$ is such that $G(E)$ is finite and $E \cap D^i \neq \emptyset$ then $E \supset D^i$. Indeed, considering points from $E \cap D^i$ it is easy to see that for any $B = B(\mathrm{orb}\, J, f) \in G(E)$ we have $Q(D^i) \subset \mathrm{orb}\, J$ and hence $E \supset D^i$.

Clearly, we have already described all basic and some solenoidal sets via $\succ$-ordering. However, there may exist generating intervals $\{I_j\}$ with periods $\{m_j\}$ and the corresponding solenoidal set $Q = \cap \mathrm{orb}\, I_j$ such that $Q \cap C(f) = F$ and $F \neq D^i$ $(1 \leq i \leq k)$. Then by the Decomposition Theorem $F \cap D^i = \emptyset$ $(1 \leq i \leq k)$ and $f|\mathrm{orb}\, I_N$ has no basic sets for sufficiently large N. Applying the analysis of maps with zero entropy to $f|\mathrm{orb}\, I_N$ we finally obtain the Decomposition Theorem for pm-maps.

Theorem PM4 (Decomposition Theorem for pm-maps). *Let* $f : [0, 1] \to [0, 1]$ *be a pm-map. Then there exist an at most countable family of pairs of basic and Ω-basic sets* $\{B_i \subset B_i'\}$ *and a family of triples of solenoidal sets* $\{S^{(\alpha)} \subset S_\Omega^{(\alpha)} \subset Q^{(\alpha)}\}_{\alpha \in A}$ *such that:*

1) $\Omega(f) = X_f \cup (\bigcup_\alpha S_\Omega^{(\alpha)}) \cup (\bigcup_i B_i')$;
2) $\omega(f) = \overline{\mathrm{Per}\, f} = X_f \cup (\bigcup_\alpha S^{(\alpha)}) \cup (\bigcup_i B_i)$;
3) $\mathrm{card}\, A \leq \mathrm{card}\, C(f)$;
4) $S^{(\alpha)} = S_p^{(\alpha)} = S_\omega^{(\alpha)}$ *for any* $\alpha \in A$;
5) *intersections in this decomposition are only possible between different basic or Ω-basic sets, intersection of any three sets is empty and intersection of any two sets is finite;*
6) *there exist a finite number of pairwise disjoint subsets* $\{D^i\}_{i=1}^k$, $\{F_j\}_{j=1}^l$ *of* $C(f)$, *a finite collection of basic sets* $\{\widetilde{B}_r\}_{r=1}^R$ *and a finite collection of cycles of intervals* $\{\mathrm{orb}\, K_j\}_{j=1}^l$ *with the following properties:*
   a) *for any* i, $1 \leq i \leq k$ *the family* $G(D^i)$ *is an infinite chain* $B_1^i = B(\mathrm{orb}\, I_1^i, f) \succ B_2^i = B(\mathrm{orb}\, I_2^i, f) \succ \ldots$ *of basic sets with periods* $p_1^i < p_2^i < \ldots$ *and* $Q(D^i) = \bigcap_n \mathrm{orb}\, I_n^i$ *is a solenoidal set with the corresponding group* $H(p_1^i, p_2^i, \ldots)$;
   b) $f|B_n^i$ *is semiconjugate to a one-sided shift in a topological Markov chain and the semiconjugation is at most* 2*-to-*1;
   c) *if* $i \neq j$, $B \in G(D^i)$, $\widehat{B} \in G(D^j)$ *then neither* $B \succ \widehat{B}$ *nor* $\widehat{B} \succ B$;
   d) *all basic sets of* f *are* $\{\widetilde{B}_r\}_{r=1}^R \cup \bigcup_{i=1}^k \{B_n^i\}_{n=1}^\infty$;
   e) *for any* j, $1 \leq j \leq l$ *the cycle of intervals* $\mathrm{orb}\, K_j$ *has period* N_j, *there exists a unique solenoidal set* $Q_j \subset \mathrm{orb}\, K_j$, $h(f|\mathrm{orb}\, K_j) = 0$, $\mathrm{orb}\, K_j \cap C(f) = F_j \subset Q_j \subset \mathrm{orb}\, K_j$ *and the group corresponding to* Q_j *is* $H(N_j, 2N_j, 4N_j, \ldots)$;
   f) $\{Q(D^i)\}_{i=1}^k \cup \{Q_j\}_{j=1}^l = \{Q^{(\alpha)}\}_{\alpha \in A}$;

g) *there exists a countable set of pairwise disjoint cycles of intervals $\{orb\, L_j\}$ (perhaps some of them are degenerate) such that $C(f) \cap int\,(orb\, L_j) = \emptyset\ (\forall j)$ and $X_f \subset \bigcup_j orb\, L_j$.*

Remark that we have not included the proofs of statements 3) and 6.g) which are left to the reader.

Let us make several historical remarks. Jonker and Rand [JR1,JR2] constructed the "spectral" decomposition of $\Omega(f)$ for a map with a unique turning point (*a unimodal map*); they used the kneading theory of Milnor and Thurston [MilT]. The unimodal case was studied also in [Str]. The decomposition was extended to an arbitrary pm-map by Nitecki [N2] and Preston [P1–P2]. For piecewise-monotone maps with discontinuities the decomposition is due to Hofbauer [H2,H3].

Our Decomposition Theorem for a pm-map is related to those of Nitecki and Preston. However, we would like to note some differences: 1) we deduce the Decomposition Theorem for a a pm-map from the general Decomposition Theorem for a continuous map of the interval; 2) we investigate the properties of basic sets using an approach which seems to be new.

1.12. Properties of Piecewise-Monotone Maps of Specific Kinds

To conclude the part of this Introduction concerning pm-maps we discuss some specific kinds of pm-maps. First we need some definitions. A pm-map f is said to be *topologically expanding* or simply *expanding* if there exists $\gamma > 1$ such that $\lambda(fI) \geq \gamma \cdot \lambda(I)$ for any interval I provided $f|I$ is monotone (here $\lambda(\cdot)$ is Lebesgue measure). Let g be a continuous interval map, J be a non-degenerate interval such that $g^n|J$ is monotone for any $n \geq 0$ (recall that by monotone we mean non-strictly monotone); following Misiurewicz we call *J a homterval*. Remark also that one can define the topological entropy of $f|K$ without assuming K to be an invariant or even compact set [B3]. Now we are able to formulate

Lemma PM5[Bl3]. *The following properties of f are equivalent:*
1) f is topologically conjugate to an expanding map;
2) if $d < b$ then $f|[d, b]$ is non-degenerate and if $\{c_1, \ldots, c_k\} = C(f) \cap int\,(\overline{Per\, f})$ then $\bigcup_{i=1}^{k}(\bigcup_{n\geq 0} f^{-n}c_i)$ is a dense subset of $[0, 1]$;
3) there exists $\delta > 0$ such that $h(f|J) \geq \delta$ for any non-degenerate interval J;
4) f has neither homtervals nor solenoidal sets.

Proof. We give here only a sketch of the proof. It is based on the Decomposition Theorem and the following important theorem of Milnor and Thurston, proved in [MilT].

Theorem MT. *Let f be a pm-map with $h(f) > 0$. Then there exists an expanding map g with two properties:*
1) $\lambda(g[d, b]) = e^{h(f)} \cdot \lambda([d, b])$ for any $d < b$ provided that $g|[d, b]$ is monotone;
2) f is topologically semiconjugate to g by a monotone map.

An expanding map g with the properties from Theorem MT is called *a map of a constant slope*.

Suppose that statement 1) from Lemma PM5 holds for a map f. Then the properties of solenoidal sets and the definition of a homterval imply that statement 4) holds. Indeed, we may assume f itself to be an expanding map with the constant of expansion $\gamma > 1$. Let n be such that $\gamma^n > 2$. Consider the map f^n and and let ε be the length of the shortest lap of f^n. Then any interval J with $\lambda(J) \leq \varepsilon$ covers no more than than 1 turning point of f and hence $\lambda(f^n J) > 2\lambda(J)$. So there exists a number k such that for any $i > k$ we have $\lambda(f^{in} J) > \varepsilon$; now properties of continuous maps easily imply that for the corresponding $\delta > 0$ and any $j > kn$ we have $\lambda(f^j J) > \delta$. Clearly it implies that f has neither homtervals nor solenoidal sets, i.e. that statement 4) holds.

Now let statement 4) hold. Then by the Decomposition Theorem we see that because of the non-existence of solenoidal sets there are only finitely many basic sets. Besides it follows from the non-existence of solenoidal sets and homtervals that there are no periodic intervals on which f has zero entropy; this implies that all $\succ$-minimal basic sets are cycles of intervals on which the map f is transitive.

Let us show that it implies statement 2) of Lemma PM5. Indeed, the non-existence of homtervals implies that the map f is non-degenerate on every non-degenerate interval, so 2a) holds. Now let us prove that 2b) holds too. Let J be an interval; consider the orbit of J under iterations of f. The non-existence of homtervals implies that there are numbers $n < m$ such that $f^n J \cap f^m J \neq \emptyset$. It is easy to see now that there is a weak cycle of intervals $I, fI, \ldots, f^{k-1}I, f^k I \subset I$ and a number n such that $\bigcup_{i \geq n} f^i J = \bigcup_{r=0}^{k-1} f^r I$. But by what has already been proved in the previous paragraph $\bigcup_{r=0}^{k-1} f^r I$ should contain a cycle of intervals M on which f is transitive (otherwise it would have contained either homtervals or solenoidal sets which is impossible). On the other hand by the properties of basic sets $M \subset \overline{Per\,f}$;hence there is $c \in C(f)$ such that $c \in int\,(M) \subset int(Per\,f)$ which means that that there exists $c \in int\,(\overline{Per\,f})$ with preimages in the interval J for any J. It proves statement 2) of Lemma PM5.

Now suppose that statement 2) holds. Let us show that there exist cycles of intervals on which the map is transitive. Indeed, otherwise every basic or solenoidal set has an empty interior. Suppose that for any n the the set $\{x : f^n x = x\}$ has an empty interior too. Then the set $A = \bigcup_{n \geq 0} \{x : f^n x = x\} \cup (\bigcup_\alpha S^{(\alpha)}) \cup (\bigcup_i B_i)$ is of the first category and so has an empty interior. At the same time by the Decomposition Theorem $A = \overline{Per\,f}$ and by the statement 2) $A = \overline{Per\,f}$ has a non-empty interior; this contradiction shows that there exists n such that $int\{x : f^n x = x\} \neq \emptyset$. However this in its turn contradicts statement 2) and shows that our first assumption was false and there exist cycles of intervals on which the map is transitive.

Now take all cycles of intervals on which f is transitive. On each cycle there exists a semiconjugation to a constant slope map (Theorem MT); indeed, the semiconjugation exists because transitive interval maps have positive entropy (see e.g. Lemma 9.3; this fact also may be easily deduced from the Decomposition Theorem). Moreover, in fact it must be a conjugation because otherwise "expanding" properties of transitive maps (see Lemma 8.3) imply that the constant slope map in question is degenerate. Using some technical arguments and statement 2) itself one can now construct a conjugation between the map f and some expanding map, i.e. statement 1) holds. The equivalence of all these statements and statement 3) may be proved by similar methods. It completes the sketch of the proof of Lemma PM5. □

For a map with constant slope the Decomposition Theorem may be refined. Namely, in [Bl10] the following theorem is proved.

Theorem PM6[Bl10]. *Let f be a map of constant slope and $\{B_i\}_{i=1}^N$ be the family of all basic sets of f. Then $N \leq \operatorname{card} C(f)$, the family of limit sets of genus 0 is finite and there is no solenoidal sets.*

Let us apply Theorem PM6 and investigate the continuity of topological entropy for pm-maps. Let M_n be the class of pm-maps f such that $\operatorname{card} C(f) \leq n$. For any $c \in C(f)$ let $q(c, f)$ be the number of basic sets $B = B(\operatorname{orb} I, f)$ such that $c \in \operatorname{orb} I$ if $c \in \operatorname{Per} f$ or ∞ otherwise.

The machinery of discontinuity of the entropy as a function h from M_n to the set of real numbers was investigated in [Mi5] and in different way in [Bl12]; in [MiŚl] the analogous result was obtained for piecewise-monotone maps with discontinuities. Roughly speaking if h is not continuous at $f \in M_n$ (where $\operatorname{card} C(f) = n$) then there exists $c \in C(f) \cap \operatorname{Per} f$ which can "blow up" turning into a periodic interval J such that for a new map $g \in M_n$ we have $h(g|\operatorname{orb}_g J) > h(f)$. However, this is impossible if $q(c, f) \geq n$. Namely, the following theorem holds.

Theorem PM7[Bl12]. *Let $f \in M_n$, $\operatorname{card} C(f) = n$ and $q(c, f) \geq n$ for any $c \in C(f)$. Then the entropy function h as a function from M_n to the set of real numbers is continuous at f.*

As a corollary we obtain in [Bl12] a new proof of the following result of M.Misiurewicz [Mi5].

Corollary PM8[Mi5],[Bl12]. *Let $f \in M_1$, $C(f) = \{c\}$ and either $h(f) = 0$ and $c \notin \operatorname{Per} f$ or $h(f) > 0$. Then the entropy function h as a function from M_1 to the set of real numbers is continuous at f.*

The most important example of a pm-map is perhaps *a smooth map of an interval*, by which we mean a C^∞-map $f : [0, 1] \to [0, 1]$ with a finite number of non-flat critical points. We denote the set of all smooth maps with n critical points by Sm_n; $Sm \equiv \cup Sm_n$. Let us define *the Schwarzian derivative* as $Sf \equiv f'''/f' - 3/2 \cdot (f''/f')^2$. If for $f \in Sm_n$ we have $Sf < 0$ outside the critical points of f then we say that f is a map with *negative Schwarzian*. The family of all such f is denoted by NS_n; $NS \equiv \bigcup_{n\geq 0} NS_n$.

Does there exist a smooth map with a wandering interval? This question is related to Denjoy theorem [D] and since 1970's (namely, since the appearence of G. Hall's example which we discuss later) it has been attracting great attention. The main conjecture was that the answer is negative. Let us describe the history of the verification of this conjecture.

0) [D] for a diffeomorphism of a circle with the first derivative of bounded variation;
1) [Mi1] for a map $f \in NS_1$ with a 2-adic solenoid;
2) [Gu] for a map $f \in NS_1$;
3) [MSt] for a map $f \in Sm_1$;
4) [Yo] for a smooth homeomorphism of the circle with a finite number of non-flat critical points;
5) [L] for a map $f \in NS$ with critical points which are turning points (the principal step towards the polymodal case);

6) [BL1] for a map $f \in Sm$ with critical points which are turning points;
7) [MMSt] for a map $f \in Sm$.

Remark also that in [MMSt] the following nice theorem was proved.

Theorem MMS. *Let $f \in Sm$. Then there exist a positive integer N and a number $\xi > 0$ such that for any periodic point p of minimal period $n > N$ the following inequality holds:* $|Df^n(p)| \geq 1 + \xi$.

Remark. G.Hall constructed an example of a C^∞-piecewise-monotone map with finitely many critical points (among them there are flat critical points) which has a homterval. It shows that C^∞-property alone is not sufficient for the conjecture in question to be true.

Together with Theorem 6.2′ and the Decomposition Theorem for pm-maps these results imply the following

Corollary PM9. *Let $f \in Sm$. Then there exist k cycles of intervals $\{orb\, I_j\}_{j=1}^i$, q solenoids $\{Q_j\}_{j=1}^q$ and l cycles of intervals $\{L_j\}_{j=1}^l$ such that $i + q \leq C(f)$ and the following statements are true:*
1) $f|orb\, I_j$ is transitive $(1 \leq j \leq i)$;
2) $int\,(orb\, L_j) \cap C(f) = \emptyset\,(1 \leq j \leq l)$;
3) there exists a residual subset $G \subset [0, 1]$ such that for $x \in G$ either $\omega(x) \subset orb\, L_j$ is a cycle for some $1 \leq j \leq l$, or $\omega(x) = Q_j$ for some $1 \leq j \leq q$, or $\omega(x) = orb\, I_j$ and $V_f(x) = M_{f|orb\, I_j}$ for some $1 \leq j \leq i$.

Remark. In [Bl1] we describe generic limit sets for pm-maps without wandering intervals.

1.13. Further Generalizations

Now we would like to discuss possible generalizations of these results. First note that we consider a pm-map as a particular case of a continuous map of the interval; at the same time one can consider a continuous map as a generalization of a pm-map. It is natural to ask whether there are other generalizations and here a pm-map with a finite number of discontinuities is another important example.

This class of maps was investigated by F.Hofbauer in his papers [H2–H4] where he constructed and studied the corresponding "spectral" decomposition. It is necessary to mention also the paper [HR] where components of Hofbauer's decomposition with zero entropy are investigated and the paper [W] where topologically generic limit behavior of pm-maps with finite number of discontinuities is studied.

However, we are mostly interested in continuous maps; this leads to the generalization of our results to continuous maps $f : M \to M$ of a one-dimensional branched manifold ("graph") into itself. It turns out that the "spectral" decomposition and the classification of its components can be generalized for a continuous map of a "graph" with slight modifications.

More precisely, let $f : M \to M$ be a continuous map of a "graph". Let $K = \bigcup_{i=1}^n K_i$ be a submanifold with connected components $K_1, \ldots, K_n$; we call K a *cyclical submanifold* if K is invariant and f cyclically permutes the components $K_1, \ldots, K_n$. A cyclical submanifold R can generate a maximal limit set; the definition is analogous to that for the interval. Namely, let $L = \{x \in R :$ for any relatively open neighborhood U of x in R we have $\overline{orb\, U} = R\}$ be an infinite set. Then there are two possibilities.

1) $f|R$ *has no cycles*. Then $f|L$ acts essentially as an irrational rotation of the circle. In this case we denote L by $Ci(R, f)$ and call $Ci(R, f)$ a *circle-like* set. For instance, if $g : S^1 \to S^1$ is the Denjoy map of the circle (i.e. the example of the circle homeomorphism with a wandering interval) then $R = S^1$ and $Ci(S^1, g)$ is the unique minimal set of g. The existence of a monotone map which semiconjugates g to the irrational rotation is in this case a well-known fact; moreover, this semiconjugation is at most 2-to-1 on $Ci(S^1, g)$, i.e. essentially $g|Ci(S^1, g)$ is similar to the corresponding irrational rotation.

In [AK] this kind of dynamics was proven to take place for any continuous maps of the circle into itself without periodic points; namely it was shown in [AK] that if $g : S^1 \to S^1$ is a map without periodic points then there is a monotone semiconjugation between g and some irrational rotation of the circle. It turns out that actually similar monotone semiconjugation exists in general case of graph maps as well. Namely if $f|R$ does not have periodic points then there is a finite union K of disjoint circles and a map $g : K \to K$ which cyclically permutes the circles, maps each circle into itself by the corresponding iteration of g as an irrational rotation and at the same time may be obtained as a factor-map of $f|R$ by a monotone semiconjugation ϕ (in other words, $f|R$ is monotonically semiconjugate to g, see [Bl5] and also [Bl9], [Bl11]). This shows that in general a map on any of its circle-like sets is similar to an irrational rotation and justifies the terminology.

Clearly, to construct an example of a graph map with no periodic points one can take an irrational circle rotation and "blow up" one or several orbits, replacing all but finitely many points in them by intervals and also the rest of the points by finite graphs; it is easy to see that this construction can be carried out so that the new graph map is continuous and it follows from the construction that it will not have periodic points. The aforementioned results in fact show that this is essentially the only way such examples may be constructed.

2) $f|R$ *has cycles*. Then we denote L by $B(R, f)$ and call $B(R, f)$ a *basic set*. The properties of a basic set of a map of a "graph" are analogous to those of a basic set of a map of the interval.

The definitions of a solenoidal set and of a limit set of genus 0 are similar to those for the interval. Limit sets of genus 0, solenoidal sets, circle-like sets and basic sets are the components of the "spectral" decomposition for a map of a "graph".

The Decomposition Theorem for a map of a "graph" and its several corollaries are proved in [Bl5,Bl9,Bl11]. For example, the generic properties of invariant measures are analogous to those for a map of the interval (clear modifications are connected with the existence of circle-like sets). It should be mentioned also that the famous Sharkovskii theorem on the co-existence of periods of cycles (Theorem Sh3) was generalized for continuous maps of the circle[Mi4], of the letter Y[ALM] and of any n-od[Ba]. There are also some recent results concerning the description of sets of periods of cycles for continuous maps of an arbitrary finite "graph" into itself [Bl13, Bl14, LM] and for continuous maps of an arbitrary finite "tree" into itself [Bl15] (here "tree" is a finite "graph" which does not contain subsets homeomorphic to the circle).

Almost all the results of this paper are contained in the author's Ph.D. Thesis (Kharkov, 1985). The preprint [Bl16] is a preliminary version of the present paper.

When I was revising the paper I learned about the nice recent Block and Coppel's book on topological dynamics of interval maps [BCo] where the authors among a lot of questions consider few problems related to those studied in the present paper. Briefly, this "overlapping" may be described as follows. First of all some classical results of A.N. Sharkovskii are proved in [BCo]; for example, Theorem Sh2 is proved there, which is apparently the first proof of this remarkable result published in English (see [BCo], Chapter 4, Proposition 6) . The authors also obtain some results, close to Theorem 7.6 and Theorem 9.4. All necessary remarks are made in the corresponding places in the text of the present paper.

I would like also to mention two books in Russian [SKSF] and [SMR] written by A.N. Sharkovskii and his collaborators in which a lot of problems of interval maps are considered; however spectral decomposition and related questions are not discussed there.

Acknowledgments. I would like to express my gratitude to the Institute for Mathematical Sciences in SUNY at Stony Brook for the kind hospitality which made possible the revising of this paper. I am also grateful to M. Lyubich and J. Milnor for providing useful comments. I would like to thank Wesleyan University where the revising of the paper was finished and E.M. Coven for helpful comments concerning entropy-minimality and the results of [AK] and [BCo]. My special thanks to the referees for careful reading of the first version of the paper and useful suggestions.

2. Technical Lemmas

From now on we will use all notions introduced in Section 1 without repeating definitions. At the same time for the sake of convenience we will repeat formulations of theorems and lemmas we are going to prove. Fix a continuous map $f : [0, 1] \to [0, 1]$. We will prove in this Section some elementary preliminary lemmas which nevertheless seem quite important. Let us start with the following easy

Lemma 2.1. *1) Let U be an interval , $f^m U \cap U \neq \emptyset$ for some m. Then there exists a weakly periodic closed interval I of period n such that $\overline{orb\, U} = \bigcup_{i=0}^{n-1} f^i I = orb\, I$ is a weak cycle of intervals and $\{orb\, I \setminus orb\, U\}$ is a finite set.*

2) Let J be a weakly l-periodic closed interval. Then $L = \bigcap_{i \geq 0} f^{il} J$ is a closed l-periodic interval and $\bigcap_i f^i(orb\, J) = orb\, L$ is a cycle of intervals.

Proof.
1) Clearly, $\bigcup_{i=0}^{\infty} f^{mi+k} U = J_k$ is an interval for $0 \leq k < m$. Thus the set $\overline{orb\, U} = \bigcup_{k=0}^{m-1} \overline{J_k}$ consists of a finite number of its components and $card\, (\overline{orb\, U} \setminus orb\, U) < \infty$. Let $I \supset U$ be a component of $\overline{orb\, U}$ and n be the minimal integer such that $f^n I \cap I \neq \emptyset$. Then $f^n I \subset I$ and the first statement is proved.
2) The proof is left to the reader. $\square$

Denote by L the left side and by R the right side of any point $x \in [0, 1]$. Now if $T = L$ or $T = R$ is a side of $x \in [0, 1]$ then denote by $W_T(x)$ a one-sided semi-neighborhood of x. Let $U = [\alpha, \beta]$ be an interval, $\alpha < \beta$, $x \in (\alpha, \beta)$. By $Si_U(x) \equiv \{L, R\}$ we denote the set of the sides of x; also let $Si_U(\alpha) \equiv \{R\}$, $Si_U(\beta) \equiv \{L\}$. We will consider a pair $(x, T)_U$ where $T \in Si_U(x)$ and call $(x, T)_U$ a U-pair or a pair in U. A set of all U-pairs is denoted by $\widehat{U}$. If $U = [0, 1]$ then we write simply $Si(x)$, (x, T) and call (x, T) a pair. If (x, T) is a pair in U then we also say that T is a side of x in U. Finally, if $f|W_T(x)$ is not degenerate for any $W_T(x)$ then we say that f is not degenerate on the side T of x.

Let us define the way f acts on pairs. Namely, say that (y, S) belongs to $f(x, T)$ if $y = fx$ and for any $W_T(x)$ there exists $W_S(y)$ such that $fW_T(x) \supset W_S(y)$.

Let us formulate without proof some properties of a continuous map of the interval.

Property C1. Let U be an interval, $x \in fU = V$ and $T \in Si_V(x)$. Then there exists $y \in U$ and $S \in Si_U(y)$ such that $(x, T) \in f(y, S)$. In particular:

1) if $x \in int\, V$ then for any side $T \in Si(x)$ there exists $y \in int\, U$ and a side S of y in U such that $(x, T) \in f(y, S)$;

2) if x is an endpoint of a non-degenerate interval V and there exists $y \in int\, U$ such that $fy = x$ then there exists $z \in int\, U$ and $S \in Si_U(z)$ such that $f(z, S) \ni (x, T)$.

Property C2. Let f be non-degenerate on the side T of x. Then $f(x, T)$ is non-empty.

Property C3. If I, J are closed intervals and $I \subset fJ$ then there exists a closed interval $K \subset J$ such that $fK = I$.

Property C4. Let U be an interval, $x \in U$ be a point, $\lambda(U) \geq \varepsilon > 0$, $n > 0$. Then there exists an interval V such that $x \in V \subset U$, $\lambda(f^i V) \leq \varepsilon$ $(0 \leq \varepsilon \leq n)$ and $\lambda(f^j V) = \varepsilon$ for some $j \leq n$.

Let us consider some examples.

Example 2.1. Let $f(x) \equiv x$. Then $f(x, L) = (x, L)$ and $f(x, R) = (x, R)$ for any $x \in [0, 1]$.

Example 2.2. Let $f(x) = 4x(1 - x)$; then $f(1/2, L) = f(1/2, R) = (1, L)$.

Example 2.3. Let f be continuous and x be a point of local strict maximum of f. Then $f(x, L) = f(x, R) = (fx, L)$.

Let I be a k-periodic interval, $M = orb\, I = \bigcup_{i=0}^{k-1} f^i I$. For every $x \in M$ we consider three sets which are similar to the well-known prolongation set. Let $\mathcal{U}$ be either the family L of all left semi-neighborhoods of x in M or the family R of all right semi-neighborhoods of x in M or the family A of all neighborhoods of x in M. For any $W \in \mathcal{U}$ and $n \geq 0$ let us consider the invariant closed set $\overline{\bigcup_{i \geq n} f^i W}$. Set $P_M^{\mathcal{U}}(x, f) \equiv P_M^{\mathcal{U}} \equiv \bigcap_{W \in \mathcal{U}} \bigcap_{n \geq 0} (\overline{\bigcup_{i \geq n} f^i W})$. Let us formulate (without proof) some properties of these sets (we will write $P_M(x)$ instead of $P_M^A(x)$ and $P^{\mathcal{U}}(x)$ instead of $P_{[0,1]}^{\mathcal{U}}$).

Property P1. $P_M^{\mathcal{U}}(x)$ is an invariant closed set and $P_M(x) = P_M^L(x) \cup P_M^R(x)$.

Property P2. Let $y \in \overline{orb\, x}$. Then $P_M(x) \subset P_M(y)$.

Property P3. *If $y = f^n x$ and $f^n(x, T) = \{(y, S_i)\}_{i=1}^t$ then $P_M^T(x) = \bigcup_{i=1}^t P_M^{S_i}(y)$.*

We say that a point y is *a limit point of orb x from the side T* or that *a side T is a limit side of $y \in \omega(x)$* if for any open semi-neighborhood $W_T(y)$ we have $W_T(y) \cap \mathrm{orb}\, x \neq \emptyset$.

Property P4. *If y is a limit point of orb x from the side T then $P_M^T(y) \supset P_M(x)$ and $P_M^T(y) \supset \omega(x)$.*

Property P5. *$f|P_M^{\mathcal{U}}(x)$ is surjective.*

Property P6. *$P_M^{\mathcal{U}}(x) = \bigcup_{i=0}^{m-1} f^i P_M^{\mathcal{U}}(x, f^m)$.*

Moreover, the following lemma is true (note that by the definition if $W \in \mathcal{U}$ then either x is an endpoint of W or $x \in W$).

Lemma 2.2. *Let I be a periodic interval, $M = \mathrm{orb}\, I$, $x \in M$. Then one of the following possibilities holds for the set $P_M^{\mathcal{U}}(x)$.*

1) There exists an interval $W \in \mathcal{U}$ with pairwise disjoint forward iterates and $P_M^{\mathcal{U}}(x) = \omega(x)$ is a 0-dimensional set.

2) There exists a periodic point p such that $P_M^{\mathcal{U}}(x) = \mathrm{orb}\, p$.

3) There exists a solenoidal set Q such that $P_M^{\mathcal{U}}(x) = Q$.

4) There exists a periodic interval J such that $P_M^{\mathcal{U}}(x) = \mathrm{orb}\, J$.

If additionally $x \in \Omega(f)$ then $x \in P(x)$.

Proof. The possibility 1) is trivial. Suppose this possibility does not hold. Clearly, it means that if $W \in \mathcal{U}$ then for some $l < n$ we have $f^l W \cap f^n W \neq \emptyset$. By Lemma 2.1 $\overline{\bigcup_{i \geq k} f^k W}$ is a weak cycle of intervals and there exists a periodic interval J_W such that $\bigcap_{k \geq l}(\overline{\bigcup_{i \geq k} f^i W}) = \mathrm{orb}\, J_W$. Let us choose a family of intervals $\{W_m\}$ so that $W_m \in \mathcal{U}$, $W_m \supset W_{m+1}$ and $\lambda(W_m) \to 0$. Denote J_{W_m} by J_m. Then $\mathrm{orb}\, J_m \supset \mathrm{orb}\, J_{m+1}$ $(\forall m)$ and $P_M^{\mathcal{U}}(x) = \bigcap_{m \geq 0} \mathrm{orb}\, J_m$. If periods of J_m tend to infinity then we get to the case 3) of the lemma. Otherwise $\mathrm{orb}\, J_m$ tend either to a cycle (the case 2)) or to a cycle of intervals (the case 4)). $\quad\square$

Let us consider some examples.

Example 2.4. Let $f : [0, 1] \to [0, 1]$ be a transitive map. Then for any pair (x, T) we have $P^T(x) = [0, 1]$.

Example 2.5. Let $f : [0, 1] \to [0, 1]$, $f(0) = 0$, $f(1) = 1$ and $fx > x$ for any $x \in (0, 1)$. Then for the pair $(0, R)$ we have $P^R(0) = [0, 1]$ and for any other pair (x, T) we have $P^T(x) = \{1\}$.

3. Solenoidal Sets

The following Theorem 3.1 is the central theorem concerning solenoidal sets.

Theorem 3.1[Bl4,Bl7]. *Let $\{I_j\}_{j=0}^{\infty}$ be generating intervals with periods $\{m_i\}_{i=0}^{\infty} = D$, $Q = \bigcap_{j \geq 0} \mathrm{orb}\, I_j$. Then there exists a continuous map $\phi : Q \to H(D)$ with the following properties:*

1) ϕ *semiconjugates* $f|Q$ *to* τ (i.e. $\tau \circ \phi = \phi \circ f$ *and* ϕ *is surjective*);
2) *there exists the unique set* $S \subset S_p$ *such that* $\omega(x) = S$ *for any* $x \in Q$ (S *is a set of all limit points of* S_Ω *and* $f|S$ *is minimal*);
3) *if* $\omega(z) \cap Q \neq \emptyset$ *then* $S \subset \omega(z) \subset S_\omega$;
4) *for any* $\mathbf{r} \in H(D)$ *the set* $J = \phi^{-1}(\mathbf{r})$ *is a connected component of* Q *and:*
   a) *if* $J = \{a\}$ *is degenerate then* $a \in S$
   b) *if* $J = [a, b], a \neq b$ *then* $\emptyset \neq S \cap J \subset S_\Omega \cap J \subset \{a, b\}$;
5) $S_\Omega \setminus S$ *is at most countable and consists of isolated points;*
6) $h(f|Q) = 0$.

Proof. If $y \in Q$ then there exists a well-defined element $\mathbf{r} = (r_0, r_1, \ldots) \in H(D)$ such that $y \in f^{r_i} I_i$ ($\forall i$). Let us define $\phi : Q \to H(D)$ as follows: $\phi(y) \equiv \mathbf{r}(y)$. Then ϕ is continuous, surjective and $\phi^{-1}(\mathbf{s}) = \bigcap_{i \geq 0} f^{s_i} I_i$ is a component of Q for any $\mathbf{s} = (s_0, s_1, \ldots) \in H(D)$. Clearly, $\tau \circ \phi = \phi \circ f$ and all the components of Q are wandering.

Now we are going to prove statement 2). Let us denote by J_z the component of Q containing z. Besides let S be the set of all limit points of S_Ω and also $x \in Q$. We will show that $\omega(x) = S$. First observe that $J_x \cap S_\Omega \neq \emptyset$; this easily implies that $\omega(x) \subset S$.

On the other hand let $y \in S$. By the definition there exists a sequence $\{U_i\}$ of intervals, where every U_i is a component of $orb\, I_i$, with the following property: $U_i \to y$, $y \notin U_i$ ($\forall i$). Since $U_i \cap Per\, f \neq \emptyset$ we have $y \in \overline{Per\, f}$. Moreover, we can choose a sequence $\{n_i\}$ such that $f^{n_i} J_x \subset U_i$ ($\forall i$). Therefore $y \in \omega(x)$ and $\omega(x) = S \subset \overline{Per\, f}$. Statement 2) is proved.

Statements 3) and 6) easily follow from what has been proved and are left to the reader (statement 3) follows from the construction and statement 6) may be deduced from statement 3) and the well-known properties of the topological entropy). Statement 4) follows from statements 1)–2) and the observation that J_z is wandering for any $z \in Q$ (indeed, $\phi(J_z)$ as a point of $H(D)$ has an infinite τ-orbit and an infinite ω-limit set which together with $\tau \circ \phi = \phi \circ f$ implies that J_z itself is a wandering interval). Statement 5) follows from statements 2) and 4). □

In the sequel it is convenient to use the following

Corollary 3.2. *Let* $\{I_j\}$ *be a family of generating intervals,* $Q = \bigcap_{j \geq 0} orb\, I_j$. *Then the following statements hold:*
1) $Q \cap Per\, f = \emptyset$;
2) *if* $J \subset int\, Q$ *is an interval then* J *is wandering;*
3) *if* $int\, Q = \emptyset$ (*i.e.* Q *is a solenoid*) *then* $f|Q$ *is conjugate to the minimal translation* τ *in* $H(D)$.

Proof. Left to the reader. □

4. Basic Sets

Now we pass to the properties of basic sets. The main role here plays the following

Theorem 4.1[Bl4,Bl7]. *Let* I *be an* n-*periodic interval,* $M = orb\, I$ *and* $B = B(M, f)$ *be a basic set. Then there exist a transitive non-strictly* n-*periodic map* $g : M' \to M'$ *and*

a monotone map $\phi : M \to M'$ *such that* ϕ *almost conjugates* $f|B$ *to* g. *Furthermore,* B
has the following properties:

a) B *is a perfect set;*

b) $f|B$ *is transitive;*

c) *if* $\omega(z) \supset B$ *then* $\omega(z) = B$ (*i.e.* B *is a maximal limit set);*

d) $h(f|B) \geq \ln 2 \cdot (2n)^{-1}$;

e) $B \subset \overline{Per\, f}$;

f) *there exist an interval* $J \subset I$, *an integer* $k = n$ *or* $k = 2n$ *and a set* $\widetilde{B} = \overline{int\, J \cap B}$ *such*
that $f^k J = J$, $f^k \widetilde{B} = \widetilde{B}$, $f^i \widetilde{B} \cap f^j \widetilde{B}$ *contains no more than* 1 *point* $(0 \leq i < j < k)$,
$\bigcup_{i=0}^{k-1} f^i \widetilde{B} = B$ *and* $f^k|\widetilde{B}$ *is almost conjugate to a mixing interval map (one can assume*
that if $k = n$ *then* $I = J$).

Let us formulate some assertions before proving Theorem 4.1; they easily follow from
Theorems 3.1 and 4.1 and show the connection between basic sets and sets of genus 1
and 2 introduced by Sharkovskii in [Sh3–Sh6].

Assertion 4.2[Bl4,Bl7]. *1) Limit sets of genus* 1 *are solenoidal sets which are maximal*
among all limit sets, and vice versa;

2) limit sets of genus 2 *are basic sets, and vice versa.*

Assertion 4.3[Bl4,Bl7]. *Two following properties of a set* $\omega(x)$ *are equivalent:*

1) for any y *the inclusion* $\omega(y) \supset \omega(x)$ *implies that* $h(f|\omega(y)) = 0$;

2) $\omega(x)$ *is either a solenoidal set or a set of genus* 0.

Now we pass to the proof of Theorem 4.1.

Proof of Theorem 4.1. We divide the proof by steps. The proofs of the first three ones
are left to the reader.

Step B1. $f|M$ *is surjective.*

Step B2. B *is an invariant closed set.*

Step B3. $B(M, f) = B(orb\, I, f) = \bigcup_{i=0}^{n-1} B(f^i I, f^n)$.

Example. Let $f : [0, 1] \to [0, 1]$ be a transitive map. Then $B([0, 1], f) = [0, 1]$.

Remark. One can make the Steps B1–B3 without the assumption $card\, B = \infty$.

In the rest of the proof we assume $I = M = [0, 1]$.

Step B4. *For any* $x \in B$ *there exists a side* T *of* x *such that* $P^T(x) = [0, 1]$ (we call
such T *a source side*).

Remark. In general case if I is an n-periodic interval, $M = orb\, I$, $x \in I$ and T is a side
of x in I such that $P_M^T(x) = M$ then we call T *a source side of* x *for* $F|M$.

Suppose that for some $x \in B$ there is no such side. Then $x \neq 0, 1$ (indeed, if, say,
$x = 0$ then the fact that $x \in B$ implies that $P^R(x) = [0, 1]$ which proves Step B4).

Furthermore, the assumption implies that $P^L(x) \neq [0, 1]$ and $P^R(x) \neq [0, 1]$. On the other hand $x \in B$, i.e. by the definition $P(x) = P^L(x) \cup P^R(x) = [0, 1]$ (the fact that $P(x) = P^L(x) \cup P^R(x)$ follows from Property P1 in Section 2). By Lemma 2.2 it implies that $P^L(x)$ and $P^R(x)$ are cycles of intervals. But the set B is infinite; hence there exist a point $y \in B$ and a side S such that $y \in int\, P^S(x)$ and so necessarily $P^S(x) = [0, 1]$ which is a contradiction.

Step B5. *Let U be an interval and $x \in B \cap int\, (fU)$. Then there exists $y \in (int\, U) \cap B$.*

Indeed, first let us choose the side S of x in $int\, U$ such that $P^S(x) = [0, 1]$ (it is possible by Step B4 and because $int\, (fU)$ is open). Then by Property C1.1 from Section 2 we can find a point $y \in int\, (U)$ such that $fy = x$ and, moreover, $(x, S) \in f(y, T)$ for some side T of y in U. Now by the definition of a basic set we see that $y \in int\, (U) \cap B$.

Let us denote by $\mathfrak{B}$ the set of all maximal intervals complementary to B.

Step B6. *If $U \in \mathfrak{B}$ then $(int\, fU) \cap B = \emptyset$ and either $\overline{U}$ has pairwise disjoint forward iterates or for some m, n we have $f^{m+n}\overline{U} \subset f^m \overline{U}$.*

Follows from Step B5.

Step B7. *Let $x \in B$ and T be a source side of x. Then for any $V_T(x)$ we have $(int\, V_T(x)) \cap B \neq \emptyset$ (and so B is a perfect set).*

Suppose that there exists $V_T(x)$ such that $(int\, V_T(x)) \cap B = \emptyset$. We may assume that $V_T(x) \in \mathfrak{B}$. By Step B6 and the definition of a source side it is easy to see that $f^n \overline{V_T(x)} \subset \overline{V_T(x)}$ for some n and $\bigcup_{i=0}^{n-1} f^i \overline{V_T(x)} = [0, 1]$. But B is infinite which implies that $(int\, f^i V_T(x)) \cap B \neq \emptyset$ for some i. Clearly, it contradicts Step B6.

Step B8. *Let $\phi : [0, 1] \to [0, 1]$ be the standard continuous monotone increasing surjective map such that for any interval U the set $\phi(U)$ is degenerate if and only if $(int\, U) \cap B = \emptyset$. Then ϕ almost conjugates $f|B$ to a transitive continuous map $g : [0, 1] \to [0, 1]$.*

The existence of the needed map ϕ is a well-known fact. Moreover, by Steps B6 and B7 one can easily see that there exists the continuous map g with $g \circ \phi = \phi \circ f$. Now let us take any open interval $W \subset [0, 1]$ and prove that its g-orbit is dense in $[0, 1]$. Indeed, by the construction $\phi^{-1}W$ is an open interval containing points from B, so the f-orbit of $\phi^{-1}W$ is dense in $[0, 1]$ which implies that g-orbit of W is dense in $[0, 1]$ as well. So g-orbit of any open set is dense and g is transitive.

Step B9. *$f|B$ is transitive.*

Follows from Step B8.

Statements a)–c) of Theorem 4.1 are proved. Statements d)–f) follow from the lemmas which will be proved later. Namely in Lemma 9.3 we will prove that $h(g) \geq 1/2 \cdot \ln 2$ provided $g : [0, 1] \to [0, 1]$ is transitive. Clearly, it implies statement d). In Lemma 8.3 we establish the connection between transitive and mixing maps of the interval into itself

and show that $\overline{Per\,g} = [0, 1]$ provided $g : [0, 1] \to [0, 1]$ is transitive; statements e) and f) will follow from Lemma 8.3. These remarks complete the proof of the theorem.
□

Corollary 4.4. *Let B be a basic set. Then B is either a cycle of intervals or a Cantor set.*

Proof. Follows from the fact that B is a perfect set.   □

Now we may construct the "spectral" decomposition for the sets $\overline{Per\,f}$ and $\omega(f)$. However to extend the decomposition to the set $\Omega(f)$ we need some more facts. Let I be a k-periodic interval, $M = orb\,I$. Set $E(M, f) \equiv \{x \in M$: there exists a side T of x in M such that $P_M^T(x) = M\}$ (in the case of a basic set we call such side *a source side*). By Theorem 4.1 if there exists the set $B = B(orb\,I, f)$ then $E(M, f) = B$. In particular, if $card\,E(M, f) = \infty$ then $E(M, f) = B(M, f)$. The other possibilities are described in the following

Lemma 4.5. *Let $N = [a, b]$ be an s-periodic interval, $M = orb\,N$, $E = E(M, f)$ is finite and non-empty. Then $E = orb\,x$ is a cycle of period k, $M \setminus E$ is an invariant set and one of the following possibilities holds:*
1) $k = s$, $f^s[a, x] = [x, b]$, $f^s[x, b] = [a, x]$;
2) $k = s$ and either $x = a$ or $x = b$;
3) $k = 2s$ and we may assume $x = a$, $f^s = b$.

Remark. Note that by Theorem 4.1 and Lemma 4.5 $E(M, f) \subset \overline{Per\,f}$.

Proof. Let us assume $N = M = [0, 1]$. Clearly, f is surjective. Let $\mathcal{B}$ be the family of all intervals complementary to B. As in Steps B5–B6 of the proof of Theorem 4.1 we have that

(E1) *for any $U \in \mathcal{B}$ there exists $V \in \mathcal{B}$ such that $f\overline{U} \subset \overline{V}$.*

Surjectivity of f implies that

(E2) $\mathcal{B}$ *consists of several cycles of intervals; moreover, if $U, V \in \mathcal{B}$ and $f\overline{U} \subset \overline{V}$ then $f\overline{U} = \overline{V}$.*

Let us now consider some cases.

Case 1. *There are no fixed points $a \in (0, 1)$.*

Clearly, Case 1 corresponds to the possibility 2) of the lemma.

Case 2. *There is a fixed point $a \in (0, 1) \setminus E$.*

Let $a \in U = (\alpha, \beta) \in \mathcal{B}$. First assume that $U \not\supset (0, 1)$. Then by E1 we see that $\overline{U}$ is f-invariant and by E2 we see that $\overline{[0, 1] \setminus U}$ is f-invariant. Clearly, it implies that neither α nor β have a source side which is a contradiction.
So we may assume that $U \supset (0, 1)$. First suppose that $0 \in E$ and there exists $x \in (0, 1)$ such that $fx = 0$. Then by Property C1 from Section 2 we see that $(0, 1) \cap E \neq \emptyset$ which

is a contradiction. The similar statement holds for 1. We conclude that E is invariant and $M \setminus E$ is invariant.

It remains to show that the possibility "$E = \{0, 1\}$, $f(0) = 0$, $f(1) = 1$" is excluded (the other possibilities correspond to the possibilities 2) and 3) of Lemma 4.5); note that we will prove it without making use of the fact that there is a fixed point $a \in (0, 1) \setminus E$. Suppose that $E = \{0, 1\}$, $f(0) = 0$, $f(1) = 1$. Then for any $b \in (0, 1)$ neither $[0, b]$ nor $[b, 1]$ are invariant. Choose $\eta < 1$ such that $|x - y| \le 1 - \eta$ implies that $|fx - fy| \le \eta$ for any x, y.

Let us show that if $[c, d] \ne [0, 1]$ is invariant then $d - c \le \eta$. Indeed, otherwise $[0, d]$ and $[c, 1]$ are invariant which is a contradiction. Thus if $J = [c, d]$ is a maximal by inclusion invariant proper subinterval containing the fixed point a then $\lambda(J) \le \eta$. Suppose that $c \ne 0$. Then by the maximality of J for any $\gamma \in [0, c)$ we get $[0, c] \supset \bigcup_{i \ge 0} f^i[\gamma, c]$ and hence $[0, 1] = \overline{\bigcup_{i \ge 0} f^i[\gamma, c]}$ which contradicts the fact that $c \notin E$.

Case 3. *There is a fixed point $a \in (0, 1) \cap E$ and there is no fixed point in $(0, 1) \setminus E$.*

Let $U = (c, a)$ and $V = (a, d)$ be the components of $\mathcal{B}$. At least one of them is not invariant (because of the fact that $a \in E$). By E1–E2 we have $f[c, a] = [a, d]$ and $f[a, d] = [c, a]$; so $c = 0$ and $d = 1$. But by what has been proved in the end of consideration of Case 2 we have $0 \notin E([0, a], f^2)$ and $1 \notin E([a, 1], f^2)$. Hence $c, d \notin E$, i.e. $E = \{a\}$. Now it is easy to see that $M \setminus E$ is invariant which completes the proof. $\square$

Now let us describe the properties of Ω-basic sets and the set $\Omega(f) \setminus \omega(f)$ (more detailed investigation of the properties of this set one can find in Section 7). To this end we will need the following theorem of Coven and Nitecki.

Theorem CN [CN]. *Let $f : [0, 1] \to [0, 1]$ be an arbitrary continuous map of the interval $[0, 1]$ into itself. Then the following statements hold:*
1) $\Omega(f) = \Omega(f^n)$ for any odd n;
2) if x has an infinite orbit and $x \in \Omega(f)$ then $x \in \Omega(f^n)$ $(\forall n)$;
3) if $x \in \Omega(f)$ then $x \in \overline{\bigcup_{n > 0} f^{-n} x}$;
4) if $0 \in \Omega(f)$ then $0 \in \overline{Per f}$ and if $1 \in \Omega(f)$ then $1 \in \overline{Per f}$.

We will also need Theorem Sh2 which was formulated in Subsection 1.1 of Introduction.

Lemma 4.6. *Let $x \in \Omega(f) \setminus \omega(f)$. Then there exist a number m and an m-periodic interval I such that the following statements are true:*
1) $x \in \Omega(f^m)$;
2) x is one of the endpoints of I;
3) if x does not belong to a solenoidal set then the following additional facts hold: a) $x \in B'(\text{orb } I, f)$; b) $f^k x \in B(\text{orb } I, f)$ provided $f^k x \in \text{int}(\text{orb } I)$; c) $f^{2m} x \in B(\text{orb } I, f)$.

Proof. First of all note that since by Theorem Sh2 $\omega(f) \subset \overline{Per f}$ we have $x \notin \overline{Per f}$. By Theorem CN.4) and Theorem Sh2 we have $x \ne 0, 1$. By Theorem CN.3) we may assume that there exist sequences $n_i \nearrow \infty$ and $x_i \nearrow x$ such that $f^{n_i} x_i = x$ $(\forall i)$. Finally by Theorem Sh2 we may assume that there exists $\eta > 0$ such that the interval $(x - \eta, x)$ has disjoint from itself forward iterates and the same is true for the interval $(x, x + \eta)$.

Fix j such that $x_j \in (x - \eta, x)$ and consider the set $\overline{\bigcup_{i \geq n_j} f^i[x_j, x]}$. By Lemma 2.1 there exists a weakly periodic interval $J = [x, z]$ of period u such that $\overline{\bigcup_{i \geq n_j} f^i[x_j, x]} = orb J$ and $orb J \cap [x_j, x) = \emptyset$. Moreover, $\bigcap_{r \geq 0} f^{ru} J = N$ is a u-periodic interval such that $x \in N$ is its endpoint. In other words, we have proved the existence of a periodic interval having x as its endpoint.

Remark that $f^k|[x - \delta, x]$ is not degenerate for any $\delta > 0$ and any positive integer k (otherwise $x \in Per f$). Moreover, *for any positive integer k and any side T of $f^k x$ such that $T \in f^k(L, x)$ we have $x \in P^T(f^k x)$ and if x does not belong to a solenoidal set then $P^T(f^k x)$ is a cycle of intervals.* Now if x belongs to a solenoidal set then $orb x$ is infinite and by Theorem CN.2) x belongs to $\Omega(f^n)$ for any n. So in case when x belongs to a solenoidal set we are done and it remains to consider the case when x does not belong to a solenoidal set.

Note that if $M = [x, \zeta]$ is a periodic interval then $x \notin E(orb M, f)$. Indeed, by Lemma 4.5 and Theorem 4.1 (see Remark after the formulation of Lemma 4.5) $E(M, f) \subset \overline{Per f}$ and at the same time $x \notin \overline{Per f}$ so $x \in E(M, f)$ is impossible. Let us assume that $I = [x, y]$ is the minimal by inclusion periodic interval among all periodic intervals having x as an endpoint. Let I have a period m. Let us consider two possibilities.

A) *There exists a positive integer k and a side T of $f^k x$ in $f^k I$ such that $(f^k x, T) \in f^k(x, L)$* (for example this holds provided that $f^k x \in int (f^k I)$).

Choose the minimal integer k among those existing by the supposition and prove that $f^k x \in E(orb I, f)$ and $E(orb I, f) = B(orb I, f) = B$ is infinite. Indeed, by the minimality of the interval I for any semi-neighborhood $V_T(f^k x)$ we easily have that $\overline{orb V_T(f^k x)} = orb I$ and so $f^k x \in E(orb I, f)$. Now we see that $E(orb I, f)$ is not an f^{-1}-invariant set; so by Lemma 4.5 the set $E(orb I, f) = B(orb I, f) = B$ is infinite. Hence $f^k x \in B$ and by the choice of k we see that $f^v x \notin int (orb I)$ for any $0 \leq v < k$. It proves that $x \in B'(orb I, f)$; moreover, we have also proved statement 3b) of Lemma 4.6.

In the preceding paragraph we have shown that $\overline{orb V_T(f^k x)} = orb I$ where T is a side of $f^k x$ in $f^k I$ such that $(f^k x, T) \in f^k(x, L)$; clearly, it implies that $x \in \Omega(f^m)$. Furthermore, if $f^m x \in int I$ or $f^{2m} x \in int I$ then $f^{2m} x \in B$. Otherwise we may assume that $f^m x = f^{2m} x = y$; now the fact that $f^k x \in B$ and the choice of k easily imply that $y = f^{2m} x \in B$ which completes the consideration of the possibility A).

B) *There are no positive integers k and side T of $f^k x$ in $f^k I$ such that $T \in f^k(x, L)$.*

Clearly, we see that $f^m x = f^{2m} x = y$ and $f^{km}(x, L) = (y, R)$ for any $k \geq 1$. Let us consider the set $P^R(y)$. By Lemma 2.2 $P^R(y) = orb K \ni y$ is a cycle of intervals; we may assume that $y \in K$. Clearly, the fact that $f^m(x, L) = (y, R)$ implies that $x \in orb K$ and $[x - \eta, x) \cap orb K = \emptyset$, so x is an endpoint of one of the intervals of $orb K$. Now by the minimality of I we see that x is an endpoint of K and $I \subset K$. Moreover, it is easy to see that $I \neq K$ (otherwise the possibility B) is excluded) which implies that $y \in int K$. At the same time $y \in E(orb K, f)$ by the definition. Repeating now the arguments from the previously considered possibility A) we obtain the conclusion. $\square$

5. The Decomposition

The aim of this section is to prove the Decomposition Theorem. First let us describe intersections between basic sets, solenoidal sets and sets of genus 0.

Lemma 5.1. *1) Let* $B_1 = B(orb\,I_1, f)$ *and* $B_2 = B(orb\,I_2, f)$ *be basic sets,* B_1' *and* B_2' *be the corresponding* Ω-*basic sets. Let* $B_1 \neq B_2$ *and* $B_1 \cap B_2 \neq \emptyset$. *Finally let* A *be the union of endpoints of intervals from* $orb\,I_1$ *and endpoints of intervals from* $orb\,I_2$. *Then* $B_1 \cap B_2 \subset B_1' \cap B_2' \subset A$ *and so* $B_1 \cap B_2$ *and* $B_1' \cap B_2'$ *are finite. Moreover, if* $x \in B_1' \cap B_2'$ *then* x *is not a limit point for both* B_1 *and* B_2 *from the same side.*

2) Intersection of any three Ω-*basic sets is empty and intersection of any two basic sets is finite.*

Proof. 1) Obviously it is enough to consider the case when $x \in B_1 \cap B_2$. It is easy to see that there is no side T of x such that T is a source side for both $f|orb\,I_1$ and $f|orb\,I_2$. For the definiteness let L be the only source side of x for $f|orb\,I_1$ and R be the only source side of x for $f|orb\,I_2$. Let us suppose that $x \in int\,(orb\,I_2)$ and prove that x is an endpoint of one of the intervals from $orb\,I_1$. Indeed, otherwise for open U such that $int\,(orb\,I_2) \cap int\,(orb\,I_1) \supset U \ni x$ we have $\overline{orb\,U} = orb\,I_1 = orb\,I_2$ which is a contradiction.

2) Follows from 1).   □

Example. Suppose that $g : [0, 1] \to [0, 1]$ has the following properties:
1) $g[0, 1/2] = [0, 1/2]$, $g|[0, 1/2]$ is transitive;
2) $g[1/2, 1] = [1/2, 1]$, $g|[1/2, 1]$ is transitive.
Then $B_1 = [0, 1/2]$ and $B_2 = [1/2, 1]$ are basic sets and $B_1 \cap B_2 = \{1/2\}$.

Lemma 5.2. *The family of all basic sets of* f *is at most countable.*

Proof. First consider basic sets B with non-empty interiors. Properties of basic sets easily imply that these interiors are pairwise disjoint so the family of such sets is at most countable.

Now let us consider a basic set $B = B(M, f)$ with an empty interior; then by Corollary 4.4 B is a Cantor set. We will show that there exists an interval $W \equiv W(B)$ in M complementary to B and such that its forward iterates are disjoint from itself and its endpoints belong to B and do not coincide with the endpoints of intervals from M. Indeed, denote by $\mathcal{B}$ the family of all complementary to B in M intervals; by Theorem 4.1 they are mapped one into another by the map f. Choose two small intervals $I \in \mathcal{B}$ and $J \in \mathcal{B}$ belonging to the same interval $K \in M$. If one of them is not periodic then it has the required properties. Otherwise we may suppose that $f^N I \subset I$, $f^N J \subset J$ for some N; moreover, denoting by L the interval lying between I and J we may assume that L is non-degenerate and there are no intervals from $orb\,I$ or $orb\,J$ in L. If for some n we have $f^n L \cap I \neq \emptyset$ then one may take as the required interval the subinterval of L which is complementary to B in M and is mapped by f^n in I. The same argument shows that one can find the required interval if $f^m L \cap J \neq \emptyset$. On the other hand if for any i we have $f^i L \cap (I \cup J) = \emptyset$ then we get to the contradiction with the fact that by the definition of a basic set $\overline{orb\,L} = M \supset K \supset (I \cup J)$.

Now suppose that there are two basic sets $B_1 \neq B_2$; then it is easy to see that $W(B_1) \cap W(B_2) = \emptyset$. Indeed, $\overline{W(B_1)}$ and $\overline{W(B_2)}$ have no common endpoints (otherwise by Lemma 5.1 these points are endpoints of intervals from generating B_1 and B_2 cycles of intervals which contradicts the choice of $W(B_1)$ and $W(B_2)$). On the other hand by of the choice of $W(B_2)$ no endpoints of $\overline{W(B_1)}$ can belong to $W(B_2)$ because the endpoints of $\overline{W(B_1)}$ are non-wandering. Similarly no endpoints of $\overline{W(B_2)}$ belong to $W(B_1)$. Hence $W(B_1) \cap W(B_2) = \emptyset$ which implies that the family of intervals $W(B)$ and the family of all basic sets are at most countable.   □

Lemma 5.3 *1) Let $I_0 \supset I_1 \supset \ldots$ be generating intervals and $Q = \bigcap_{j \geq 0} orb\, I_j$. Then $Q \cap B = \emptyset$ for any basic set B and if $J_0 \supset J_1 \supset \ldots$ are generating intervals and $Z = \bigcap_{i \geq 0} orb\, J_i$ then either $Z \cap Q = \emptyset$ or $Z = Q$.*
2) There is at most countable family of those solenoidal sets $Q = \bigcap_{j \geq 0} orb\, I_j$ which have non-empty interiors.

Proof. The proof easily follows from the properties of solenoidal sets (Theorem 3.1) and is left to the reader.   □

Now we can prove the Decomposition Theorem. Recall that by X_f we denote the union of all limit sets of genus 0 of a map f.

Theorem 5.4 (Decomposition Theorem)[B14,B17]. *Let $f : [0, 1] \to [0, 1]$ be a continuous map. Then there exist an at most countable family of pairs of basic and Ω-basic sets $\{B_i \subset B_i'\}$ and a family of collections of solenoidal sets of corresponding types $\{S^{(\alpha)} \subset S_p^{(\alpha)} \subset S_\omega^{(\alpha)} \subset S_\Omega^{(\alpha)} \subset Q^{(\alpha)}\}_{\alpha \in A}$ with the following properties:*
1) $\Omega(f) = X_f \cup (\bigcup_\alpha S_\Omega^{(\alpha)}) \cup (\bigcup_i B_i')$;
2) $\omega(f) = X_f \cup (\bigcup_\alpha S_\omega^{(\alpha)}) \cup (\bigcup_i B_i)$;
3) $\overline{Per\, f} = X_f \cup (\bigcup_\alpha S_p^{(\alpha)}) \cup (\bigcup_i B_i)$;
4) the set $S_\Omega^{(\alpha)} \setminus S^{(\alpha)}$ is at most countable set of isolated points, the set $\{\alpha : int\, Q^{(\alpha)} \neq \emptyset\}$ is at most countable and $S^{(\alpha)} = Q^{(\alpha)}$ for all other $\alpha \in A$;
5) intersections in this decomposition are possible only between different basic or Ω-basic sets, each three of them have an empty intersection and the intersection of two basic sets or two Ω-basic sets is finite.

Remark. Note that in statement 5) of the Decomposition Theorem we do not take into account intersections between a basic set and an Ω-basic set with the same subscript or between different solenoidal sets with the same superscripts.

Proof of the Decomposition Theorem. We start with statement 2). Let us consider some cases assuming that $x \in \omega(f)$. If $x \in X_f$ then we have nothing to prove. If $x \in Q$ for some solenoidal set Q then by Theorem 3.1 $x \in S_\omega^{(\alpha)}$ for the corresponding solenoidal set $S_\omega^{(\alpha)}$. Thus we may assume that $x \notin X_f \cup (\bigcup_\alpha Q_\omega^{(\alpha)})$. Hence there exists $\omega(z) \ni x$ such that $\omega(z)$ is neither a cycle nor a solenoidal set. Clearly, we may assume that $\omega(z)$ is infinite.

Let us construct a special cycle of intervals $orb\, I$ such that $x \in B(orb\, I, f)$. Recall that we say that a point y is *a limit point of orb ξ from the side T* or that *a side T is a limit side of $y \in \omega(\xi)$* if for any open semi-neighborhood $W_T(y)$ we have $W_T(y) \cap orb\, \xi \neq \emptyset$. If T is a limit side of $x \in \omega(z)$ then by Property P4 $P^T(x) \supset \omega(z)$ and hence $P^T(x) = orb\, I$ is a cycle of intervals. Moreover, the fact that $\omega(z) \subset P^T(x)$ is infinite implies that if $\zeta \in \omega(z)$ and N is a limit side of $\zeta \in \omega(z)$ then N is a side of ζ in $P^T(x)$. Thus $P^N(\zeta) \subset P^T(x)$; the converse is also true and thus $P^N(\zeta) = P^T(x) = orb\, I$ for any $\zeta \in \omega(z)$ and any limit side N of ζ. By the definition we have $\omega(z) \subset E(orb\, I, f) = B(orb\, I, f)$ which proves statement 2).

It remains to note that now statement 1) follows from Lemma 4.6, statement 3) follows from Theorem 3.1 and Theorem 4.1, statement 4) follows from Theorem 3.1 and Corollary 3.2 and statement 5) follows from Lemma 5.1 and Lemma 5.3. Moreover, the family of all basic sets is at most countable by Lemma 5.2. It completes the proof. $\square$

Corollary 5.5. *For an arbitrary $x \in [0, 1]$ one of the following possibilities holds:*
1) $\omega(x)$ is a set of genus 0;
2) $\omega(x)$ is a solenoidal set;
3) $\omega(x) \subset orb\, I$ where $orb\, I$ is cycle of intervals and $f|orb\, I$ is transitive;
4) $\omega(x) \subset B$ for some basic set B, B is a Cantor set and if x does not belong to a wandering interval then $\omega(x)$ is a cycle or $f^n x \in B$ for some n.

Proof. The proof is left to the reader. $\square$

6. Limit Behavior for Maps Without Wandering Intervals

In this section we describe topologically generic limit sets for maps without wandering intervals. We will need the following notions: $Z_f \equiv \{x : \omega(x) \text{ is a cycle }\}$, $Y_f \equiv int\, Z_f$.

Lemma 6.1. *A map $f : [0, 1] \to [0, 1]$ has no wandering intervals if and only if the set Z_f is dense.*

Proof. By the definition if a map f has a wandering interval J then Z_f is not dense because $int\, J \cap Z_f = \emptyset$. Now suppose that f has no wandering intervals and at the same time there is an interval I such that $I \cap Z_f = \emptyset$ (and so Z_f is not dense). Let us show that I is a wandering interval. Suppose that there exist n and m such that $f^n I \cap f^{n+m} I \neq \emptyset$. Hence the set $\bigcup_{i=0}^{\infty} f^{n+im} I = K$ is an interval, $f^m K \subset K$ and on the other hand K contains no cycle of f. It is easy to see now that all points from $int\, K$ tend under iterations of f^m to one of the endpoints of $\overline{K}$ which is a periodic point of f. In other words all points from $int\, K$ belong to Z_f which is a contradiction. So I has pairwise disjoint forward iterates. But $I \cap Z_f = \emptyset$ and so I is a wandering interval which is a contradiction. This completes the proof. $\square$

Remark that all solenoidal sets of f are in fact solenoids provided f has no wandering intervals.

Theorem 6.2[Bl8]. *Let f have no wandering intervals. Then there is a residual set $G \subset [0, 1]$ such that for any $x \in G$ one of three possibilities holds:*
1) $\omega(x)$ is a cycle;
2) $\omega(x)$ is a solenoid;
3) $\omega(x) = orb\, I$ is a cycle of intervals.

Proof. Let us investigate the set $\Gamma_f = [0, 1] \setminus Y_f$. It is easy to see that Γ_f has the following properties:

NW1. Γ_f is closed and invariant;

NW2. $f|K$ is non-degenerate for any interval $K \subset \Gamma_f$;

NW3. for any non-degenerate component I of Γ_f there exist a non-degenerate component J of Γ and integers m, n such that J is a weakly m-periodic interval and $f^n I \subset J$.

Clearly, property NW2 easily implies the following property:

NW4. if there are two intervals $L, M, fL \subset M$ and, moreover, W is a residual subset of M then $f^{-1}W \cap L$ is a residual subset of L.

Thus it remains to show that if an interval J is a weakly periodic component of Γ_f then Theorem 6.2 holds for $f|orb\, J$. We may assume that $J = [0, 1]$. Then $Y_f = \emptyset$ and $f|K$ is non-degenerate for any open interval K. Let B be a nowhere dense basic set. Then $f^{-n}B$ is nowhere dense for any n. On the other hand by Lemma 5.2 the family of all basic sets is at most countable. Let $D_f = \{x :$ there is no nowhere dense basic set B such that $f^l x \in B$ for some positive integer l. Clearly, it follows from what we have shown that D_f is residual in $[0, 1]$ and by Corollary 4.5 for $x \in D_f$ one of the following three possibilities holds:

i) $\omega(x)$ is a cycle;

ii) $\omega(x)$ is a solenoid;

iii) there is a cycle of intervals $orb\, I$ such that $f|orb\, I$ is transitive and $\omega(x) \subset orb\, I$.

Denote by $\mathcal{T}$ the family of all cycles of intervals $orb\, I$ such that $f|orb\, I$ is transitive. Suppose that there are chosen residual invariant subsets $\Pi_{orb\, I}$ of any cycle of intervals $orb\, I \in \mathcal{T}$. Now instead of condition iii) let us consider the following condition:

iii*) there is a cycle of intervals $orb\, I \in \mathcal{T}$ such that $orb\, x$ eventually enters the set $\Pi_{orb\, I}$.

Then it is easy to show that the set G_Π of all the points for which one of the conditions i), ii) and iii*) is fulfilled is a residual subset of $[0, 1]$. Indeed, since D_f is residual in $[0, 1]$ we may assume that $D_f = \bigcap_{i=0}^\infty H_i$ where H_i is an open dense in $[0, 1]$ set for any i. Consider the set $R = \{x : orb\, x$ enters an interior of some cycle of intervals $orb\, I \in \mathcal{T}\}$. Then R is an open subset of D_f. Now set $T_i \equiv int\,(H_i \setminus R)$ and replace every H_i by $H_i' = R \cup T_i$. Then T_i is an open set, $T_i \cap R = \emptyset$ for any i and H_i' is dense in $[0, 1]$ so that $D_f' = \cap H_i'$ is a residual in $[0, 1]$ set.

At the same time by the choice of sets $\Pi_{orb\, I}$ and by the property NW4 we may conclude that preimages of points from the set $\bigcup_{orb\, I \in \mathcal{T}} \Pi_{orb\, I}$ form a residual subset of R. So by what we have proved in the previous paragraph it implies that the set of the points for which one of the conditions i), ii), and iii*) is fulfilled is a residual subset of $[0, 1]$. To prove Theorem 6.2 it is enough to observe now that one can choose as $\Pi_{orb\, I}$ the set of all points in $orb\, I$ with dense in $orb\, I$ orbit. However, in Section 10 we will show that choosing sets $\Pi_{orb\, I}$ in a different way one can further specify the limit behavior of generic points whose orbits are dense in cycles of intervals.   □

7. Topological Properties of the Sets $Per\, f$, $\omega(f)$ and $\Omega(f)$

In this section we are going mostly to investigate the properties of the set $\Omega(f) \setminus \overline{Per\, f}$. Set $A(x) \equiv (\bigcup_{n \geq 0} f^{-n}x) \cap \Omega(f)$.

Lemma 7.1.

1) If $x \notin \Omega(f)$ then $A(x) = \emptyset$.

2) *Let* $x \in \Omega(f)$, $I \ni x$ *be a weakly periodic interval and* $f^n x \in int\,(orb\,I)$ *for some* n. *Then* $A(x) \subset orb\,I$.

3) *Let* $x \in \Omega(f) \setminus \overline{Per\,f}$, I *be periodic interval such that* x *is an endpoint of* I. *Then* $A(x) \cap int\,(orb\,I) = \emptyset$.

Proof. The proof is left to the reader; note only that statement 3) follows from Theorem CN.4) (Theorem CN was formulated in Section 4).   □

Corollary 7.2. *Let* $x \in \Omega(f) \setminus \overline{Per\,f}$, I *be periodic interval such that* x *is an endpoint of* I *and* $f^n x \in int\,(orb\,I)$ *for some* n. *Then* $A(x) \subset \partial(orb\,I)$.

Proof. Follows immediately from Lemma 7.1, statements 2) and 3).   □

Lemma 7.3. *If* $x \in \Omega(f) \setminus \omega(f)$ *then there exists a periodic interval* J *such that* x *is an endpoint of* J *and* $A(x) \subset \partial(orb\,J)$; *if* x *does not belong to a solenoidal set then we may also assume that* $x \in B'(orb\,J, f)$.

Proof. By Lemma 4.6 we may assume that there exists a periodic interval $I = [x, y]$ having x as one of its endpoints and if, moreover, x does not belong to a solenoidal set then $x \in B'(orb\,I, f)$. Let us consider two possibilities.

1) *There exists* k *such that* $f^k x \in int\,(orb\,I)$.

Then by Corollary 7.2 $A(x) \subset \partial(orb\,I)$ which proves Lemma 7.3 in this case.

2) *For any* k *we have* $f^k x \notin int\,(orb\,I)$.

Then by Lemma 4.6 we may assume that $x \in B'(orb\,I, f)$, $I = [x, y]$ has a period m and $f^m x = y = f^m y$. By Lemma 7.1.3) $A(x) \cap int\,(orb\,I) = \emptyset$. Suppose $A(x) \not\subset \partial(orb\,I)$ and show that there exists a periodic interval J such that $x \in B'(orb\,J, f)$ and $A(x) \subset \partial(orb\,J)$.

Indeed, if $A(x) \not\subset \partial(orb\,I)$ then there exists $z \in A(x) \setminus orb\,I$. By Lemma 4.6 $z \in B'(orb\,J, f)$ for some n-periodic interval J. We may assume $x \in J$; then $f^{nm} x = y \in f^{nm} J = J$ and thus $I = [x, y] \subset J$, $I \neq J$. Clearly, $x \in B'(orb\,J, f) \setminus B(orb\,J, f)$ because $z \in B'(orb\,J, f)$ is a preimage of x under the corresponding iteration of f and at the same time $x \notin \omega(f)$; so x is an endpoint of $J = [x, \zeta]$. Hence $f^m x = y \in int\,J$ and as in case 1) we see that by Corollary 7.2 $A(x) \subset \partial(orb\,J)$. This completes the proof.
□

To formulate the next corollary connected with the results from [Y] and [N] we need some definitions. Let c be a local extremum of f. It is said to be *an o-extremum* in the following cases:

1) c is an endpoint of an interval $[c, b]$ such that i) $f|[c, b]$ is degenerate, ii) f is not degenerate in any neighborhood of each c and b, iii) c and b are either both local minima or both local maxima;

2) there is no open interval (c, b) such that $f|(c, b)$ is degenerate (note that neither in case 1) nor in case 2) we assume that $c < b$).

In [Y] the following theorem was proved.

Theorem Y. *Let* $f : [0, 1] \to [0, 1]$ *be a pm-map and* $x \in \Omega(f) \setminus \overline{Per\,f}$. *Then there exists* $n > 0$ *and a turning point* c *such that* $f^n c = x$.

On the other hand [N1] contains the following

Theorem N. *If f is a pm-map then $\overline{Per\,f} = \omega(f)$.*

Remark. Note, that Theorem N may be also easily deduced from Lemma PM2 (see Subsection 1.11 of Introduction) and the Decomposition Theorem.

So the following Corollary 7.4 generalizes Theorem Y.

Corollary 7.4. *If $x \in \Omega(f) \setminus \omega(f)$ then there exist an o-extremum c and $n > 0$ such that $f^n c = x$.*

Proof. Take the interval J existing for the point x by Lemma 7.3. Then $f|orb\,J$ is a surjective map and at the same time x is not a periodic point. Hence we may choose the largest n such that there exists an endpoint y of an interval from $orb\,J$ with the following properties: $f^n y = x$ and $y, fy, \ldots, f^n y$ are endpoints of intervals from $orb\,J$. Then by the choice of y there exists a point z and an interval $[a, b]$ from $orb\,J$ such that $z \in (a, b)$, $fz = y$, $fa \neq y$, $fb \neq y$. Now it is easy to see that we may assume z to be an o-extremum. $\square$

Remark. Corollary 7.4 was also proved in the recent paper [Li] (see Theorem 2 there). Theorem 7.5 describes another sort of connection between the sets $\omega(f)$ and $\Omega(f)$.

Theorem 7.5[Bl1],[Bl7]. $\omega(f) = \bigcap_{n \geq 0} f^n \Omega(f)$.

Remark. Another proof of Theorem 7.5 may be found in [BCo].

Proof. By the properties of limit sets for any z we have $f\omega(z) = \omega(z)$. It implies that $\omega(f) = f\omega(f) \subset \bigcap_{n \geq 0} f^n \Omega(f)$. At the same time by Lemma 7.3 the set $A(x)$ is finite for any $x \in \Omega(f) \setminus \omega(f)$. So by the definition of $A(x)$ we see that if $x \in \Omega(f) \setminus \omega(f)$ then $x \notin \bigcap_{n \geq 0} f^n \Omega(f)$ which implies the conclusion. $\square$

Finally in the following Theorem 7.6 we study the structure of the set $\Omega(f) \setminus \overline{Per\,f}$.

Theorem 7.6. *Let $U = (a, b)$ be an interval complementary to $\overline{Per\,f}$. Then up to the orientation one of the following four possibilities holds.*
1) $\Omega(f) \cap U = \emptyset$.
*2) $\Omega(f) \cap U = \{x_1 < x_2 < \ldots < x_n\}$ is a finite set, $card(orb\,x_1) < \infty, \ldots,$
$card(orb\,x_{n-1}) < \infty$, $(\bigcup_{i=1}^{n-1} x_i) \cap \omega(f) = \emptyset$ and there exist periodic intervals $J_i = [x_i, y_i]$ such that $x_i \in B'(orb\,J_i, f)$ for $1 \leq i \leq n - 1$ and $J_i \supset J_{i+1}$ for $1 \leq i \leq n - 2$. Moreover, for x_n there exist two possibilities: a) x_n belongs to a solenoidal set; b) x_n belongs to an Ω-basic set $B'(orb\,J_n, f)$ where $J_n = [x_n, y_n] \subset J_{n-1}$.*
3) $\Omega(f) \cap U = (\bigcup_{i=1}^{\infty} x_i) \cup x$, $x_1 < x_2 < \ldots$, $x = \lim x_i$, and there exist generating intervals $J_i = [x_i, y_i]$ such that:
a) $x_i \in B'(orb\,J_i, f)$, $card(orb\,x_i) < \infty$ $(\forall i)$ and $(\bigcup_{i=1}^{\infty} x_i) \cap \omega(f) = \emptyset$;

b) $x \in S_\omega(\{orb\, J_i\}_{i=1}^\infty) = \omega(f) \cap (\bigcap_{i=1}^\infty orb\, J_i)$.
4) $\Omega(f) \cap U = \bigcup_{i=1}^\infty x_i$, $x_1 < x_2 < \ldots$, $\lim x_i = b$, $card(orb\, x_i) < \infty$ $(\forall i)$, $(\bigcup_{i=1}^\infty x_i) \cap \omega(f) = \emptyset$ *and there exist periodic intervals* $J_i = [x_i, y_i]$ *such that* $x_i \in B'(orb\, J_i, f), J_i \supset J_{i+1}$ $(\forall i)$ *and* $\bigcap_{i=1}^\infty J_i = \{b\}$. *Moreover, either periods of* J_i *tend to infinity,* $\{J_i\}$ *are generating intervals and b belongs to the corresponding solenoidal set, or periods of* J_i *do not tend to infinity and b is a periodic point. In any case* $card\{\omega(f) \cap U\} \le 1$.

Remark. Some results related to Theorem 7.6 are proved in [BCo] (see the corresponding remark in Subsection 1.7).

Proof. We divide the proof into steps.

Step A. *$card\{\omega(f) \cap U\} \le 1$ and if $x \in \omega(f) \cap U$ then x belongs to a solenoidal set.*

If $x \in \omega(f) \cap U$ then by the Decomposition Theorem there exists a solenoidal set $Q \ni x$; so if $x \in \omega(f) \cap U$ then $card(orb\, x) = \infty$. Now it follows from Theorem 3.1 and Corollary 3.2 that if J is the component of Q containing x then up to the orientation we may assume that $J = [x, b]$ and, moreover, J is a wandering interval.

Suppose that there exists $y \in \omega(f) \cap U$, $y \ne x$. Then the fact that J is a wandering interval implies that $a < y < x$. Moreover, similarly to what we have seen in the previous paragraph it is easy to see now that there exists a solenoidal set $\tilde{Q}$ such that $K = [a, y]$ is its component. By the properties of solenoidal sets there exist intervals $M = [\tilde{a}, \tilde{y}]$ and $N = [\tilde{x}, \tilde{b}]$ such that $y < \tilde{y} < \tilde{x} < x$ and $f^n M = M, f^n N = N$ for some n. Clearly, it implies that $f^n[\tilde{y}, \tilde{x}] \supset [\tilde{y}, \tilde{x}]$ and so there exists a point $z \in [\tilde{y}, \tilde{x}]$ such that $f^n z = z$ which is a contradiction. So $card\{\omega(f) \cap U\} \le 1$.

Step B. *$\Omega(f) \cap U$ has in U at most one limit point, which necessarily belongs to some solenoidal set S_ω.*

By Theorem Sh2 limit points of $\Omega(f)$ belong to $\omega(f)$. Thus Step B follows from Step A.

Let J be a periodic interval and suppose that one of the endpoints of J belongs to U. Then the endpoint of U belonging to J is uniquely determined; we denote this endpoint of U by $e = e(J)$.

Step C. *The point e is uniquely defined and does not depend on J.*

Clearly, it is sufficient to show that there is no pair of periodic intervals $I = (a', y)$ and $J = (x, b')$ where $x, y \in U, a' < a, b' > b$. To prove this fact observe that if these intervals existed then the interval K with endpoints x, y would have the property $f^n K \supset K$ for some n which is impossible.

In the rest of the proof we assume that $e = b$.

Step D. *If $z \in \Omega(f) \cap U$ and orb z is infinite then $[z, b]$ has pairwise disjoint forward iterates.*

If z belongs to a solenoidal set then Step D is trivial by the properties of solenoidal sets (see Theorem 3.1 and Corollary 3.2). So we may assume that there exists a periodic interval $J = [z, c]$ such that $z \in B'(orb\, J, f) = B'$. Hence there exists an interval $[z, d]$, $d \ge b$, which is a complementary to $B(orb\, J, f) = B$ in $orb\, J$ interval. If $[z, d]$

does not have pairwise disjoint iterates then there exists a weakly periodic interval K which is a complementary to B in $orb\,J$ interval and, moreover, $f^m z \in int(orb\,K)$ for large m. At the same time by the definition of an Ω-basic set $f^m z \in B$ for a large m. Clearly, this is a contradiction.

Step E. *If $x, y \in \Omega(f) \cap U$, $x < y$, then $card(orb\,x) < \infty$ and x belongs to an Ω-basic set $B'(orb\,[x, d], f)$ for some periodic interval $[x, d]$.*

If $card\,(orb\,x) = \infty$ then $[x, b]$ has pairwise disjoint forward iterates which is impossible because $y \in \Omega(f) \cap (x, b)$. Hence $card\,(orb\,x) < \infty$; by the Decomposition Theorem it implies that x belongs to an Ω-basic set $B'(orb\,[x, d], f)$ for some periodic interval $[x, d]$.

Step F. *Let $x, y \in \Omega(f) \cap U$, $x < y$, $x \in B'(orb\,J, f)$ and $y \in B'(orb\,I, f)$, where $J = [x, c]$ and $I = [y, d]$. Then $d < c$.*

Suppose that $c \leq d$. Then by Step C those iterations of I which do not coincide with I have empty intersections with U. Thus by the definition of a basic set we have $B(orb\,J, f) \cap (y, c) = \emptyset$. Moreover, by the Decomposition Theorem $B(orb\,J, f) \subset \overline{Per\,f}$ and so $[x, y] \cap B(orb\,J, f) = \emptyset$. Hence $B(orb\,J, f) \cap J \subset \{c\}$ which contradicts the definition of a basic set.

Step G. *The point a is not a limit point of $\Omega(f) \cap U$.*

Suppose that a is a limit point of $\Omega(f) \cap U$. We may assume that $x_{-i} \searrow a$ while $i \to \infty$ and (by Step E) that $card(orb\,x_{-i}) < \infty$ $(\forall i > 0)$. By Step F we may assume also that for any $i > 0$ there exists an n_i-periodic interval $J_i = [x_{-i}, d_i]$ such that $x_{-i} \in B'(orb\,J_i, f)$ and $J_{i+1} \supset J_i$ $(\forall i > 0)$. Clearly, we may assume that $n_i = 1$ $(\forall i > 0)$. Indeed, as we have just shown $J_{i+1} \supset J_i$, so periods of J_i decrease and hence become equal to some constant; we will consider the case when this constant is 1, the arguments in the general case are similar.

By the definition and Theorem 4.1 $B(orb\,J_{i+1}, f) = B_{i+1} \subset [d_i, d_{i+1}]$ for any $i > 0$. Indeed, basic sets belong to $\overline{Per\,f}$, so $B_{i+1} \cap U = \emptyset$ and $B \subset [b, d_{i+1}]$. But by the definition of a basic set and the fact that $[x_{-i}, d_i]$ is invariant we see that there are no points of B_{i+1} in $[b, d_i)$ which implies that $B_{i+1} \subset [d_i, d_{i+1}]$.

Let us choose $i > 0$ such that for any y, z we have $|fz - fy| < |d_1 - x_{-1}|$ provided $|z - y| < d_{i+1} - d_i$; clearly, it is possible because $d_{i+1} - d_i \to 0$ while $i \to \infty$. We are going to show that the interval $[x_{-i}, d_{i+1}]$ is invariant. Indeed, let $z \in [x_{-i}, d_{i+1}]$. If in fact $z \in [x_{-i}, d_i]$ then $fz \in [x_{-i}, d_i] \subset [x_{-i}, d_{i+1}]$. If $z \in [d_i, d_{i+1}]$ then by the choice of i we see that $|fz - f\zeta| < |d_1 - x_{-1}|$ for any $\zeta \in [d_i, d_{i+1}]$. Choose any $\zeta \in B_{i+1} \subset [d_i, d_{i+1}]$; then $f\zeta \in B_{i+1} \subset [d_i, d_{i+1}]$ as well and so $|d_1 - x_{-1}| > |fz - f\zeta| > |fz - d_i|$ which implies that $fz \in [x_{-i}, d_{i+1}]$. Hence $[x_{-i}, d_{i+1}]$ is invariant which contradicts the definition of a basic set and the existence of the basic set B_{i+1}.

Recall that by Steps B and G the set $\Omega(f) \cap U$ has at most one limit point which we denote by x. By Step G $x \neq a$. Now if $x = b$ then $\Omega(f) \cap U < b$. If $x \in U$ then by Step B x belongs to some solenoidal set S_ω and the fact that $e = b$ (see Step C) implies that $[x, b]$ is a wandering interval and so all non-limit points of $\Omega(f) \cap U$ are less then x. This observation shows that the formulation of Step H is correct.

Step H. *Let $\Omega(f) \cap U \supset \{x_i\}_{i=0}^\infty$ where $\{x_i\}_{i=0}^\infty$ is the whole set of non-limit points of $\Omega(f) \cap U$; moreover, let $x_0 < x_1 < \ldots, x_n \to x$. Then there exist periodic intervals*

$J_i = [x_i, d_i]$, $J_0 \supset J_1 \supset \ldots$ *such that* $x_i \in B'(orb\, J_i, f)$ $(\forall i)$ *and* $\cap J_i = [x, b]$.
Moreover, if periods of the intervals J_i *tend to infinity then* $[x, b]$ *belongs to a solenoidal*
set and either $x = b$ *and* $\Omega(f) \cap U = \{x_i\}_{i=0}^{\infty}$ *or* $x < b$ *and* $\Omega(f) \cap U = \{x_i\}_{i=0}^{\infty} \cup \{x\}$.
On the other hand, if periods of J_i *do not tend to infinity then* $x = b$ *and so* $\cap J_i = \{b\}$.

The existence of the intervals $J_i = [x_i, d_i]$ such that $x_i \in B'(orb\, J_i, f)$ and $J_i \supset$
J_{i+1} $(\forall i)$ follows from Steps E and F. If periods of J_i tend to infinity then the required
property follows from the properties of solenoidal sets (Theorem 3.1, Corollary 3.2).
Now suppose that periods of J_i do not tend to infinity; consider the case when all J_i are
invariant (i.e. have period 1), the general case may be considered in the similar way.

We are going to show that $\cap J_i = \{b\}$. Indeed, let $\cap J_i = [b', d']$, $b' < d'$; then clearly
$\lim x_i = b' \le b$. Choose i such that for any y, z we have $|fz - fy| < d' - b'$ provided
$|z - y| < |d_i - d'|$. Now repeating all the arguments from Step G we get the same
contradiction. Indeed, for any i the set $B(orb\, J_i, f) = B_i$ has an empty intersection with
$[x_i, b')$ because $B_i \subset \overline{Per\, f}$ by Theorem 4.1 and at the same time there is no points of
$\overline{Per\, f}$ in $[x_i, b')$. On the other hand the choice of i and the fact that B_i is invariant imply
(as in Step G) that $[b', d_i]$ is an invariant interval which contradicts the definition of a
basic set and the existence of the set B_i. This contradiction shows that $b' = d' = b$
which completes Step H.

Now let us consider different cases depending on the properties of the set $\Omega(f) \cap U$.
First of all let us note that the properties of points $x \in \Omega(f) \cap U$ such that $(x, b) \cap \Omega(f) \neq \emptyset$
are fully described in Steps E and F; together with the definitions it completes the
consideration of case 2) and proves the corresponding statements from the other cases.
Furthermore, by Step B we see that $\Omega(f) \cap U$ has at most one limit point in U and if
so then Steps E–H imply case 3) and also the first part of case 4) of Theorem 7.4. The
second part of case 4) follows from Step H. This completes the proof of Theorem 7.4.
□

8. Transitive and Mixing Maps

In this section we will investigate the properties of transitive and mixing interval maps
which are closely related to the properties of maps on their basic sets as it follows from
Theorem 4.1. Let us start with the following simple

Lemma 8.1[Bl7]. *Let* $f : [0, 1] \to [0, 1]$ *be a transitive map,* $x \in (0, 1)$ *be a fixed point,*
$\eta > 0$. *Then there exists* $y \in (x, x + \eta)$ *such that* $f^2 y > y$ *or* $y \in (x - \eta, x)$ *such that*
$f^2 y < y$.

Proof. First suppose there is a point $z \in (x, x + \eta)$ such that $fz > z$. Then choose the
maximal fixed point ζ among fixed points which are smaller than z. Clearly, if we take
$y > \zeta$ close enough to ζ we will see that $f^2 y > y$. Moreover, we can similarly consider
the case when there is a point $z \in (x - \eta, x)$ such that $fz < z$. So we may assume that
for points from $(x - \eta, x + \eta)$ we have $fz < z$ if $x < z$ and $fz > z$ if $x > z$.

Now choose $\delta > 0$ such that $\delta < \eta$, $f[x, x + \delta] \subset (x - \eta, x + \eta)$. The map f is
transitive so $f[x, x + \delta] = [a, b]$ where $a < x$ and $b \ge x$. Moreover, by the transitivity
of f one can easily see that there is a point $d \in [a, x]$ such that $fd > x + \delta$ (otherwise

$[a, x + \delta]$ is an invariant interval). Take $y \in [x, x + \delta]$ such that $fy = d$; clearly, y is the required point.   □

Lemma 8.2. *Let* $f : [0, 1] \to [0, 1]$ *be a transitive map,* $\eta > 0$. *Then there exist a fixed point* $x \in (0, 1)$, *a periodic point* $y \in (0, 1)$, $y \neq x$ *with minimal period 2 and an interval* $U \subset [x - \eta, x + \eta]$ *such that* $x \in U \subset fU$.

Proof. The existence of a fixed point in $(0, 1)$ easily follows from the transitivity of f. Let us show that there exists a point y of minimal period 2. We may assume that 1 is not a periodic point of minimal period 2. Suppose that x is a fixed point and there exists $\varepsilon > 0$ such that for points from $(x - \varepsilon, x + \varepsilon)$ we have $fz < z$ if $x < z$ and $fz > z$ if $x > z$. By Lemma 8.1 there exists, say, $\zeta \in (x, x + \varepsilon)$ such that $f^2\zeta > \zeta$. Now if there are no fixed points in $(x, 1]$ then set $\xi = 1$; otherwise let ξ be the nearest to ζ fixed point which is greater than ζ. By the construction for any $\alpha \in (\zeta, \xi)$ we have $f\alpha < \alpha$; it easily implies that if a point $\beta \in (\zeta, \xi)$ is sufficiently close to ξ then $f^2\beta < \beta$. Together with $f^2\zeta > \zeta$ it shows that there is a periodic point $y \in (\zeta, \beta)$ such that $f^2y = y$; at the same time by the choice of ξ we have $fy \neq y$, so the minimal period of y is 2.

Now suppose that there is no fixed point x for which there exists $\varepsilon > 0$ such that for points from $(x - \varepsilon, x + \varepsilon)$ we have $fz < z$ if $x < z$ and $fz > z$ if $x > z$. Then clearly, there are at least two fixed points, say, a and b, and we may assume that $a < b$ and $z < fz$ for $z \in (a, b)$. Let us show that $a \in f[b, 1]$. Indeed, otherwise $I = [b, 1] \cup f[b, 1] \neq [0, 1]$ is an f-invariant interval which contradicts the transitivity. Choose the smallest $c \in [b, 1]$ such that $fc = a$; then $b < c$. It is easy to see that again by the transitivity there exists $d \in (a, c)$ such that $fd = c$. Choose the fixed point a' in such a way that the interval (a', d) does not contain fixed points. Then for z sufficiently close to a' we have $f^2z > z$ which together with the fact that $f^2d = a < d$ implies that there is a periodic point $y \in (z, d)$ of minimal period 2.

The proof of the existence of the interval $U \subset (x - \eta, x + \eta)$ with $x \in U \subset fU$ uses arguments similar to those from Lemma 8.1. Indeed, if there is a point $z \in (x, x + \delta)$ such that $fz > z$ or $z \in (x - \delta, x)$ such that $fz < z$ then it is sufficient to take $U = (x, z)$. So we may assume that for points from $(x - \eta, x + \eta)$ we have $fz < z$ if $x < z$ and $fz > z$ if $x > z$. Now take a point $y \in (x - \eta, x + \eta)$ which exists by Lemma 8.1; we may assume that $y \in (x, x + \delta)$, $f(x, x + \delta) \subset (x - \eta, x + \eta)$ and $f^2y > y$. Then it is easy to see that $U = [x, y] \cup f[x, y]$ is the required interval.   □

Lemma 8.3[Bl7]. *Let* $f : [0, 1] \to [0, 1]$ *be a transitive map. Then one of the following possibilities holds:*

1) the map f *is mixing and, moreover, for any* $\eta > 0$ *and any non-degenerate interval* U *there exists* n_0 *such that* $f^nU \supset [\eta, 1 - \eta]$ *for any* $n > n_0$;

2) the map f *is not mixing and, moreover, there exists a fixed point* $a \in (0, 1)$ *such that* $f[0, a] = [a, 1]$, $f[a, 1] = [0, a]$, $f^2|[0, a]$ *and* $f^2|[a, 1]$ *are mixing.*

In any case $\overline{\text{Per } f} = [0, 1]$.

Proof.

1) First suppose there exists a fixed point $x \in (0, 1)$ such that $x \in int\, f[0, x]$ or $x \in int\, f[x, 1]$. To be definite suppose that $x \in int\, f[0, x]$ and prove that f is mixing and has all the properties from statement 1). Clearly, we may assume that $x \in int\, f[b, x]$

for some $0 < b < x$. By Lemma 8.2 there exists a closed interval $U \subset f[0, x]$ such that $x \in U \subset fU$. Let V be any open interval. By Lemma 2.1 the set $[0, 1] \setminus \bigcup_{m \geq 0} f^m V$ is finite. On the other hand, the set $\bigcup_{n \geq 0} f^{-n} x$ is infinite. So $x \in f^k V$ for some k. Now the transitivity implies that $f^l V \supset [b, x]$ for some l and so $U \subset f^{l+1} V$. At the same time the inclusion $U \subset fU$ and the transitivity imply that for any $\varepsilon > 0$ there exists $N = N(\varepsilon)$ such that $f^n U \supset [\varepsilon, 1 - \varepsilon]$ for $n \geq N$. Thus $f^m V \supset [\varepsilon, 1 - \varepsilon]$ for $m > N + l$. It completes the consideration of the case 1).

2) Suppose there exists a fixed point $a \in (0, 1)$ such that $a \notin int\, f[0, a]$ and $a \notin int\, f[a, 1]$. By the transitivity $f[0, a] = [a, 1]$, $f[a, 1] = [0, a]$; moreover, $f^2|[0, a]$ and $f^2|[a, 1]$ are transitive and hence by the case 1) $f^2|[0, a]$ and $f^2|[a, 1]$ are mixing. The fact that $\overline{Per\, f} = [0, 1]$ easily follows from what we have proved. $\square$

In the proof of Theorem 4.1 we announced that statements e) and f) of it would follow from Lemma 8.3. Let us prove the statements now; for the sake of convenience we will recall their formulations.

e) *If $B = B(orb\, I, f)$ is a basic set then $B \subset \overline{Per\, f}$.*

Proof. Clearly, it is enough to consider the case when the period of the interval I is 1. In this case by the preceding statements of Theorem 4.1 $f|B$ is almost conjugate by a monotone map $\phi : I \to [0, 1]$ to a transitive map $g : [0, 1] \to [0, 1]$. By Lemma 8.3 $\overline{Per\, g} = [0, 1]$. Now the fact that B is perfect (statement a) of Theorem 4.1) and monotonicity of ϕ easily imply that $B \subset \overline{Per\, f}$.

f) *there exist an interval $J \subset I$, an integer $k = n$ or $k = 2n$ and a set $\widetilde{B} = \overline{int\, J \cap B}$ such that $f^k J = J$, $f^k \widetilde{B} = \widetilde{B}$, $f^i \widetilde{B} \cap f^j \widetilde{B}$ contains no more than 1 point $(0 \leq i < j < k)$, $\bigcup_{i=0}^{k-1} f^i \widetilde{B} = B$ and $f^k|\widetilde{B}$ is almost conjugate to a mixing interval map (one can assume that if $k = n$ then $I = J$).*

Proof. Again consider the case when the period of the interval I is 1 and $f|B$ is almost conjugate by a monotone map $\phi : I \to [0, 1]$ to a transitive map $g : [0, 1] \to [0, 1]$. If g is in fact mixing then set $k = n = 1$, $J = I$; clearly then all the properties from statement b) hold. If g is not mixing then by Lemma 8.3 there exist such $a \in (0, 1)$ that $g[0, a] = [a, 1]$, $g[a, 1] = [0, a]$, $g^2|[0, a]$ and $g^2|[a, 1]$ are mixing. Set $k = 2, J = \phi^{-1}[0, a]$; again it is easy to see that all the properties from statement f) hold which completes the proof.

Corollary 8.4[B17]. *If $f : [0, 1] \to [0, 1]$ is mixing then there exist a fixed point $a \in (0, 1)$ and a sequence of intervals $\{U_i\}_{i=-\infty}^{\infty}$ with the following properties:*

1) $U_i \subset U_{i+1} = fU_i$ $(\forall i)$;

2) $\cap U_i = \{a\}$;

3) *for any open V there exists $n = n(V)$ such that $f^n V \supset U_0$;*

4) $\bigcup_{i=-\infty}^{\infty} U_i \supset (0, 1)$.

Proof. Follows from Lemmas 8.2 and 8.3. $\square$

Let $A(f) \equiv A$ be the set of those from points $0, 1$ which have no preimages in $(0, 1)$.

Lemma 8.5[B17]. *If $f : [0, 1] \to [0, 1]$ is mixing then there are the following possibilities for A:*

1) $A = \emptyset$;
2) $A = \{0\}$, $f(0) = 0$;
3) $A = \{1\}$, $f(1) = 1$;
4) $A = \{0, 1\}$, $f(0) = 0$, $f(1) = 1$;
5) $A = \{0, 1\}$, $f(0) = 1$, $f(1) = 0$.

Moreover, if I is a closed interval, $I \cap A = \emptyset$, then for any open U there exists n such that $f^m U \supset I$ for $m > n$ (in particular, if $A = \emptyset$ then for any open U there exists n such that $f^n U = [0, 1]$).

Remark. Results closely related to Lemmas 8.3–8.5 were also obtained in [BM1–BM2].

Proof. The map f is surjective; thus A is f^{-1}-invariant set which together with Lemma 8.3 implies the conclusion. $\square$

Lemma 8.6[Bl7].
1) Let $A \neq \emptyset$, $a \in A$, $f(a) = a$. If f is mixing then there exists a sequence $c_n \to a$, $c_n \neq a$ of fixed points.
2) Let $A = \{0, 1\}$, $f(0) = 1$, $f(1) = 0$. If f is mixing then there exists a sequence of periodic points $\{c_n\}$ of minimal period 2 such that $c_n \to 0$, $c_n \neq 0$.

Proof. It is sufficient to consider the case $0 \in A$, $f(0) = 0$. Suppose that 0 is an isolated fixed point. Then by the transitivity $fx > x$ for some $\eta > 0$ and any $x \in (0, \eta)$. At the same time $0 \in A$ and so $0 \notin f[\eta, 1]$. Let $z = \inf f|[\eta, 1]$; by the transitivity $z < \eta$. Then because of the properties of $f|[0, \eta]$ we see that in fact $z = \inf_k f^k|[\eta, 1]$ and so $[z, 1] \subset (0, 1]$ is an invariant interval which is a contradiction. $\square$

Let us prove that a mixing map of the interval has the specification property. In fact we introduce a property which is slightly stronger than the usual specification property (we call it *the i-specification property*) and then prove that mixing maps of the interval have the i-specification property. Actually, we need this variant of the specification property to make possible the consideration of interval maps on their basic sets; they are closely related to mixing maps (see Theorem 4.1).

We will not repeat the definition of the specification property (see Section 1); instead let us introduce the notion of the i-specification property. To this end we first need the following definition. Let $z \in Per\ f$ have a period m. Moreover, let $f^m[z, z + \eta]$ lie to the left of z and $f^m[z - \eta, z]$ lie to the right of z for some $\eta > 0$. Then we say that *the map f^m at the point z (of period m) is reversing*; otherwise we say that *the map f^m at the point z (of period m) is non-reversing*.

Now let $f : I \to I$ be a continuous interval map. The map f is said to have *the i-specification property* or simply *i-specification* if for any $\varepsilon > 0$ there exists an integer $M = M(\varepsilon)$ such that for any $k > 1$, any k points $x_1, x_2, \ldots, x_k \in I$, any semi-neighborhoods $U_i \ni x_i$ with $\lambda(U_i) = \varepsilon$, any integers $a_1 \leq b_1 < a_2 \leq b_2 < \ldots < a_k \leq b_k$ with $a_i - b_{i-1} \geq M$, $2 \leq i \leq k$ and any integer p with $p \geq M + b_k - a_1$ there exists a point $x \in I$ of period p such that f^p is non-reversing at the point x and, moreover, $d(f^n x, f^n x_i) \leq \varepsilon$ for $a_i \leq n \leq b_i, 1 \leq i \leq k$ and $f^{a_i} z \in U_i$, $1 \leq i \leq k$. The additional properties which are required by the i-specification property compare to the usual specification property give us the possibility to lift some properties of mixing interval maps (which as we are going to prove have i-specification) to interval maps on basic sets.

Theorem 8.7[Bl7]. *If a map $f : [0, 1] \to [0, 1]$ is mixing then it has the i-specification property.*

Proof. We will consider some cases depending on the structure of the set $A(f)$ (see Lemma 8.5).

First we consider the case $A(f) = \emptyset$. Suppose that $\eta > 0$. Choose $M = M(\eta)$ such that for any interval U we have $f^M U = [0, 1]$ provided $\lambda(U) > \eta/2$ (which is possible by Lemma 8.5). Let us consider points $x_1, \ldots, x_n$ with semi-neighborhoods $U_i \ni x_i$ of length η and integers $a_1 \le b_1 < a_2 \le b_2 < \ldots < a_n \le b_n$, p such that $b_i - a_{i-1} \ge M$ ($2 \le i \le n$), $p \ge M + b_n - a_1$. From now on without loss of generality we will suppose that $a_1 = 0$. We have to find a periodic point z of period p such that f^p is non-reversing at z and, moreover, $|f^t z - f^t x_i| \le \eta$ for $a_i \le t \le b_i$ and $f^{a_i} z \in U_i$ ($1 \le i \le n$).

First let us find an interval W with an orbit which approximates pieces of orbits $\{f^t x_i : a_i \le t \le b_i\}_{i=1}^n$ quite well; we show that one can find W in such a way that $f^p W = [0, 1]$. Recall the following

Property C4(see Section 2). *Let U be an interval, $x \in U$ be a point, $\lambda(U) \ge \eta > 0$, $n > 0$. Then there exists an interval V such that $x \in V \subset U$, $\lambda(f^i V) \le \eta$ ($0 \le i \le n$) and $\lambda(f^j V) = \eta$ for some $j \le n$.*

By Property C4 there exists an interval V_1 such that $x_1 \in V_1 \subset U_1$, $\lambda(f^i V_1) \le \eta$ ($a_1 \le i \le b_1$) and $\lambda(f^{t_1} V_1) = \eta$ for some t_1, $0 = a_1 \le t_1 \le b_1$. Clearly, $[0, 1] = f^{a_2 - b_1}(f^{b_1} V_1) = f^{a_2 - t_1}(f^{t_1} V_1)$ since $a_2 - t_1 \ge a_2 - b_1 \ge M$. Then we can find an interval $W_1 \subset V_1$ such that $f^{a_2} W_1 = U_2$. Repeating this argument we get an interval $W = [\alpha, \beta]$ such that for any $1 \le i \le n$ and $a_i \le t \le b_i$ we have $f^t W \subset [f^t x_i - \eta, f^t x_i + \eta]$, $f^{a_i} W \subset U_i$ and for some $a_n \le l \le b_n$ we have $\lambda(f^l W) = \eta$. Since $p \ge M + b_n - a_1 = M + b_n$ we see that $f^p W = f^{p-l}(f^l W) = [0, 1]$.

It remains to show that there exists a periodic point $z \in W$ of period p such that f^p is non-reversing at z. Suppose that f^p is reversing at all p-periodic points in W. Then it is easy to see that there is only one p-periodic point $z \in W$ and $z \in int\ W = (\alpha, \beta)$. At the same time $\lambda(f^l W) \ge \eta$ and so we may assume that, say, $\lambda([f^l z, f^l \beta]) \ge \eta/2$; by the choice of M it implies that $f^p[z, \beta] = f^{p-l}(f^l[z, \beta]) = [0, 1]$ and hence there is another p-periodic point in $(z, \beta]$ which is a contradiction. It completes the consideration of the case $A(f) = \emptyset$.

Consider the case $A(f) = \{0\}$, $f(0) = 0$; the other cases which are left may be considered similarly. Again suppose that $\eta > 0$. We will say that a point y δ-*approximates* a point x if $|f^n x - f^n y| \le \delta$ ($\forall n$). Let us prove the following

Assertion 1. *There exists a closed interval I such that $I \cap A(f) = \emptyset$ and for any $x \in [0, 1]$ there exists $y \in I$ which $\eta/3$-approximates x; moreover, if $x \in I$ then we can set $y = x$.*

Indeed, by Lemma 8.6 we can find two fixed points $0 < e < d$ such that $d < \eta/3$, $f[0, e] \subset [0, d]$. Let us show that $I = [e, 1]$ has the required property.

We may assume that $x \in [0, e]$ (otherwise we can set $y = x$). If $orb\ x \subset [0, e]$ then set $y = e$. If $orb\ x \not\subset [0, e]$ then first let us choose the smallest n such that $f^n x \notin [0, e]$.

Clearly, $f^n x \in (e, d]$. Now it is easy to see that there exists $y \in (e, d]$ such that $f^i y \in (e, d]$ for $0 \le i \le n - 1$ and $f^n y = f^n x$. Obviously y is the required point which completes the proof of Assertion 1.

Let $M = M(\eta)$ be an integer such that for any interval U longer than $\eta/6$ we have $f^m U \supset I$ for any $m \ge M$. To show that f has the i-specification property let us consider points $x_1, \ldots, x_n$ with semi-neighborhoods $U_i \ni x_i$ of length η and integers $0 = a_1 \le b_1 < a_2 \le b_2 < \ldots < a_n \le b_n$, p such that $b_i - a_{i-1} \ge M$ $(2 \le i \le n)$, $p \ge M + b_n - a_1$. We have to find a periodic point z of period p such that f^p is non-reversing at z and, moreover, $|f^t z - f^t x_i| \le \eta$ for $a_i \le t \le b_i$ and $f^{a_i} z \in U_i$ $(1 \le i \le n)$.

First let us find points $y_i \in I$ which $\eta/3$-approximate points x_i and belong to U_i (it is possible by Assertion 1 and the fact that if $x_i \notin I$ then the only semi-neighborhood of x_i of length η is $U_i = [x_i, x_i + \eta)$). Then choose one-sided semi-neighborhoods V_i of y_i such that $V_i \subset U_i$, $\lambda(V_i) = \eta/3$, $V_i \subset I$ $(1 \le i \le n)$. Now it is easy to see that one can replace U_i by V_i, then repeat the arguments from the case $A(f) = \emptyset$ and get a point z with the required properties. This completes the proof. $\square$

9. Corollaries Concerning Periods of Cycles

Let us pass to the corollaries concerning periods of cycles of continuous maps of the interval. Theorem Sh1 and well-known properties of the topological entropy imply that $h(f) = h(f|\overline{Per\, f})$. However, it is possible to get a set D such that $h(f) = h(f|D)$ using essentially fewer periodic points of f. Indeed, let A be some set of positive integers and define the set $K_f(A)$ as follows: $\{y \in Per\, f$: minimal period of y belongs to $A\}$.

Theorem 9.1[Bl4,Bl7]. *The following two properties of A are equivalent:*
1) $h(f) = h(f|\overline{K_f(A)})$ for any f;
2) for any k there exists $n \in A$ which is a multiple of k.

Proof. First suppose that statement 2) holds and prove that it implies statement 1). By the Decomposition Theorem it is enough to show that $\overline{\cup B_i} \subset \overline{K_f(A)}$ where $\cup B_i$ is the union of all basic sets of f. Fix a basic set $B = B(orb\, I, f)$; then by Theorem 4.1.f) we see that there is an interval $J \subset I$, a number m such that $f^m J = J$, a set $\tilde{B} = \overline{int\, J \cap B}$ and a monotone map $\phi : J \to [0, 1]$ such that $\bigcup_{i=0}^{m-1} f^i \tilde{B} = B$ and $f^m|\tilde{B}$ is almost conjugate by ϕ to a mixing map $g : [0, 1] \to [0, 1]$. By Theorem 8.7 the map g has the specification property. Now we need the following easy property of maps with specification.

Property X. *If $\psi : X \to X$ is a map with specification and H is some infinite set of positive integers then $\overline{K_\psi(H)} = X$.*

To prove Property X it is necessary to observe first that there exist at least two different ψ-periodic orbits. Now we need to show that for any $z \in X$ there is a point from $K_\psi(H)$ in any open $U \ni z$. To this end we may apply the specification property and pick up a point $y \in U$ which first approximates the orbit of z for a lot of time, then approximates one of the previously chosen periodic orbits for only one iteration of f and also has the property $\psi^N y = y$ where $N \in H$ is a large number (the periodic orbit we consider here should not contain z; that is why first we needed to find two distinct periodic orbits). Clearly, taking the appropriate constants and large enough number N from H we can see that the minimal period of y is exactly N which completes the proof of Property X.

Let us return to the proof of Theorem 9.1. Consider the set $A' = \{n : mn \in A\}$. Then by statement 2) from Theorem 9.1 we see that A' is infinite; so by Property X we have that $K_g(A')$ is dense in $[0, 1]$. Now by the properties of almost conjugations we see that $\widetilde{B} \subset \overline{K_f(A)}$ and hence $B \subset \overline{K_f(A)}$. It completes the proof of the fact that statement 2) implies statement 1) of Theorem 9.1.

To show that statement 1) implies statement 2) suppose that A is a set of positive integers such that for some k there are no multiples of k in A. We need to construct a map f such that $h(f) > h(f|\overline{K_f(A)}$. To this end consider some pm-map g with a periodic interval I of period k. Let us construct a new map f which coincides with g on the set $[0, 1] \setminus orb_g I$ and may be obtained by changing of the map g only on the set $orb_g I$ in such a way that $orb_g I = orb_f I$ remains the cycle of intervals for the map f as well as for the map g and

$$h(f|orb_f I) > h(f|\{x : f^n x \notin orb_f I \ (\forall n)\}) = h(f|\{x : f^n x \notin orb_f I \ (\forall n)\}).$$

Clearly, it is possible and this way we will get a map f such that $h(f) > h(f|\overline{K_f(A)})$. It completes the proof of Theorem 9.1. $\quad\square$

Now we are going to study how the sets $\Omega(f), \Omega(f^2), \ldots$ vary for maps with a fixed set of periods of cycles. In what follows by a period of a periodic point we always mean the minimal period of the point. In [Sh1] A.N. Sharkovskii introduced the notion of L-scheme.

L-scheme. If there exist a fixed point x and a point y such that either $f^2 \le x < y < fy$ or $fy < y < x \le f^2 y$ then it is said that f has L-scheme and points x, y form L-scheme.

Theorem Sh4[Sh1]. If f has L-scheme then f has cycles of all periods.

Lemma 9.2. If f has L-scheme then $h(f) \ge \ln 2$.

Proof. It follows from the well-known results on the connection between symbolic dynamics and one-dimensional dynamical systems (see, for example, [BGMY]). $\quad\square$

Lemma 9.3[Bl2,Bl7]. Let $f : [0, 1] \to [0, 1]$ be a transitive continuous map. Then:
1) f^2 has L-scheme;
2) $h(f) \ge 1/2 \cdot \ln 2$;
3) f has cycles of all even periods.

Proof. By Theorem Sh4 and Lemma 9.2 it is sufficient to prove statement 1). Consider some cases.

Case 1. There exist $0 \le a < b \le 1$ such that $fa = a, fb = b$.

Assuming $z < fz$ for $z \in (a, b)$ let us prove that $a \in f[b, 1]$. Indeed, otherwise $I = [b, 1] \cup f[b, 1] \ne [0, 1]$ is an f-invariant interval which contradicts the transitivity. Choose the smallest $c \in [b, 1]$ such that $fc = a$; then $b < c$. It is easy to see that there exists $d \in (a, c)$ such that $fd = c$ and points a, d form L-scheme. In other words, we have shown that in this case the map f itself has L-scheme; in particular, if $f(0) = 0$ or $f(1) = 1$ then f has L-scheme.

Case 2. *There exists a fixed point $t \in (0, 1)$ such that $fy > y$ for any $y \in [0, t)$ and $fy < y$ for any $y \in (t, 1]$.*

If $f[0, t] = [t, 1]$, $f[t, 1] = [0, t]$ then by Case 1 we may conclude that f^2 has L-scheme. Hence it is enough to consider the maps for which these equalities do not hold. Then by Lemma 8.3 we may assume that f is mixing which implies that f^2 is transitive and has an f^2-fixed point $y \neq t$ (by Lemma 8.2). Now Case 1 implies the conclusion. □

Note that Lemma 9.3 implies statement d) of Theorem 4.1.

Theorem 9.4[B14,B17,B18]. *Let $n \geq 0, k \geq 0$ be fixed, f have no cycles of period $2^n(2k + 1)$. Then:*
1) if $B = B(orb\, I, f)$ is a basic set and I has a period m then $2^n(2k + 1) \prec m \prec 2^{n-1}$;
2) $\Omega(f) = \Omega(f^{2^n})$;
3) if f is of type 2^l, $l \leq \infty$ then $\Omega(f) = \Omega(f^r)$ ($\forall r$).

Remark. Another proof of statements 2) and 3) of Theorem 9.4 is given in Chapter 4 of [BCo]. The statement 3) was also proved in [N3] and [Zh].

Proof.
1) By the Sharkovskii theorem about the coexistence of periods of cycles for interval maps and by the definition of a periodic interval we have $2^n(2k + 1) \prec m$. Suppose $m = 2^i$, $i \leq n - 1$. Then by Lemma 9.3 and Theorem 4.1 f has a cycle of period $2^i \cdot 2(2k + 1) \prec 2^n(2k + 1)$ which is a contradiction.
2) It is sufficient to prove that if $x \in \Omega(f) \setminus \omega(f)$ then $x \in \Omega(f^{2^n})$; indeed, obviously $\omega(f) \in \Omega(f^r)$ for any r and so $\omega(f) \subset \Omega(f^{2^n})$. By Theorem 3.1 if x belongs to a solenoidal set then $orb\, x$ is infinite and so by Theorem CN $x \in \Omega(f^{2^n})$ (remind that Theorem CN was formulated in Section 4). Now let $x \in B'(orb\, I, f)$ where I is chosen by Lemma 4.6; then I has a period m and $x \in \Omega(f^m)$. On the other hand by statement 1) $m = 2^n j$, $1 \leq j$ and so $x \in \Omega(f^m) \subset \Omega(f^{2^n})$.
3) Follows from statement 2) and Theorem CN.1).   □

10. Invariant Measures

It is well-known that the specification property has a lot of consequences concerning invariant measures (see, for example, [DGS]). We summarized some of them in Theorem DGS in Section 1. In the rest of Section 10 we rely on the results of Sections 2–5 to make use of Theorem 8.7 and Theorem DGS. First we need the following

Lemma 10.1. *Let $f : [0, 1] \to [0, 1]$ be continuous, $B = B([0, 1], f) \neq \emptyset$ and $f|B$ be mixing. Let also $\eta > 0$ and $x_1, x_2, \ldots, x_m \in Per(f|B)$. Then one can find $M = M(\{x_i\}_{i=1}^m, \eta)$ such that for any integers $a_1 \leq b_1 < a_2 \leq b_2 < \ldots < a_m \leq b_m$, p with $a_{i+1} - b_i \geq M$ ($1 \leq i \leq m - 1$), $p \geq M + b_m - a_1$ there exists a periodic point $z \in B$ of period p such that f^p is non-reversing at z and, moreover, $|f^n z - f^n x_i| \leq \eta$ for $a_i \leq n \leq b_i$ ($1 \leq i \leq m$).*

Proof. First consider the case when $m = 2$; let $x_2 = y$. For the sake of convenience let us reformulate our lemma in this situation. Namely, $x, y \in Per(f|B)$ and we have to find $M = M(\{x, y\}, \eta)$ such that for any $a_1 \le b_1 < a_2 \le b_2$, p with $a_2 - b_1 \ge M$, $p \ge M + b_2 - a_1$ there exists a periodic point $z \in B$ of period p such that f^p is non-reversing at z and, moreover, $|f^n z - f^n x| \le \eta$ for $a_1 \le n \le b_1$ and $|f^n z - f^n y| \le \eta$ for $a_2 \le n \le b_2$.

Let us assume that x and y are fixed points; the result in the general situation may be deduced from this case or may be proved by the similar arguments. Choose a semi-neighborhood V of x in the following way. First choose a side T of x such that x is a limit point for B from the side T. If x is not an endpoint of some interval complementary to B then let $V = V_T(x)$ be a semi-neighborhood of x of length smaller than η. If, for example, (x, α) is an interval complementary to B then let $V = V_T(x)$ have the properties $f\overline{V} \not\ni \alpha$ and $\lambda(V) < \eta$. Similarly we find a semi-neighborhood W of y.

By Theorem 4.1 there exist a mixing map $g : [0, 1] \to [0, 1]$ and a non-strictly increasing map $\phi : [0, 1] \to [0, 1]$ such that ϕ almost conjugates f to g. We may assume that $\phi(W) = W'$ and $\phi(V) = V'$ have the same length δ and, moreover, $W = \phi^{-1}(W')$ and $V = \phi^{-1}V'$; by the construction V' and W' are semi-neighborhoods of $\phi(x) = x'$ and $\phi(y) = y'$ respectively. Furthermore, we may assume that if x is not an endpoint of an interval complementary to B then $[x' - \delta, x' + \delta] \subset int(\phi[x - \eta, x + \eta])$ and the similar property holds for y.

By Theorem 8.7 there exists $M = M(\delta)$ corresponding to the constant δ in the i-specification property for g. Again we may assume without loss of generality that $a_1 = 0$. Now let $0 = a_1 \le b_1 < a_2 \le b_2$, p be integers with the properties from Lemma 10.1 with this number M. Applying Theorem 8.7 to the points x', y' with the semi-neighborhoods V', W' and the integers $0 = a_1 \le b_1 < a_2 \le b_2$, p we can find a periodic point z' such that g^p is non-reversing at z' and, moreover, $|g^n z' - g^n x'| \le \delta$ for $a_1 \le n \le b_1$, $|g^n z' - g^n y'| \le \delta$ for $a_2 \le n \le b_2$ and $z' = g^{a_1} z' \in V'$, $g^{a_2} z' \in W'$.

Properties of ϕ imply that $\phi^{-1}(z')$ is either a point or a closure of an interval complementary to B. In the first case set $z = \phi^{-1}(z')$. In the second case it is easy to see that since z' is a g-periodic point of period p at which g^p is non-reversing then there exists an endpoint z of the interval $\phi^{-1}(z')$ such that $f^p z = z$. In any case we get a f-periodic point $z \in B$ of period p such that f^p is non-reversing at z and $\phi(z) = z'$.

Let us show that z is the required point. Suppose that x is not an endpoint of an interval complementary to B. Then $|g^n z' - g^n x'| = |g^n z' - x'| \le \delta$ implies $|f^n z - f^n x| = |f^n z - x| < \eta$ by the choice of δ. So we may assume that (x, α) is an interval complementary to B. By the construction $z' = g^{a_1} z' \in V'$ and so $z = f^{a_1} z \in V$. Suppose that there exist numbers $r \le b_1$ such that $f^r z \notin V$ and let n be the smallest such number. If $f^n z$ lies to the left of V then $|\phi(f^n z) - x'| = |g^n z' - x'| > \delta$ although by the i-specification property $|g^n z' - x'| \le \delta$ (since $n \le b_1$). Thus $f^n z$ lies to the right of V which means that it lies to the right of α. At the same time $f^{n-1} z \in V$, $fx = x$ and by the choice of V we have $f\overline{V} \not\ni \alpha$. Clearly, we get to the contradiction and so $f^r z \in V, a_1 \le r \le b_1$. Applying the similar arguments to the point y we obtain the conclusion.

The proof in case when $m > 2$ is similar and left to the reader. $\qquad\square$

Corollary 10.2. *Let $d_1, \ldots, d_n$ be periodic points belonging to a basic set B, l be a positive integer and $\mu = \sum_{i=1}^n \alpha_i \cdot \nu(d_i)$ be an invariant measure. Then μ can be approximated by CO-measures with supports in B and minimal periods greater than l.*

Proof. We only outline here the proof which is very is similar to that of Proposition 21.8 from [DGS] (note that we are going to apply Lemma 10.1 instead of the specification property).

Namely, suppose that a neighborhood of μ is given. We may assume that $n > 1$ and orbits of $d_1, \ldots, d_n$ are pairwise distinct. Choose η such that $dist(orb\, d_i,\ orb\, d_j) > 10\eta$ $(i \neq j)$. Then approximate the measure μ by a measure of the same type, i.e. by a measure $\mu' = \sum_{i=1}^{n} \beta_i \cdot \nu(d_i)$, where β_i are properly chosen and very close to α_i rationals. The next step is to construct a collection of integers $a_1 = 0 < b_1 < a_2 < b_2 < \ldots < a_n < b_n$, p which are required in Lemma 10.1 in such a way that for any $1 \leq i \leq n$ we have $(b_i - a_i)/p = \beta_i$, $b_i - a_i \gg M = M(\{d_i\}_{j=1}^{n}, \eta)$ and $a_{i+1} = b_i + M$; furthermore, we may assume that $p \gg l$. Take the periodic point z of period p which exists for this collection of integers and periodic points by Lemma 10.1 and approximates pieces of orbits of $d_1, \ldots, d_n$. Then because of the choice of η it is easy to see that p is the minimal period of z. At the same time similarly to the proof of Proposition 21.8 from [DGS] it is easy to see that in fact the constants may be chosen in such a way that the point z generates the required CO-measure $\nu(z)$; in other words, we may assume that $\nu(z)$ approximates μ, lying in the previously given neighborhood of μ. It completes the proof. $\square$

Theorem 10.3 (cf. Theorem DGS). *Let B be a basic set. Then the following statements are true.*
1) *For any positive integer l the set $\bigcup_{p \geq l} P_f(p)$ is dense in $M_{f|B}$.*
2) *The set of ergodic non-atomic invariant measures μ with $supp\, \mu = B$ is residual in $M_{f|B}$.*
3) *The set of all invariant measures which are not strongly mixing is a residual subset of $M_{f|B}$.*
4) *Let $V \subset M_{f|B}$ be a non-empty closed connected set. Then the set of all points $x \in B$ such that $V_f(x) = V$ is dense in B (in particular, every measure $\mu \in M_{f|B}$ has generic points).*
5) *The set of points with maximal oscillation for $f|B$ is residual in B.*

Proof. First observe that if g is a transitive non-strictly periodic map then it is easy to see that Theorem 10.3 holds for g by Theorem DGS, Theorem 8.7 and Lemma 8.3. Now let us pass to the proof of statement 1) assuming that B is a Cantor set.

Let $B = B(orb\, I, f)$, g be a transitive non-strictly periodic map and ϕ almost conjugate $f|orb\, I$ to g (maps ϕ and g exist by Theorem 4.1). Let $\mu \in M_{f|B}$ and l be a positive integer. We have to prove that μ belongs to the closure of $\bigcup_{p \geq l} P_f(p)$ in $M_{f|B}$.

The case when μ is non-atomic is quite clear and we leave it to the reader (indeed, it is enough to consider the measure $\mu' \in M_g$ which is the ϕ-image of μ, apply Theorem DGS to the measure μ' and then lift the approximation we found for the measure μ' to the approximation of the measure μ which is possible since μ is non-atomic). On the other hand it is easy to see that any invariant measure from $M_{f|B}$ may be approximated by a measure μ of type $\mu = \alpha_0 \cdot \tilde{\mu} + \sum_{i=1}^{N} \alpha_i \cdot \nu(e_i)$ where $\tilde{\mu}$ is non-atomic and $N < \infty$. By the non-atomic case we can approximate $\tilde{\mu}$ by a CO-measure $\nu(e_0)$. Applying Corollary 10.2 we can approximate the measure $\sum_{i=0}^{N} \alpha_i \cdot \nu(e_i)$ by a CO-measure $\nu(c)$ where c is a periodic point with a minimal period $m \geq l$. This completes the proof of statement 1).

Looking through the proofs of Propositions 21.9–21.21 from [DGS, Section 21] which correspond to statements 2)–5) of Theorem DGS one can check that they are based on statement 1) of Theorem DGS and the property of invariant measures which is proved in Corollary 10.2. Hence repeating the arguments from [DGS, Section 21] one can prove statements 2)–5) of Theorem 10.3. □

Property 5) from Theorem 10.3 shows that if f is a transitive interval map then points with maximal oscillation form a residual subset of the interval. Applying this result we can easily specify Theorem 6.2 as it was explained in the proof of this theorem. Namely, in the proof of Theorem 6.2 we need to choose a residual subset $\Pi_{orb\,I}$ of any cycle of intervals $orb\,I$ such that $f|orb\,I$ is transitive and in the previous version of this theorem we chose $\Pi_{orb\,I}$ to be the set of all points with dense orbits in $orb\,I$. Now to specify Theorem 6.2 one can now choose the set of points with maximal oscillation as the set $\Pi_{orb\,I}$. It leads to the following

Theorem 6.2′ (cf.[Bl1],[Bl8]). *Let $f : [0,1] \to [0,1]$ be a continuous map without wandering intervals. Then there exists a residual subset $G \subset [0,1]$ such that for any $x \in G$ one of the following possibilities holds:*
1) $\omega(x)$ is a cycle;
2) $\omega(x)$ is a solenoid;
3) $\omega(x) = orb\,I$ is a cycle of intervals and $V_f(x) = M_{f|orb\,I}$.

Theorem 10.4. *Let μ be an invariant measure. Then the following properties of μ are equivalent:*
1) there exists $x \in [0,1]$ such that $supp\,\mu \subset \omega(x)$;
2) the measure μ has a generic point;
3) the measure μ can be approximated by CO-measures.

Remark. For non-atomic measures Theorem 10.4 was proved in [Bl4,BL7].

Proof. Clearly, 2)⇒1). If $\omega(x)$ is a cycle then the implications 1)⇒2) and 1)⇒3) are trivial. If $\omega(x)$ is a basic set then the implications 1)⇒2) and 1)⇒3) follow from Theorem 10.3. The case when $\omega(x)$ is a solenoidal set may be easily deduced from Theorem 3.1; this case is left to the reader.

It remains to prove that 3)⇒1). Let $\{e_i\}$ be a sequence of periodic points such that $\nu(e_i) \to \mu$. Set $L \equiv \{z : \text{for any open } U \ni z \text{ there exists a sequence } n_k \to \infty \text{ such that } orb\,e_{n_k} \cap U \neq \emptyset \ (\forall k)\}$. Obviously, L is compact, $supp\,\mu \subset L$, $fL = L$. We may assume that $e_i \searrow e$. Consider the set $P^R(e) = P^R$; then $L \subset P^R$. Finally we have $supp\,\mu \subset L \subset P^R$. By Lemma 2.2 there are the following possibilities for P^R.
1) P^R *is a cycle.* This case is trivial since $supp\,\mu$ belongs to a cycle and hence statement 1) holds.
2) P^R *is a solenoidal set.* Then by Theorem 3.1 the fact that $supp\,\mu \subset L \subset P^R$ implies that $supp\,\mu = S$ where S is the unique minimal subset if P^R. This completes the consideration of the case 2).
3) $\{P^R\}$ *is a cycle of intervals.* Consider two subcases.
3a) e *is the right endpoint of a component $[d,e]$ of P^R.* Then $orb\,e_i \cap P^R = \emptyset$ and hence $L \subset \partial(P^R)$. Surjectivity of $f|L$ implies that $e \in Per\,f$ and we may assume that

$fe = e$. Clearly, it implies that $\{L\} = \{e\}$ and completes the consideration of the subcase 3a).

3b) $e \in [z, y)$ *where* $[z, y]$ *is a component of* P^R. Then it is easy to see that $L \subset E(P^R, f)$ (the definition of the set $E(orb\, I, f)$ for cycle of intervals $orb\, I$ may be found in Section 4 before Lemma 4.5). Indeed, we may assume that $orb\, e_i \subset P^R$ for any i. Let $\zeta \in L$ and T is a side of ζ from which points of $orb\, e_{n_k}$ approach the point ζ. Then T is a side of ζ in the corresponding component of P^R.

Consider $P^T(\zeta)$; clearly, $P^T(\zeta) \subset P^R$. At the same time it is easy to see that any iterate of any semi-neighborhood $W_T(\zeta)$ is not wandering as a set and so by Lemma 2.1 the set $\overline{\bigcup_{i>n} f^i W_T(\zeta)}$ is a weak cycle of intervals for any n. Since this set contains infinitely many periodic orbits $orb\, e_i$ we can conclude that for any n the set $\overline{\bigcup_{i>n} f^i W_T(\zeta)}$ contains some right semi-neighborhood of e which implies that $P^T(\zeta) \supset P^R$. Finally $P^T(\zeta) = P^R$ and so $L \subset E(P^R, f)$ by the definition. Hence by Theorem 4.1 and Lemma 4.5 either L is a cycle or $L \subset B(P^R, f)$. In both cases statement 1) holds so this completes the proof of Theorem 10.4. $\square$

Corollary 10.5[Bl4,Bl7]. *CO-measures are dense in all ergodic measures.*

Remark. In [Bl4,Bl7] Corollary 10.5 was deduced from the version of Theorem 10.4 for non-atomic measures proved in [Bl4,Bl7].

Proof. Follows immediately from Theorem 10.4 and the fact that every ergodic measure has a generic point. $\square$

11. Discussion of Some Recent Results of Block and Coven and Xiong Jincheng

There are some recent papers ([BC], [X]) in which the authors investigate the sets $\omega(f) \setminus \overline{Per\, f}$ and $\Omega(f) \setminus \overline{Per\, f}$. Let us discuss some of their results.

First observe that by the Decomposition Theorem if $x \in \omega(f) \setminus \overline{Per\, f}$ then $x \in S_\omega$ for some solenoidal set S_ω and thus by Theorem 3.1 $\omega(x) = S$ is a minimal solenoidal set. It implies the following theorem proved in [BC].

Theorem BC. *If* $x \in \omega(f) \setminus \overline{Per\, f}$ *then* $\omega(x)$ *is an infinite minimal set.*

In [X] some new notions were introduced. Let us recall them. For a set $Y \subset [0, 1]$ by $\Lambda(Y)$ we denote the set $\bigcup_{x \in Y} \omega(x)$; let $\Lambda^1 = \Lambda([0, 1]) = \omega(f)$, $\Lambda^2 = \Lambda(\Lambda^1)$ etc. Obviously $\Lambda^1 \supset \Lambda^2 \supset \ldots$; let $\Lambda^\infty \equiv \bigcap_{n=1}^\infty \Lambda^n$.

By $\alpha(x)$ we denote the set of all α-limit points of x; in other words, $y \in \alpha(x)$ if and only if there exist sequences $x_{-i} \to y$ and $n_i \to \infty$ such that $f^{n_i} x_{-i} = x$ for any i. A point y is called a γ-limit point of x if $y \in \omega(x) \cap \alpha(x)$. Let $\gamma(x) \equiv \omega(x) \cap \alpha(x)$ and $\Gamma(f) \equiv \Gamma \equiv \bigcup_{x \in [0,1]} \gamma(x)$.

In the following lemma we use the notation from the Decomposition Theorem.

Lemma 11.1. $\Gamma = (\bigcup_i B_i) \cup (\bigcup_{\beta \in \mathscr{A}} S^{(\beta)}) \cup X_f$.

Proof. First let us prove that $\Gamma \supset (\bigcup_i B_i) \cup (\bigcup_{\beta \in \mathscr{A}} S^{(\beta)}) \cup X_f$. Clearly, $X_f \cup (\bigcup_{\beta \in \mathscr{A}} S^{(\beta)}) \subset \Gamma$ (for $S^{(\beta)}$ it follows for example from the fact that $f|S^{(\beta)}$ is minimal by Theorem 3.1). By Theorem 4.1 to prove that $B_i \subset \Gamma$ ($\forall i$) it is sufficient to show that $\Gamma(g) = [0, 1]$ provided $g : [0, 1] \to [0, 1]$ is a transitive map. Consider this case. If $x \in (0, 1)$ then by Lemma 8.3 $\alpha(x) = [0, 1]$. Thus if $x \in (0, 1)$ has a dense orbit in $[0, 1]$ then $\gamma(x) = [0, 1]$ and so $\Gamma(g) = [0, 1]$. Hence finally we may conclude that $\Gamma \supset (\bigcup_i B_i) \cup (\bigcup_{\beta \in \mathscr{A}} S^{(\beta)}) \cup X_f$.

Now let us prove that $\Gamma \subset (\bigcup_i B_i) \cup (\bigcup_{\beta \in \mathscr{A}} S^{(\beta)}) \cup X_f$. Indeed, $\Gamma \subset \omega(f) = (\bigcup_i B_i) \cup (\bigcup_{\beta \in \mathscr{A}} S^{(\beta)}_\omega) \cup X_f$ by the definition of Γ. So to prove Lemma 11.1 it remains to show that if $x \in S^{(\beta)}_\omega \setminus S^{(\beta)}$ then $x \notin \Gamma$ (here $\beta \in \mathscr{A}$). Suppose there exists z such that $x \in \omega(z) \cap \alpha(z)$. Then the fact that $x \notin S^{(\beta)}$ implies that $z \notin Q^{(\beta)}$ because otherwise $x \in \omega(z) = S^{(\beta)}$ by Theorem 3.1. Hence $\alpha(z) \cap Q^{(\beta)} = \emptyset$ which contradicts the fact that $x \in \alpha(z) \cap Q^{(\beta)}$. It completes the proof of Lemma 11.1. $\square$

Let us show how to deduce some of the results of [X] from our results.

Theorem X1[X]. *1)* $\Omega(f) \setminus \Gamma$ *is at most countable.*
2) $\Lambda^1 \setminus \Gamma$ *is either empty or countable.*
3) $\overline{Per\, f} \setminus \Gamma$ *is either empty or countable.*

Proof. 1) By the Decomposition Theorem $\Omega(f) \setminus \overline{Per\, f}$ is at most countable. By Theorem 3.1 $S^{(\beta)}_p \neq S^{(\beta)}$ for at most countable family of solenoidal sets and $S^{(\beta)}_p \setminus S^{(\beta)}$ is at most countable. This implies statement 1).

2) First recall that $\Lambda^1 = \omega(f)$. If $\Lambda^1 \setminus \Gamma \neq \emptyset$ then by the Decomposition Theorem and Lemma 11.1 there exist a solenoidal set $\omega(z)$ and a point $x \in \omega(z) \setminus S$ where S is the unique minimal set belonging to $\omega(z)$ (see Theorem 3.1); actually $\omega(z) \setminus S \subset \Lambda^1 \setminus \Gamma$. Now the fact that $f|\omega(z)$ is surjective implies that $\omega(z) \setminus S$ is countable and the inclusion $\omega(z) \setminus S \subset \Lambda^1 \setminus \Gamma$ implies the conclusion.

3) Consider the case when $\overline{Per\, f} \setminus \Gamma \neq \emptyset$. Similarly to the proof of statement 2) we see that then there exists a solenoidal set $Q = \bigcap_i orb\, J_i$ such that $(\overline{Per\, f} \cap Q) \setminus S \neq \emptyset$ where S is the unique minimal set belonging to Q. Denote $(\overline{Per\, f} \cap Q)$ by R.

We are going to prove the fact that $R \setminus S$ is a countable set by repeating the arguments from the proof of statement 2) replacing $\omega(z)$ by R. However, to this end we need to show that $f|R$ is surjective. Consider a point $y \in R$ and show that it has f-preimages in R. The fact that $y \in R \subset \overline{Per\, f}$ implies that there is a point $z \in \overline{Per\, f}$ such that $fz = y$. Let us prove that $z \in Q^{(\beta)}$. Suppose that $z \notin Q^{(\beta)}$. Then the fact that $fz = y$ and Theorem 3.1 imply that there exist an open $U \ni z$ and a number N such that $U \cap Q^{(\beta)} = \emptyset$ and for any $n > N$ we have $f^n U \subset Q^{(\beta)}$. Clearly, it contradicts the fact that $z \in \overline{Per\, f}$ and shows that actually $z \in Q^{(\beta)}$; hence $z \in Q^{(\beta)} \cap \overline{Per\, f} = R$ and so $f|R$ is surjective. Now the fact that $R \setminus S$ is a countable set may be proved similarly to statement 2). $\square$

Theorem X2[X]. $\Lambda^\infty = \ldots = \Lambda^3 = \Lambda^2 = \Lambda(\overline{Per\, f}) = \Lambda(\Omega(f)) = \Gamma$.

Proof. By Lemma 11.1 and properties of basic sets (Theorem 4.1), solenoidal sets (Theorem 3.1) and cycles we have $\Lambda(\Gamma) = \Gamma$ and so $\Lambda(\Omega(f)) \supset \Gamma$ since $\Omega(f) \supset \Gamma$. On the other hand the Decomposition Theorem and the definition of Γ imply that $\Lambda(\Omega(f)) \subset \Gamma$; indeed, in the notation from the Decomposition Theorem we have $\Lambda(B_i') \subset B_i$ for all i, $\Lambda(X_f) \subset X_f$ and $\Lambda(Q^{(\alpha)}) \subset S^{(\alpha)}$ for any α (the last assertion follows from Theorem 3.1). So $\Lambda(\Omega(f)) = \Gamma = \Lambda(\Gamma)$ which completes the proof. $\square$

Theorem X3[X]. *The following properties of a map f are equivalent:*
1) the type of f is 2^i, $i \leq \infty$;
2) every γ-limit point of f is recurrent.

Proof. As we have shown in Section 1 the fact that f has type 2^i, $i \leq \infty$ is equivalent to the absence of basic sets (see the part of Section 1 where we discuss the connection between the Misiurewicz theorem on maps with zero entropy and the "spectral" decomposition). So in this case by Theorem X1 we see that $\Gamma = (\bigcup_{\beta \in \mathcal{A}} S^{(\beta)}) \cup X_f$. But by Theorem 3.1 every point of $S^{(\beta)}$ is recurrent and ,clearly, every point of of X_f is recurrent. Hence if the type of f is 2^i, $i \leq \infty$ then every γ-limit point of f is recurrent.

On the other hand if there is a basic set B of f then it is easy to find a non-recurrent point $z \in B$ (it follows, for example, from Theorem 4.1 and Lemma 8.3). Now by Lemma 11.1 $B \subset \Gamma$ which shows that there exist non-recurrent points in Γ and completes the proof. $\square$

References

[AKMcA] R.L. Adler, A.G. Konheim, M.H.McAndrew. *Topological entropy,* Trans. Amer. Math. Soc., **114** (1965), 309–319.

[ALM] L. Alsedá, J. Llibre, M. Misiurewicz. *Periodic orbits of maps of Y,* Trans. Amer. Math. Soc., **313** (1989), 475–538.

[AK] J. Auslender, Y. Katznelson. *Continuous maps of the circle without periodic points,* Israel J. Math., **32** (1979), 375–381.

[Ba] S. Baldwin. *An extension of Šarkovskiĭ's theorem to the n-od,* Ergod. Th. and Dynam. Syst., **11** (1991), 249–271.

[BM1] M. Barge, J. Martin. *Chaos, periodicity, and snakelike continua,* Trans. Amer. Math. Soc., **289** (1985), 355–365.

[BM2] M. Barge, J. Martin. *Dense orbits on the interval,* Michigan Math. J., **34** (1987), 3–11.

[BCo] L. S. Block, W. A. Coppel. *One-dimensional dynamics,* to appear

[BC] L. Block, E.M. Coven. *ω-limit sets for maps of the interval,* Erg. Th. and Dyn. Syst., **6** (1986), 335–344.

[BGMY] L. Block, J. Guckenheimer, M. Misiurewicz, L.-S. Young. *Periodic orbits and topological entropy of one-dimensional maps.* In: Global Theory of Dynamical Systems, Lecture Notes in Mathematics, **819**, Springer: Berlin(1980), 18–34.

[Bl1] A.M. Blokh. *On the limit behavior of one-dimensional dynamical systems,* Russ. Math. Surv., **37**, no.1 (1982), 157–158.

[Bl2] A.M. Blokh. *On sensitive mappings of the interval,* Russ. Math. Surv., **37**, no.2 (1982), 203–204.

[Bl3] A.M. Blokh. *On the "spectral" decomposition for piecewise monotone maps of segment,* Russ. Math. Surv., **37**, no.3 (1982), 198–199.

[Bl4]   A.M. Blokh *Decomposition of dynamical systems on an interval*, Russ. Math. Surv., **38**, no. 5 (1983), 133–134.

[Bl5]   A.M. Blokh. *On the connection between entropy and transitivity for one-dimensional mappings*, Russ. Math. Surv., **42**, no.5 (1987), 165–166.

[Bl6]   A.M. Blokh. *A letter to editors*, Russ. Math. Surv., **42**, no.6 (1987).

[Bl7]   A.M. Blokh. *On the limit behavior of one-dimensional dynamical systems.1* (in Russian), Preprint no.1156–82, **VINITI**, Moscow (1982).

[Bl8]   A.M. Blokh. *On the limit behavior of one-dimensional dynamical systems.2*(in Russian), Preprint no.2704–82, **VINITI**, Moscow (1982).

[Bl9]   A.M. Blokh. *On transitive maps of one-dimensional branched manifolds* (in Russian). In: Differential-difference Equations and Problems of Mathematical Physics, Kiev (1984), 3–9.

[Bl10]   A.M. Blokh. *On some properties of maps of the interval with constant slope* (in Russian). In: Mathematical Physics and Functional Analysis, Kiev (1986), 127–136.

[Bl11]   A.M. Blokh. *On dynamical systems on one-dimensional branched manifolds. 1, 2, 3* (in Russian): 1, Theory of Functions, Functional Analysis and Applications, Kharkov, **46** (1986), 8–18; 2, Theory of Functions, Functional Analysis and Applications, Kharkov, **47** (1987), 67–77; 3, Theory of Functions, Functional Analysis and Applications, Kharkov, **48** (1987), 32–46.

[Bl12]   A.M. Blokh. *On C^0-continuity of entropy* (in Russian). Preprint (1989).

[Bl13]   A.M. Blokh. *The spectral decomposition, periods of cycles and Misiurewicz conjecture for graph maps* (1990), submitted to "Proceedings of the Conference on Dynamical Systems in Güstrow" (to appear in Lecture Notes in Mathematics).

[Bl14]   A.M. Blokh. *On some properties of graph maps: spectral decomposition, Misiurewicz conjecture and abstract sets of periods*, Max-Planck-Institut für Mathematik, Preprint no.35 (June, 1991).

[Bl15]   A.M. Blokh. *Periods implying almost all periods, trees with snowflakes, and zero entropy maps*, SUNY, Institute for Mathematical Sciences, Preprint no.13/1991 (August, 1991).

[Bl16]   A.M. Blokh. *The "spectral" decomposition for one-dimensional maps*, SUNY, Institute for Mathematical Sciences, Preprint no.14/1991 (September, 1991).

[BL1]   A.M. Blokh, M.Yu. Lyubich. *Non-existence of wandering intervals and structure of topological attractors of one-dimensional dynamical systems. 2. The smooth case.*, Erg. Th. and Dyn. Syst, **9**, no.4 (1989), 751–758.

[BL2]   A.M. Blokh, M.Yu. Lyubich. *Attractors of maps of the interval*, Banach Center Publications (of the Dynamical Systems Semester held in Warsaw, 1986), **23** (1989), 427–442.

[B1]   R. Bowen. *Entropy for group endomorphisms and homogeneous spaces*, Trans. Amer. Math. Soc., **153** (1971), 401–413.

[B2]   R. Bowen. *Periodic points and measures for axiom A-diffeomorphisms*, Trans. Amer. Math. Soc., **154** (1971), 377–397.

[B3]   R. Bowen. *Topological entropy for noncompact sets*, Trans. Amer. Math. Soc., **184** (1973), 125–136.

[B4]   R. Bowen. *Equilibrium states and the ergodic theory of Anosov diffeomor-phisms*. Lecture Notes in Mathematics, **470**, Springer: Berlin (1975).

[BF]   R. Bowen, J. Franks. *The periodic points of maps of the disk and the interval,* Topology, **15** (1976), 337–342.

[CE]   P. Collet, J.-P. Eckmann. *Iterated maps on the interval as dynamical systems.* Progress in Physics, **1**, Birkhäuser: Boston (1980).

[CN]   E.M. Coven, Z. Nitecki. *Non-wandering sets of the powers of maps of the interval,* Erg. Th. and Dyn. Syst., **1** (1981), 9–31

[CS]   E.M. Coven, J. Smital. *Entropy-minimality*, preprint.

[DGS]   M. Denker, C. Grillenberger, K. Sigmund. *Ergodic theory on compact spaces.* Lecture Notes in Mathematics, **527** , Springer: Berlin (1976).

[D]   A. Denjoy. *Sur les courbes definies par les équations differentielles à la surface du tore,* J. Math. Pures et Appl., **11** (1932), 333–375.

[Di]   E.I. Dinaburg. *The relation between topological entropy and metric entropy,* Soviet Math. Dokl., **11** (1970), 13–16.

[F]   M. Feigenbaum. *Quantitative universality for a class of nonlinear transfor-mations,* J. Stat. Phys., **19** (1978), 25–52.

[Gu]   J. Guckenheimer. *Sensitive dependence to initial conditions for one dimen-sional maps,* Comm. Math. Phys., **70** (1979), 133–160.

[GJ]   J. Guckenheimer, S. Johnson. *Distortion of S-unimodal maps,* Ann. of Math., **132** (1990), 71–130.

[Go]   T.N.T. Goodman. *Relating topological entropy with measure theoretic en-tropy,* Bull. Lond. Math. Soc., **3** (1971), 176–180.

[GZ]   B.M. Gurevich, A.S. Zargaryan. *A continuous one-dimensional map without maximal measure,* Funct. Anal. and its Appl., **20**, no.2 (1986), 60–61.

[H1]   F. Hofbauer. *On intrinsic ergodicity of piecewise monotonic transformations with positive entropy,* Israel of Math., **34** (1979), 213–237; Part 2, Israel J. of Math., **38** (1981), 107–115.

[H2]   F. Hofbauer. *The structure of piecewise monotone transformations,* Erg. Th. and Dyn. Syst., **1**(1981), 135–143.

[H3]   F. Hofbauer. *Piecewise invertible dynamical systems,* Probab. Th. Rel. Fields, **72** (1986), 359–386.

[H4]   F. Hofbauer. *Generic properties of invariant measures for continuous piece-wise monotonic transformations,* Monat. Math., **106** (1988), 301–312.

[HR]   F. Hofbauer, R. Raith. *Topologically transitive subsets of piecewise monotonic maps, which contain no periodic points,* Monat. Math., **107** (1989), 217–240.

[JR1]   L. Jonker, D. Rand. *Bifurcations in one dimension.1: The non-wandering set,* Inv. Math., **62** (1981), 347–365.

[JR2]   L. Jonker, D. Rand. *Bifurcations in one dimension.2: A versal model for bifurcations,* Inv. Math., **63** (1981), 1–15.

[Ke]   G. Keller. *Exponents, attractors and Hopf decomposition for interval maps,* Erg. Th. and Dyn. Syst., **10** (1990), 717–744.

[K]   A.N. Kolmogorov. *A new metric invariant of transitive dynamical systems and automorphisms of Lebesgue spaces,* Dokl. Acad. Nauk SSSR, **119** (1958), 861–864.

[Li]   Shihai Li. *Chain recurrent set and turning points,* to appear in Proc. Amer.
            Math. Soc.

[LM]   J. Llibre, M. Misiurewicz. *Excess of gods implies chaos,* preprint (1991)

[L]   M.Yu. Lyubich. *Non-existence of wandering intervals and structure of topo-
            logical attractors of one- dimensional dynamical systems. 1. The case of
            negative Schwarzian derivative,* Erg. Th. and Dyn. Syst., **9**, no.4 (1989),
            737–750.

[Ma]   R. Mañé. *Ergodic Theory.* A Series of Modern Surveys in Mathematics, **8**,
            317p., Springer:Berlin (1987).

[MMSt]   M. Martens, W. de Melo, S.J. van Strien. *Julia-Fatou-Sullivan theory for real
            one-dimensional dynamics.* Preprint (1988).

[MSt]   W. de Melo, S.J. van Strien. *A structure theorem in one dimensional dynamics,*
            Ann. of Math., **129** (1989), 519–546.

[MilT]   J. Milnor, W. Thurston. *On iterated maps of the interval.* In: Dynamical
            systems, Lecture Notes in Mathematics, **1342**, Springer: Berlin (1988), 465–
            564.

[Mi1]   M. Misiurewicz. *Structure of mappings of an interval with zero entropy.*
            Publ. Math. IHES, **53** (1981), 5–16.

[Mi2]   M. Misiurewicz. *Horseshoes for mappings of the interval,* Bull. Acad. Pol.
            Sci., ser. sci. math., **27**, no. 2 (1979), 167–169.

[Mi3]   M. Misiurewicz. *Invariant measures for continuous transformations of* [0, 1]
            with zero topological entropy. In: Ergodic theory, Lecture Notes in Mathe-
            matics, **729**, Springer: Berlin (1979), 144–152.

[Mi4]   M. Misiurewicz. *Periodic points of maps of degree one of a circle,* Erg. Th.
            and Dyn. Syst., **2** (1982), 221–227.

[Mi5]   M. Misiurewicz. *Jumps of entropy in one dimension,* Fund. Math., **132**
            (1989), 215–226.

[MiS]   M. Misiurewicz, W. Szlenk. *Entropy of piecewise monotone mappings,* Studia
            Mathematica, **67**, no.1 (1980), 45–53.

[MiŚl]   M. Misiurewicz, S.V. Ślaćkov. *Entropy of piecewise continuous interval
            maps.* Preprint (1988).

[N1]   Z. Nitecki. *Periodic and limit orbits and the depth of the center for piecewise
            monotone interval maps,* Proc. Amer. Math. Soc., **80** (1980), 511–514.

[N2]   Z.Nitecki. *Topological dynamics on the interval.* In: Ergodic Theory and
            dynamical systems, 2. Progress in Math., **21**, Birkhäuser: Boston (1982),
            1–73.

[N3]   Z. Nitecki. *Maps of the interval with closed periodic set,* Proc. Amer. Math.
            Soc., **85** (1982), 451–456.

[P1]   C. Preston. *Iterates of maps on an interval.* Lecture Notes in Mathematics,
            999, Springer: Berlin (1983).

[P2]   C. Preston. *Iterates of piecewise monotone mappings on an interval.* Lecture
            Notes in Mathematics, **1347**, Springer, Berlin (1988).

[Sh1]   A.N. Sharkovskii. *Co-existence of cycles of continuous maps of the line into
            itself* (in Russian), Ukr. Math. J., **16** (1964), 61–71.

[Sh2]   A.N. Sharkovskii. *Non-wandering points and the center of a continuous map of the line into itself* (in Ukrainian), Dop. Acad. Nauk Ukr. RSR Ser. A (1964), 865–868.

[Sh3]   A.N. Sharkovskii. *The behavior of a map in a neighborhood of an attracting set* (in Russian), Ukr. Math. J., **18** (1966), 60–83.

[Sh4]   A.N. Sharkovskii. *The partially ordered system of attracting sets,* Soviet Math. Dokl., **7** (1966), 1384–1386.

[Sh5]   A.N. Sharkovskii. *On a theorem of G.D. Birkhoff* (in Russian), Dop. Acad. Nauk Ukr. RSR Ser. A (1967), 429–432.

[Sh6]   A.N. Sharkovskii. *Attracting sets containing no cycles* (in Russian), Ukr. Math. J., **20** (1968), 136–142.

[SKSF]   A.N. Sharkovskii, S.F. Kolyada, A.G. Sivak, V.V. Fedorenko. *Dynamics of one-dimensional mappings* (in Russian), Naukova Dumka: Kiev (1989).

[SMR]   A.N. Sharkovskii, Yu.L. Maistrenko, E.Yu. Romanenko. *Difference equations and their applications* (in Russian), Naukova Dumka: Kiev (1986).

[Si1]   K. Sigmund. *Generic properties of invariant measures for axiom-A-diffeomorphisms,* Inv. Math., **11** (1970), 99–109.

[Si2]   K. Sigmund. *On dynamical systems with the specification property,* Trans. Amer. Math. Soc., **190** (1974), 285–299.

[S]   S. Smale. *Differentiable dynamical systems,* Bull. Amer. Math. Soc., **73** (1967), 747–817.

[Str]   S.J. van Strien. *On the bifurcation creating horseshoes.* In: Lecture Notes in Mathematics, **898**, Springer: Berlin (1980), 316–351.

[W]   J. Willms. *Asymptotic behavior of iterated piecewise monotone maps,* Erg. Th. and Dyn. Syst., **8** (1988), 111–131.

[X]   J.-C. Xiong. *The attracting center of a continuous self-map of the interval,* Erg. Th. and Dyn. Syst., **8** (1988), 205–213.

[Y]   J.C. Yoccoz. *Il n'y a pas de contre-exemple de Denjoy analitiques,* C.R. Acad. Sci. Paris, ser. Math., **298**, no. 7 (1984), 141–144.

[Yo]   L.-S. Young. *A closing lemma on the interval* , Inv. Math., **54** (1979), 179–184.

[Zh]   Zhou Zuoling. *Self-mappings of the interval without homoclinic points* (Chinese), Acta Math. Sinica, **25** (1982), 633–640.

A Constructive Theory of Lagrangian Tori and Computer-assisted Applications

Alessandra Celletti[1] *and Luigi Chierchia*[2]

[1]*Dipartimento di Matematica, Università dell'Aquila, 67100–Coppito, L'Aquila (Italy)*
[2]*Dipartimento di Matematica, Terza Università di Roma, via C. Segre 2, 00146–Roma (Italy)*

1. Introduction

Perturbative techniques are among the most powerful tools in the theory of conservative dynamical systems. Besides giving finite time predictions (something well known to the astronomers of the eighteenth century), perturbation methods may be used to establish the existence of regular motions. H. Poincaré used thoroughly such methods in his investigation in Celestial Mechanics [Po], obtaining, e.g., his celebrated results on periodic orbits for Hamiltonian systems. A more recent success of perturbation ideas is the so called "KAM (Kolmogorov [Ko]-Arnold [A1]-Moser [Mo1]) theory", which ensures, under suitable smoothness assumptions, the survival under a small perturbation of "most" of the invariant maximal tori which foliate the phase-space of "integrable" conservative systems (see [B] for review and exhaustive references and [ChG] for recent developments).

One of the main themes we shall discuss here is *how small* the size of the perturbation has to be for the tori to persist.

The interest in such a problem is not only motivated by practical purposes (trying to apply KAM theory to concrete situations), but also by purely abstract questions.

Let $\mathcal{T}_\mu$ be one of the tori surviving the effect of a perturbation "of size" μ. On $\mathcal{T}_\mu$ the motion is regular and past and future are known; in low dimension ($d \leq 2$) such tori constitute obstructions for the dynamics and confine the motions in regions where exact predictions may be impossible; even in high dimensions motions starting nearby $\mathcal{T}_\mu$ will remain close to it for extremely long time.

Here are three basic questions which we shall try to answer:

(a) Is it possible to give explicit approximations of $\mathcal{T}_\mu$ keeping track of approximation errors?

(b) Can one hope to deal with sizes of parameter values coming from actual observations?

(c) Experiments (see, e.g., the discussion in [Mo5]) and some theoretical results ([Ma], [AL], [MKP]) suggest that $\mathcal{T}_\mu$ breaks down as μ is increased. Do the perturbation techniques contain the elements for explaining the break-down of the invariant tori?

In this paper we shall discuss a general theory (in the *real analytic setting*), which follows recent developments in KAM theory ([Mo4], [SZ], [CC2]), concerning constructive existence results for Lagrangian tori. This theory, based on a Lagrangian formalism, has the advantage of freeing the older formalism from infinitely many changes of variables and instead deals directly with the tori equations in a spirit close to the "hard implicit function theorems" à la Nash-Moser (see [B] for review and references therein and in particular [Mo4], [Z] and [Ha]). The method is highly quantitative and we shall work out all the estimates in full detail, keeping explicit track of the different quantities involved. We shall then apply the methods to various situations including the spin-orbit resonance problem of Celestial Mechanics.

In these applications we shall make use of computer-assisted calculations (see [L] for general informations). The need for machines comes in, e.g., for the accurate evaluation of the norm of *approximate solutions,* which are the starting point for the Newton iteration leading to the construction of true solutions.

In the final section we briefly discuss the theory of invariant curves for area-preserving (symplectic) diffeomorphisms of the cylinder, using a "direct" approach developed in [CC2], [CC3], [CC4] (see also [LR] for different techniques).

The range of applicability of the method covers parameter values of concrete interest and, in three-dimensional models, is shown to be within 70% "from optimal" (the first results in this direction, inspired by [G1], were obtained in [CC1] and [CFP]) while in symplectic map models we have been able to reach 86% of optimal [CC4].

Question (c) above is by far the most difficult to attack and we shall content ourselves by pointing out a direction ([BC], [BCCF]) which connects the disappearance of the tori with *complex singularities* in the space of the μ-parameter.

The theoretical part of this paper is partly new, while most of the applications are selected from various papers of the authors.

The purpose of our exposition is ambitiously twofold: we provide (with the help of numerous appendices) complete details so that non-specialists or graduate students could acquire a working knowledge of the main ideas and techniques in KAM theory; but, in so doing, we tried to mantain the exposition concise so that researchers active in the field can find elements of novelties before getting bored.

Acknowledgment. It is a pleasure to thank U. Kirchgraber for giving us the opportunity of attempting to achieve the just mentioned project.

2. Quasi-Periodic Solutions and Invariant Tori for Lagrangian Systems: Algebraic Structure

In this and in the following paragraph we consider the equation for *maximal* invariant tori and show how to solve it by means of a Newton-KAM method provided a "good enough" *non-degenerate approximate solution* is given.

2.1. Setup and Definitions

Let $\mathscr{L}(y, x, t)$ be a real-analytic function of $(y, x, t) \in Y \times \mathbb{T}^{d+1}$, where Y is an open set in $\mathbb{R}^d$ and $\mathbb{T}^{d+1}$ is the standard $(d + 1)$-dimensional torus with periods 2π: $\mathbb{T}^{d+1} \equiv \mathbb{R}^{d+1}/(2\pi\mathbb{Z})^{d+1}$; in other words $\mathscr{L}$ is a real-analytic function of $2d + 1$ variables, 2π-periodic in $x_1, ..., x_d, t$. The Euler-Lagrange equations for the motions $t \rightarrow x(t) = (x_1(t), ..., x_d(t))$ associated to $\mathscr{L}$ (see [A2] for generalities) are given by

$$\frac{d}{dt} \mathscr{L}_y(\dot{x}, x, t) = \mathscr{L}_x(\dot{x}, x, t) , \tag{2.1}$$

where $\mathscr{L}_y \equiv \partial_y\mathscr{L}$ and $\mathscr{L}_x \equiv \partial_x\mathscr{L}$ denote the gradients of $\mathscr{L}$ with respect to y and x.

An important class of solutions of (2.1) is given by *quasi-periodic* solutions:

Definition 2.1. *A solution* $x(t) \equiv (x_1(t), ..., x_d(t))$ *of (2.1) is said to be quasi-periodic with frequencies* $\omega \in Y \subset \mathbb{R}^d$ *if* ω *is rationally independent with 1 (i.e.* $\omega \cdot n + m = 0$ *for some* $n \in \mathbb{Z}^d$, $m \in \mathbb{Z}$ *implies* $n = 0$, $m = 0$) *and if there exists a periodic function twice differentiable* $u: (\theta, t) \in \mathbb{T}^{d+1} \to u(\theta, t) \in \mathbb{R}^d$, *such that*

$$x_i(t) \equiv \omega_i t + u_i(\omega t, t), \qquad (\text{mod } 2\pi). \tag{2.2}$$

Remark 2.2. If $\mathscr{L}$ does not depend explicitly on the time t one would replace, in the above definition, "ω rationally independent with 1" with "ω rationally independent" (*i.e.* $\omega \cdot n = 0 \Rightarrow n = 0$) and $u(\theta, t)$ with $u(\theta)$.

Remark 2.3. If $\mathscr{L}$ is independent of (x, t), i.e. $\mathscr{L} = \mathscr{L}(y)$, $\mathscr{L}$ is said to be *integrable*: the Euler-Lagrange equations are trivial and all solutions are of the form $x(t) = x_0 + \omega t$, $\omega \equiv \dot{x}_0$. Thus, up to a set of Lebesgue measure zero of initial data ($\equiv \{(x_0, \dot{x}_0) : \dot{x}_0$ is rationally dependent$\}$) all solutions of integrable Lagrangians are quasi-periodic.

To any quasi-periodic solution there is naturally associated a family of solutions parametrized by $(d + 1)$ phases $(\theta, \tau) \in \mathbb{T}^{d+1}$. In fact, since $(\omega, 1)$ is rationally independent the flow $(\omega t, t)$ is dense on $\mathbb{T}^{d+1}$; therefore it is easy to check that (2.2) is solution of (2.1) if and only if

$$\theta + \omega(t - \tau) + u(\theta + \omega(t - \tau), \ t - \tau) \tag{2.3}$$

is a solution for any $(\theta, \tau) \in \mathbb{T}^{d+1}$. This, in turn, is equivalent to require that $u(\theta, t)$ is solution of the following second-order degenerate nonlinear system of partial differential equations on $\mathbb{T}^{d+1}$:

$$D\mathscr{L}_{y_j}(\omega + Du, \ \theta + u, \ t) = \mathscr{L}_{x_j}(\omega + Du, \ \theta + u, \ t), \quad j = 1, ..., d, \tag{2.4}$$

where $D \equiv D_\omega$ denotes derivative along $(\omega, 1)$:

$$D \equiv D_\omega \equiv \omega \cdot \partial_\theta + \partial_t, \qquad Du \equiv (Du_1, ...Du_d),$$

$$Du_i \equiv \sum_{j=1}^{d} \omega_j \frac{\partial u_i}{\partial \theta_j} + \frac{\partial u_i}{\partial t}. \tag{2.5}$$

As an example, consider a planar mechanical system, made up by two interacting particles of masses m_i, constrained on concentric circles of radii r_i, whose center moves on a (coplanar) circle of radius ρ with angular velocity $\dot{\lambda}(t) = \dot{\lambda}(t + 2\pi)$, the interaction being ruled by a potential energy depending on the squared distance of the two particles. Up to an additive time-dependent function (which does not contribute to the Euler-Lagrange equation) the Lagrangian of this system is given by

$$\mathscr{L}(y_1, y_2, x_1, x_2, t) = \frac{1}{2} \sum_{i=1}^{2} m_i \left[r_i^2 y_i^2 + 2r_i y_i \rho \dot{\lambda} \cos(x_i - \lambda) \right] + \tag{2.6}$$
$$- V(\cos(x_1 - x_2)),$$

where V is related to the true potential energy U by $V(\xi) \equiv U(r_1^2 + r_2^2 - 2r_1 r_2 \xi)$. For such a system, a quasi-periodic solution with frequencies (ω_1, ω_2), $x(t) = \omega t + u(\omega t)$, satisfies the system ($i = 1, 2$):

$$m_i r_i^2 \left[D^2 u_i + \frac{\rho}{r_i} \ddot{\lambda} \cos(\theta_i + u_i - \lambda) \right] =$$
$$= - m_i r_i \rho \lambda^2 \, \sin(\theta_i + u_i - \lambda) \tag{2.7}$$
$$+ (-1)^{i-1} V' \Big(\cos(\theta_1 - \theta_2 + u_1 - u_2) \Big) \sin(\theta_1 - \theta_2 + u_1 - u_2) ,$$

where $u_i = u_i(\theta_1, \theta_2, t)$ and $Du_i \equiv \omega_1 \partial_{\theta_1} u_i + \omega_2 \partial_{\theta_2} u_i + \partial_t u_i$.

Remark 2.4. Equation (2.4), and its variational formulation, has been introduced by Percival ([Pe]).

Quasi-periodic solutions span invariant tori; to be more precise we need a definition: denote by $\mathbb{1}$ the $(d \times d)$ identity matrix and let $(u_\theta)_{ij} \equiv \frac{\partial u_i}{\partial \theta_j}$, then

Definition 2.5. *We shall say that a quasi-periodic solution is non-degenerate if* $\forall (\theta, t) \in \mathbb{T}^{d+1}$

$$\det(\mathbb{1} + u_\theta) \neq 0 . \tag{2.8}$$

If $x(t)$ is a non-degenerate quasi-periodic solution, the map $(\theta, t) \rightarrow (\theta + u(\theta, t), \ t)$ yields a *non-contractible embedding* of $\mathbb{T}^{d+1}$ into itself; in other words, non-degenerate quasi-periodic solutions correspond to homotopically non trivial invariant tori of maximal dimension $d + 1$ run by a linear flow.

2.2. Approximate Solutions and Newton Scheme

Let us begin by setting up the notations. We shall think of vectors as of *column vectors* identifying m-vectors with $m \times 1$ matrices. If f is a vector function, $f : \mathbb{R}^m \rightarrow \mathbb{R}^n$, the derivative of f is the $n \times m$ matrix $\partial_x f \equiv \frac{\partial f}{\partial x} \equiv f_x$ with entries $(f_x)_{ij} = \frac{\partial f_i}{\partial x_j}$ (so that $\frac{d}{d\varepsilon}|_{\varepsilon=0} f(x + \varepsilon \sigma) = f_x \sigma$). With these conventions the gradient of a scalar function has to be interpreted as a *row vector*, introducing a funny transpose in our basic equation (2.4) which we rewrite as

$$\mathcal{E}(u) \equiv D\mathcal{L}_y^T(\omega + Du, \theta + u, t) - \mathcal{L}_x^T(\omega + Du, \theta + u, t) = 0 . \tag{2.9}$$

Definition 2.6. *A real-analytic function $v(\theta, t)$ on $\mathbb{T}^{d+1}$ is called a non-degenerate approximate solution of (2.9) (in short approximate solution) if*

(i) *there exists a $(2d + 1)$-neighbourhood (v-dependent) $\mathcal{N} \subset Y \times \mathbb{T}^{d+1}$ of the set*

$$\Big\{ (y, x, t) = (\omega + Dv, \theta + v, t) | \ (\theta, t) \in \mathbb{T}^{d+1} \Big\} ,$$

such that, on it, the matrix $\mathcal{L}_{yy}$ is positive definite:

$$\mathcal{L}_{yy} > 0 , \quad \forall (y, x, t) \in \mathcal{N} ; \tag{2.10}$$

(ii) *for each fixed t the map $\theta \rightarrow \theta + v(\theta, t)$ is non-singular i.e.:*

$$\det(\mathbb{1} + v_\theta) \neq 0 , \quad \forall (\theta, t) \in \mathbb{T}^{d+1} \qquad ([v_\theta]_{ij} = \frac{\partial v_i}{\partial \theta_j}) . \tag{2.11}$$

To an approximate solution we will always associate an *error-function* $\varepsilon = \varepsilon(\theta, t)$ by setting

$$\varepsilon(\theta, t) \equiv \mathscr{E}(v) \equiv D\mathscr{L}_y^T(\omega + Dv, \theta + v, t) - \mathscr{L}_x^T(\omega + Dv, \theta + v, t) . \qquad (2.12)$$

Now, the basic idea is, roughly speaking, to try to solve (2.9) by *linearizing* the operator $\mathscr{E}$ at v and finding a *better approximation* v' such that as in *Newton schemes*

$$w \equiv v' - v \sim O(|\varepsilon|) , \qquad \varepsilon' \equiv \mathscr{E}(v') \sim O(|\varepsilon|^2) , \qquad (2.13)$$

in suitable norms to be defined below. Iterating this procedure one may try to get a solution of the form $v + \sum w_j$.

Under suitable number-theoretical assumptions on the frequencies ω, we shall see that this strategy is successfull *provided* one starts with an approximate solution for which the *error term is "small"* enough.

An *important example* is the following. Let $\mathscr{L}$ be a *nearly integrable* Lagrangian, i.e.

$$\mathscr{L}(y, x, t) = \mathscr{L}_0(y) + \mu\mathscr{L}_1(y, x, t), \qquad 0 < \mu \ll 1 \qquad (2.14)$$

and assume that the hessian matrix of $\mathscr{L}_0$ is positive definite:

$$\partial_y^2 \mathscr{L}_0 \equiv \left[\frac{\partial^2 \mathscr{L}_0}{\partial y_i \partial y_j}\right] > 0 ; \qquad (2.15)$$

then $v \equiv 0$ is an approximate solution ($\mathcal{N}$ will be a neighbourhood of the torus $\{\omega\} \times \mathbb{T}^{d+1}$) with error function proportional to μ:

$$\varepsilon(\theta, t) \equiv \mu\left[D\partial_y\mathscr{L}_1(\omega, \theta, t) - \partial_x\mathscr{L}_1(\omega, \theta, t)\right] . \qquad (2.16)$$

Remark 2.7. The construction we present here works for a somewhat more general class of approximate solutions satisfying

$$\det\left\{(\mathbb{1} + v_\theta)^T \mathscr{L}_{yy} (\mathbb{1} + v_\theta)\right\} \neq 0 , \qquad \det\int_{\mathbb{T}^d}\left\{[\mathbb{1} + v_\theta]^T \mathscr{L}_{yy} [\mathbb{1} + v_\theta]\right\} \neq 0 , \qquad (2.17)$$

in place of (2.10), the argument of $\mathscr{L}_{yy} = \partial_y^2\mathscr{L}$ being $(\omega + Dv, \theta + v, t)$ (see [SZ]).

The matrix appearing in (2.17) is an important quantity and it deserves a name:

Definition 2.8. *For a given non-degenerate approximate solution v we shall call the matrix*

$$\mathscr{T} \equiv \mathscr{T}_v \equiv [\mathbb{1} + v_\theta]^T \mathscr{L}_{yy}(\omega + Dv, \theta + v, t) [\mathbb{1} + v_\theta] \qquad (2.18)$$

the twist matrix *of v.*

By our non-degeneracy assumption

$$\mathscr{T} = \mathscr{T}(\theta, t) > 0 , \qquad (\theta, t) \in \mathbb{T}^{d+1} . \qquad (2.19)$$

Because of the frequent occurence of the map $(\theta, t) \in \mathbb{T}^{d+1} \rightarrow (\omega + Dv, \theta + v, t) \in \mathbb{R}^d \times \mathbb{T}^{d+1}$, we shall give to it a name too:

Definition 2.9. *Given a non-degenerate approximate solution v the map*

$$(\theta, t) \in \mathbb{T}^{d+1} \rightarrow \phi(\theta, t) \equiv \phi_v(\theta, t) \equiv (\omega + Dv(\theta, t), \; \theta + v(\theta, t), \; t)$$

will be called the v-embedding map.

2.3. The Linearized Equation

Let v be an approximate solution of (2.9) [see Definition 2.6] and let $\varepsilon(\theta, t)$ be the associated error function defined in (2.12). We want to find a (vector-) function $w(\theta, t)$ such that

$$\mathcal{E}(v + w) = \varepsilon' , \tag{2.20}$$

with ε' *quadratic* in ε: the exact meaning of "quadratic" will be clear in the next paragraph where the quantitative analysis is carried out. However, intuitively speaking, it means that if ε is replaced by $\mu\varepsilon$ ($\mu \in \mathbb{R}$), then ε' should have the form $\mu^2 \varepsilon'$.

Linearizing (2.12) at v one finds

$$\mathcal{E}(v + w) = \mathcal{E}(v) + \mathcal{E}'(v)w + q_1 \equiv \varepsilon + \mathcal{E}'(v)w + q_1 , \tag{2.21}$$

where q_1 [defined by the first equality in (2.21)] is quadratic in w and $\mathcal{E}'(v)$ is the second order linear-differential operator:

$$\begin{aligned} \mathcal{E}'(v) &\equiv D\Big[\mathcal{L}_{yy}D + \mathcal{L}_{yx}\Big] - \mathcal{L}_{xy}D - \mathcal{L}_{xx} , \qquad\qquad \text{i.e. ,} \\ \mathcal{E}'(v)g &\equiv D\Big[\mathcal{L}_{yy}Dg + \mathcal{L}_{yx}g\Big] - \mathcal{L}_{xy}Dg - \mathcal{L}_{xx}g , \qquad \forall g : \mathbb{T}^{d+1} \rightarrow \mathbb{R}^d , \end{aligned} \tag{2.22}$$

where

$$[\mathcal{L}_{yx}]_{ij} = \frac{\partial^2 \mathcal{L}}{\partial y_i \partial x_j} \qquad \text{and} \qquad \mathcal{L}_{xy} = \mathcal{L}_{yx}^T , \tag{2.23}$$

all derivatives (with respect to x, y) being evaluated at $(y, x, t) = \phi_v(\theta, t) \equiv (\omega + Dv, \theta + v, t)$. The explicit expression for q_1 is:

$$\begin{aligned} q_1 \equiv D\Big[\mathcal{L}_y^T \circ \phi_{v'} - \mathcal{L}_y^T \circ \phi_v - \mathcal{L}_{yy} \circ \phi_v Dw - \mathcal{L}_{yx} \circ \phi_v w\Big] \\ - \Big[\mathcal{L}_x^T \circ \phi_{v'} - \mathcal{L}_x^T \circ \phi_v - \mathcal{L}_{xy} \circ \phi_v Dw - \mathcal{L}_{xx} \circ \phi_v w\Big] , \end{aligned} \tag{2.24}$$

where $\phi_{v'} \equiv \phi_{v+w}$.

Thus, the linear equation to be solved is

$$\mathcal{E}'(v)w + \varepsilon = q \tag{2.25}$$

for some (real-analytic) function q quadratic in ε (or, what is the same, in w) and in this way we would have

$$\mathcal{E}(v + w) = q_1 + q \equiv \varepsilon' . \tag{2.26}$$

Remarks 2.10.

(*i*) At first sight, equation (2.25) does not look very promising, the operator $\mathscr{E}'$ being a non-constant coefficient, degenerate second order operator on $\mathbb{T}^{d+1}$.

(*ii*) It is important to introduce the "extra error" q; in fact the equation $\mathscr{E}'(v)w + \varepsilon = 0$, in general, does *not* admit any solution.

The most delicate part of the whole method is the reduction of (2.25) to a *constant coefficient* equation "explicitly" solvable.

2.4. Solution of the Linearized Equation

Taking the θ-gradient of (2.12) brings in naturally the operator $\mathscr{E}'(v)$: denoting by $\mathcal{M}$ the (invertible) matrix $\mathbb{1} + v_\theta$ (see (2.11)) one finds

$$\varepsilon_\theta = \mathscr{E}'(v)\mathcal{M} , \qquad \mathcal{M} \equiv \mathbb{1} + v_\theta , \tag{2.27}$$

where as usual $[\varepsilon_\theta]_{ij} = \frac{\partial \varepsilon_i}{\partial \theta_j}$. This suggests to look for w in the form

$$w \equiv \mathcal{M}z \tag{2.28}$$

for some vector-function $z = z(\theta, t)$ to be determined. Thus

$$\begin{aligned}
\mathscr{E}'(v)w + \varepsilon &\equiv \mathscr{E}'(v)(\mathcal{M}z) + \varepsilon \\
&= (\mathscr{E}'(v)\mathcal{M})z + \mathscr{L}_{yy}D\mathcal{M}Dz + D(\mathscr{L}_{yy}\mathcal{M}Dz) \\
&\quad + (\mathscr{L}_{yx} - \mathscr{L}_{xy})\mathcal{M}Dz + \varepsilon \\
&= \varepsilon_\theta z + \mathscr{L}_{yy}D\mathcal{M}Dz + D(\mathscr{L}_{yy}\mathcal{M}Dz) \\
&\quad + (\mathscr{L}_{yx} - \mathscr{L}_{xy})\mathcal{M}Dz + \varepsilon \\
&\equiv q_2 + \mathscr{L}_{yy}D\mathcal{M}Dz + D(\mathscr{L}_{yy}\mathcal{M}Dz) + \mathscr{L}_{yx}^A\mathcal{M}Dz + \varepsilon ,
\end{aligned} \tag{2.29}$$

where the superscript A denotes the antisymmetric part of a matrix $[B^A \equiv B - B^T]$ and

$$q_2 \equiv \varepsilon_\theta z \tag{2.30}$$

is quadratic in ε, w.

Some more algebra is needed: denoting by $\mathcal{A}$ the antisymmetric part of $\mathcal{M}^T \partial_\theta \mathscr{L}_y^T \equiv \mathcal{M}^T \partial_\theta (\mathscr{L}_y^T(\omega + Dv, \theta + v, t))$:

$$\begin{aligned}
\mathcal{A} &\equiv (\mathcal{M}^T \partial_\theta \mathscr{L}_y^T)^A \equiv \mathcal{M}^T \partial_\theta \mathscr{L}_y^T - (\partial_\theta \mathscr{L}_y^T)^T \mathcal{M} \\
&= \mathcal{M}^T \mathscr{L}_{yy}D\mathcal{M} + \mathcal{M}^T \mathscr{L}_{yx}\mathcal{M} - D\mathcal{M}^T \mathscr{L}_{yy}\mathcal{M} - \mathcal{M}^T \mathscr{L}_{xy}\mathcal{M}
\end{aligned} \tag{2.31}$$

and recalling the definition of twist matrix $\mathcal{T}$ (cfr. (2.18)) we see that (2.29) can be rewritten as:

$$\mathscr{E}'(v)w + \varepsilon = q_2 + \mathcal{M}^{-T}\Big[D(\mathcal{T}Dz) + \mathcal{A}Dz\Big] + \varepsilon , \tag{2.32}$$

where $\mathcal{M}^{-T} \equiv (\mathcal{M}^T)^{-1}$.

To proceed, we have to bring in a key element: the operator $D \equiv \omega \cdot \partial_\theta + \partial_t$ acting on the space of real-analytic functions of $(\theta, t) \in \mathbb{T}^{d+1}$ is *invertible* on its range *provided* the $(d+1)$-vector $(\omega, 1)$ satisfies the "Diophantine condition":

$$|\omega \cdot n + m| \geq \frac{1}{\gamma |n|^\tau} , \qquad \forall n \in \mathbb{Z}^d \backslash \{0\}, \ m \in \mathbb{Z} , \tag{2.33}$$

for some $\gamma, \tau > 0$.

Assumption 2.11. *We assume that the vector ω entering in (2.9) through $D \equiv \omega \cdot \partial_\theta + \partial_t$ is a Diophantine vector, i.e. satisfies (2.33).*

Remark 2.12.

(i) Because of the rational independence of $(\omega, 1)$ the range of D consists of functions with vanishing average on $\mathbb{T}^{d+1}$.

(ii) We shall assume that in (2.33) it is:

$$\tau \geq d \quad \text{and} \quad \gamma \geq \frac{\sqrt{5}+3}{2} ; \tag{2.34}$$

$\tau \geq d$ is implied by a classical theorem by Liouville; the second inequality is assumed for simplicity ($\gamma = (\sqrt{5}+3)/2$ is the diophantine constant for the golden mean $\omega = \omega_g \equiv (\sqrt{5}-1)/2$). In the case $\tau > d$ almost all (in the sense of Lebesgue measure) ω's in $\mathbb{R}^d$ are Diophantine.

(iii) In the time independent case one would just suppress m in (2.33) and assume that $\tau \geq d - 1$; for $\tau > d - 1$ one has a set of full measure.

Using Fourier expansions we see immediately that the *unique solution with zero average* of

$$Dg = h(\theta, t) , \qquad h \equiv \sum_{(n,m) \neq 0} \hat{h}_{n,m} \, e^{i(n \cdot \theta + mt)} \tag{2.35}$$

for a given analytic function h with zero average is given by

$$g = \sum_{(n,m) \neq 0} \frac{\hat{h}_{n,m}}{i(\omega \cdot n + m)} \, e^{i(n \cdot \theta + mt)} \equiv D^{-1} h \tag{2.36}$$

and in general all solutions of (2.35) are given by $c + D^{-1}h$ for a constant c. Analyticity of h implies that the Fourier coefficients decay exponentially fast in $|n|+|m|$ and, therefore, by (2.36) $D^{-1}h$ is also analytic (and real-analytic if so is g). On the other hand, there exist Liouville vectors ω which can be approximated by rational vectors arbitrarily fast [e.g. $\exists \omega$: $0 < |\omega \cdot n + m| \leq \exp [\exp - (|n| + |m|)] \ \forall (n, m) \neq 0]$; for such vectors the expansion (in (2.36)) may not make sense and we see that the assumption (2.33) is essential.

There is one more step in order to describe explicitly the solution of the linearized equation and it consists in recognizing that $\mathscr{A}$ is given by the formula

$$\mathscr{A} = D^{-1}(\mathcal{M}^T \varepsilon_\theta)^A , \tag{2.37}$$

showing that the term $\mathscr{A}Dz$ is quadratic in ε, w so that

$$\begin{aligned} \mathscr{E}'(v)w + \varepsilon &= \mathcal{M}^{-T} \left[D(\mathcal{J}Dz) + \mathcal{M}^T \varepsilon \right] + q_2 + q_3 , \\ q_3 &\equiv \mathcal{M}^{-T} D^{-1} (\mathcal{M}^T \varepsilon_\theta)^A \, Dz . \end{aligned} \tag{2.38}$$

The proof of (2.37) is given in Appendix 1; notice however that (2.37) implies in particular that the entries of $\mathscr{A}$ and of $(\mathscr{M}^T \varepsilon_\theta)^A$ are functions with zero average:

$$\langle \mathscr{A} \rangle = 0 , \qquad \langle \mathscr{M}^T \varepsilon_\theta - \varepsilon_\theta^T \mathscr{M} \rangle = 0 , \qquad \langle \cdot \rangle \equiv \int_{\mathbb{T}^{d+1}} \cdot \frac{d\theta dt}{(2\pi)^{d+1}} . \qquad (2.39)$$

Now, since

$$\int \mathscr{M}^T \varepsilon \, d\theta dt = \int \Big[(D\mathscr{L}_y - \mathscr{L}_x)\mathscr{M} \Big]^T$$
$$= -\int \Big[\mathscr{L}_y D\mathscr{M} + \mathscr{L}_x \mathscr{M} \Big]^T = -\int \Big[\partial_\theta \mathscr{L} \Big]^T = 0 , \qquad (2.40)$$

we see that by our assumption on v [see (i), (ii), Definition 2.6] and on ω [Assumption 2.11], the equation

$$D(\mathscr{T}Dz) = -\mathscr{M}^T \varepsilon , \qquad \mathscr{T} \equiv \mathscr{M}^T \mathscr{L}_{yy} \mathscr{M} , \qquad (2.41)$$

can be solved and admits the *general solution*

$$z = D^{-1} \Big\{ \mathscr{T}^{-1} \Big[c_0 - D^{-1}(\mathscr{M}^T \varepsilon) \Big] \Big\} + c_1 , \qquad (2.42)$$

with

$$c_0 \equiv \langle \mathscr{T}^{-1} \rangle^{-1} \, \langle \mathscr{T}^{-1} D^{-1}(\mathscr{M}^T \varepsilon) \rangle , \qquad (2.43)$$

so that (cfr. (2.32))

$$\mathscr{E}'(v)w + \varepsilon = q_2 + q_3 \equiv q . \qquad (2.44)$$

Notice that the choice of the "integration constant" c_0 is enforced by the fact that $\langle Dz \rangle = 0$, while c_1 is arbitrary. We normalize $w = \mathscr{M}z$ by requiring that

$$\langle w \rangle \equiv \langle \mathscr{M}z \rangle = 0 \quad \Leftrightarrow \quad c_1 \equiv -\langle \mathscr{M} D^{-1} \Big\{ \mathscr{T}^{-1} \Big[c_0 - D^{-1}(\mathscr{M}^T \varepsilon) \Big] \Big\} \rangle . \qquad (2.45)$$

We collect the results of this section in the following

Proposition 2.13. *Let ω satisfy Assumption 2.11, let v be a (non-degenerate) approximate solution of the equation (2.9) and let $\varepsilon(\theta, t)$ be the associated error function: $\varepsilon = \mathscr{E}(v)$ (see (2.12)). If we set $w \equiv \mathscr{M}z \equiv (\mathbb{1} + v_\theta)z$ with z defined in (2.42), (2.43), (2.45) (see (2.41) for the definition of $\mathscr{T}$), then $w(\theta, t)$ is a real-analytic function with zero average and setting $v' \equiv v + w$, one has*

$$\mathscr{E}(v') = q_1 + q_2 + q_3 \equiv \varepsilon' , \qquad (2.46)$$

with

$$q_1 \equiv \mathscr{E}(v + w) - \mathscr{E}(v) - \mathscr{E}'(v)w , \quad q_2 \equiv \varepsilon_\theta z , \quad q_3 \equiv (\mathscr{M}^T)^{-1} D^{-1} \Big[(\mathscr{M}^T \varepsilon_\theta)^A \Big] Dz , \qquad (2.47)$$

$\mathscr{E}'(v)$ *being defined in (2.22).*

3. Quasi-Periodic Solutions and Invariant Tori for Lagrangian Systems: Quantitative Analysis

Here we introduce the scale of function spaces necessary to carry out the quantitative analysis and prove the main estimates.

3.1. Spaces of Analytic Functions and Norms

The linearized operator $\mathscr{E}'$ (see (2.22)) involves the degenerate vector field $D \equiv \omega \cdot \partial_\theta + \partial_t$ and, as we already noticed, in order for the inverse D^{-1} to make sense in general the rationally independent vector $(\omega, 1)$ has to satisfy suitable number-theoretical requirements (see Assumption 2.11). However, even in such a case, the fact that it may happen that $|\omega \cdot n_i + m_i| = O(\frac{1}{|n_i|^\tau})$ for suitable sequences $\{(n_i, m_i)\}$, shows that $D^{-1}h$ is *less differentiable* than h [not of course in the direction $(\omega, 1)$ where $D^{-1}h$ *gains* differentiability].

This problem is as old as the modern foundation of mechanics ([A2] and references therein) and is known as the *small divisors problem*. It was only with Carl Siegel [S] in 1942 (in the simpler context of linearization of complex maps around a fixed point) and later with Kolmogorov, Arnold and Moser, that it was possible to overcome technically this problem for the first time (see also Eliasson [El] for a remarkable proof avoiding the Newton method and [Her1], [Her2] for "non local" methods).

The basic technical idea of KAM theory is the following (see [Mo1], [Mo2], [Z], [B], [G1], [G2] for other introductive discussions). One picks a monotone family of Banach spaces of periodic functions on $\mathbb{T}^{d+1}$

$$\mathscr{B}_\xi \subset \mathscr{B}_{\xi'} \quad \text{if} \quad \xi' \leq \xi , \tag{3.1}$$

where the real parameter ξ measures the regularity of the functions, so that if $\langle h \rangle = 0$ and $h \in \mathscr{B}_{\xi'}$, then

$$|D^{-1}h|_{\xi'} \leq K|h|_\xi \tag{3.2}$$

for a suitable constant K depending on ξ'. The unboundedness of D^{-1} reflects in $K \uparrow \infty$ as $\xi' \uparrow \xi$. In a Newton scheme like the one described in the preceding section, the constant K, which will necessarily appear in estimating the new error ε', will be compared with $|\varepsilon|^2$ and one hopes that this square will under iteration eventually beat the divergence due to K.

Let us begin the concrete work.

Denote by $\mathscr{B}_\xi$ the Banach space of real-analytic (periodic) functions on $\mathbb{T}^{d+1}$ which admits an analytic continuation to a domain containing the complex neighbourhood

$$\Delta_\xi \equiv \left\{ (\theta, t) \in \mathbb{C}^{d+1} : |\operatorname{Im} \theta_i| \leq \xi, \ |\operatorname{Im} t| \leq \xi \right\}$$

and let $|\cdot|_\xi$ denote the sup norm on Δ_ξ:

$$|h|_\xi \equiv \sup_{|\operatorname{Im}\theta_i|\leq\xi, |\operatorname{Im}t|\leq\xi} |h(\theta, t)| . \tag{3.3}$$

Let $\mathcal{B}_\xi^d$ denote the space of real-analytic vector valued functions $u : \Delta_\xi \to \mathbb{C}^d$ with the norm

$$|u|_\xi \equiv \sum_{i=1}^d |u_i|_\xi \equiv \sum_{i=1}^d \sup_{\Delta_\xi} |u_i| \,. \tag{3.4}$$

Remark 3.1. It will be important for us to consider functional equations containing one or more parameters μ belonging to some compact subset $\mathcal{P}$ of $\mathbb{C}^m$, e.g. $\mathcal{P} = \{\mu \in \mathbb{C}, |\mu| \le \mu_0\}$ for some $\mu_0 > 0$. In such a case, the above "$\sup_{\Delta_\xi}$" will be replaced by "$\sup_{\Delta_\xi \times \mathcal{P}}$" and the uniformity in the estimates will yield, as byproduct, regular (e.g. analytic) dependence upon the parameter(s) $\mu \in \mathcal{P}$. However, since the set $\mathcal{P}$ will not change in our iterations we shall often denote indifferently

$$\sup_{\Delta_\xi \times \mathcal{P}} |\cdot| \equiv |\cdot|_\xi \equiv |\cdot|_{\xi,\mathcal{P}} \,. \tag{3.5}$$

The norm of matrix/tensor-valued functions will then be defined by the standard operator norm: if

$$\mathcal{M} : \Delta_\xi \supset \mathbb{T}^{d+1} \to L(\mathbb{C}^d) \equiv L(\mathbb{C}^d, \mathbb{C}^d) \equiv \{d \times d - \text{matrices}\} \,, \tag{3.6}$$

then we set

$$|\mathcal{M}|_\xi = \sup_{|c|=1} |\mathcal{M}c|_\xi \qquad (c \in \mathbb{C}^d \,, \quad |c| \equiv \sum_{i=1}^d |c_i|) \tag{3.7}$$

($\mathcal{M}c$ is a $\mathbb{C}^d$ valued function in Δ_ξ and therefore the $|\cdot|_\xi$ is defined in (3.4) above) and in general by induction if

$$\mathcal{T} : \Delta_\xi \to L(\underbrace{\mathbb{C}^d, L(\mathbb{C}^d, ..., L(\mathbb{C}^d, \mathbb{C}^d), \mathbb{C}^d), ...)}_{p \text{ times}} \equiv L^p(\mathbb{C}^d) \tag{3.8}$$

then, for $c \in \mathbb{C}^d$, $\mathcal{T}c \in L^{p-1}(\mathbb{C}^d)$ and we set

$$|\mathcal{T}|_\xi \equiv \sup_{|c|=1} |\mathcal{T}c| \qquad \left(c \in \mathbb{C}^d, |c| = \sum_{i=1}^d |c_i| \right) \,. \tag{3.9}$$

For example if $u : \Delta_\xi \to \mathbb{C}^d$, $u_\theta : \Delta_\xi \to L(\mathbb{C}^d)$ and

$$|u_\theta|_\xi \equiv \sup_{|c|=1} |u_\theta c|_\xi \equiv \sup_{|c|=1} \sum_{i=1}^d |\sum_{j=1}^d \frac{\partial u_i}{\partial \theta_j} c_j|_\xi$$

$$\equiv \sup_{|c|=1} \sum_{i=1}^d \sup_{\Delta_\xi} |\sum_{j=1}^d \frac{\partial u_i}{\partial \theta_j} c_j| \tag{3.10}$$

or if ϕ is a map of Δ_ξ into the domain of $\mathcal{L}(y, x, t)$ then

$$|\mathcal{L}_{yxy} \circ \phi|_\xi = \sup_{|c|=|b|=1} \sum_{i=1}^d \sup_{\Delta_\xi} |\sum_{j,k=1}^d \mathcal{L}_{y_i x_j y_k} \circ \phi \, b_j c_k| \,. \tag{3.11}$$

3.2. Analytic Tools

The basic technical tools go back to Cauchy and give the possibility of estimating the derivative of a holomorphic function in a domain Ω by the supremum of the function in a bigger domain Ω' divided by the distance between the boundaries. In formulae:

Lemma 3.2. *Let h be an analytic map from $\Omega \times \mathcal{P} \to \mathbb{C}$, where Ω is a (smooth) domain in $\mathbb{C}^d$ and $\mathcal{P} \subset \mathbb{C}^k$ a space of parameters. Then for any subdomain $\Omega' \subset \Omega$ with $\mathrm{dist}(\Omega', \partial\Omega) \equiv \delta > 0$ and for any multi index $m = (m_1, ..., m_d) \in \mathbb{N}^d$ one has*

$$\sup_{\Omega' \times \mathcal{P}} |\partial_z^m h| \equiv \sup_{\Omega' \times \mathcal{P}} |\frac{\partial^{|m|} h}{\partial z_1^{m_1} ... \partial z_d^{m_d}}| \leq m! \; \delta^{-|m|} \sup_{\Omega \times \mathcal{P}} |h| \qquad (3.12)$$

$(|m| = m_1 + ... + m_d)$.

Let h be an analytic map $h : \Omega \times \mathcal{P} \to L^p(\mathbb{C}^d)$ for some $p \in \mathbb{N}$ $(L^0(\mathbb{C}^d) \equiv \mathbb{C}^d)$; then, $\forall l \in \mathbb{Z}_+$, $\partial_z^l h \in L^{p+l}(\mathbb{C}^d)$ and

$$\sup_{\Omega' \times \mathcal{P}} |\partial_z^l h| \leq l! \; \delta^{-l} \sup_{\Omega \times \mathcal{P}} |h| . \qquad (3.13)$$

The proof of this simple lemma, which is based on Cauchy's integral formula, is given in Appendix 2. A consequence of this lemma is that if $h : \Delta_\xi \to \mathbb{C}^d$, then

$$|\partial_\theta^l h|_{\xi - \delta} \leq l! \delta^{-l} |h|_\xi ; \qquad (3.14)$$

notice that in the last inequality it would not be necessary to reduce the domain of t, which is simply playing the role of a parameter.

Similar statements hold also for the operator D^{-1}.

Lemma 3.3. *Let $h = h(\theta, t; \mu)$ be a real-analytic map of $\Delta_\xi \times \mathcal{P}$ into $\mathcal{H}$, where $\mathcal{H}$ is either $\mathbb{C}$, or $\mathbb{C}^d$ or $L^p(\mathbb{C}^d)$ and let $l \geq 1$. Then $(|\cdot|_{\xi,\mathcal{P}} \equiv \sup_{\Delta_\xi \times \mathcal{P}} |\cdot|)$*

$$|D^{-1} \partial_\theta^l h|_{\xi - \delta, \mathcal{P}} \leq \sigma_l(2\delta) \, |h|_{\xi,\mathcal{P}} , \qquad (3.15)$$

where

$$\sigma_l(\rho) = \left[2^{d+1} \sum_{(n,m) \in \mathbb{Z}^{d+1} \backslash (0,0)} (\frac{\|n\|^l}{\omega \cdot n + m})^2 \, e^{-\rho(|n|+|m|)} \right]^{\frac{1}{2}} , \qquad (3.16)$$

$(\|n\| \equiv (\sum_{i=1}^d |n_i|^2)^{1/2}, |n| \equiv \sum_{i=1}^d |n_i|)$. *The same estimate holds for $l = 0$ provided h has vanishing mean value over $\mathbb{T}^{d+1}$. If $(\omega, 1)$ verifies Assumption 2.11 then*

$$\sigma_l(\rho) < K_l \gamma \delta^{-(\tau+l)} , \qquad K_l \equiv 2^{d+2-(\tau+l)} \sqrt{\Gamma(2(\tau+l)+1)} , \qquad (3.17)$$

Γ *being Euler's gamma function.*

The proof is given in Appendix 3. Notice that if $(\omega, 1)$ satisfies (2.33) it is very easy to check (3.15) with σ replaced by $K\gamma\delta^{-(\tau+l+d)}$. For example, in the case $l = 0$ and $\mathcal{H} = \mathbb{C}$, recalling that the Fourier coefficients, $h_{(n,m)}$, of an analytic function h decay exponentially:

$$|h_{(n,m)}| \leq e^{-(|n|+|m|)\xi} \, |h|_\xi , \qquad (3.18)$$

one finds immediately $(h_{(0,0)} \equiv \langle h \rangle = 0)$:

$$
\begin{aligned}
|D^{-1}h|_{\xi-\delta} &= | \sum_{(n,m)\neq 0} \frac{h_{(n,m)}}{i(\omega \cdot n + m)} \, e^{i(n\cdot\theta+mt)}|_{\xi-\delta} \\
&\leq |h|_{\xi} \, \gamma \sum_{(n,m)\neq 0} |n|^{\tau} e^{-\xi} e^{(|n|+|m|)(\xi-\delta)} \\
&= \gamma|h|_{\xi} \sum_{(n,m)\neq 0} |n|^{\tau} e^{-\delta(|n|+|m|)} \\
&\leq K\gamma|h|_{\xi}\delta^{-(\tau+d)} ,
\end{aligned}
\tag{3.19}
$$

for some positive constant K depending on τ, d.

The fact that $\tau + d$ in (3.19) can be actually replaced by τ is a quite remarkable fact due to Rüssmann ([R1], [R2], [R3]) (see also Appendix 3).

There is also another reason for leaving the explicit expression of σ_l in (3.16) and is related to our computer-assisted technique: in trying to establish "sharp" numerical bounds there will be delicated points where we shall estimate σ "accurately" with the aid of the computer rather than using the simple (and necessarily "non sharp") bound in (3.17).

3.3. Norm-Parameters

Let us go back to (2.9) and assume that a non degenerate solution v is given (see Definition 2.6). In this section we introduce several positive numbers controlling the norms of the relevant objects. We need to define the following complex domains of $\mathbb{C}^{d+1}$ (recall that $\Delta_{\xi} \equiv \{(\theta, t) \in \mathbb{C}^{d+1} : |\mathrm{Im}\,(\theta_i)| \leq \xi, \, |\mathrm{Im}\,(t)| \leq \xi\}$)

$$
\mathcal{D}^{\xi} \equiv \Phi_v \Delta_{\xi} \equiv \{(y, x, t) \in \mathbb{C}^{2d+1} \mid y = \omega + Dv(\theta, t), \, x = \theta + v(\theta, t), \, (\theta, t) \in \Delta_{\xi}\}
\tag{3.20}
$$

and for any $\rho = (\rho_1, \rho_2) \in \mathbb{R}^2_+$ $(\Omega = |(\omega, 1)| = \sum_{i=1}^{d} |\omega_i| + 1)$:

$$
\mathcal{D}^{\xi}_{\rho} \equiv \{(y, x, t) = (y_0 + y_1, x_0 + x_1, t_0) \mid (y_0, x_0, t_0) \in \mathcal{D}^{\xi}, \, |y_1| \leq \Omega\rho_1, \, |x_1| \leq \rho_2\} .
\tag{3.21}
$$

Now, because of Definition 2.6 and the analyticity of $\mathcal{L}$ we can assume that there exist ξ, $\alpha > 0$ such that

$$
\begin{aligned}
&(i) \quad \mathcal{D}^{\xi}_{(\alpha,\alpha)} \subset \text{ analyticity domain of } \mathcal{L} \\
&(ii) \quad \sup_{\mathcal{D}^{\xi}_{(\alpha,\alpha)}} |\mathcal{L}^{-1}_{yy}| < \infty \\
&(iii) \quad \sup_{\Delta_{\xi}} |(I + v_{\theta})^{-1}| \equiv \sup_{\Delta_{\xi}} |\mathcal{M}^{-1}| < \infty .
\end{aligned}
\tag{3.22}
$$

We then set $(|\cdot|_\xi \equiv \sup_{\Delta_\xi} |\cdot|$ and notice that $\mathcal{D}^\xi \subset \mathcal{D}_\alpha^\xi)$

$$
\begin{aligned}
&|\mathcal{L}_{yy}|_{\mathcal{D}^\xi} \leq L , &&|\mathcal{L}_{yy}^{-1}|_{\mathcal{D}^\xi} \leq \overline{L} , \\
&|\mathbb{1} + v_\theta|_\xi \equiv |\mathcal{M}|_\xi \leq M , &&|\mathcal{M}^{-1}|_\xi \leq \overline{M} , \\
&|v_t|_\xi \leq S , \\
&|\varepsilon|_\xi \equiv |\mathscr{E}(v)|_\xi \leq E .
\end{aligned}
\tag{3.23}
$$

We need more norms relative to the derivatives of $\mathscr{L}$ (in the following formula $|\cdot|$ is short for $|\cdot|_{\mathcal{D}^\xi}$)

$$
\begin{aligned}
&\max\{\, \Omega^2|\mathscr{L}_{xyy}|, \ \Omega|\mathscr{L}_{xxy}|, \ |\mathscr{L}_{xxx}| \,\} \leq L_3 , \\
&\max\{\Omega^2|\mathscr{L}_{yyy}|, \ \Omega|\mathscr{L}_{yyx}|, \ |\mathscr{L}_{yxx}| \,\} \,\Omega \ \leq \ L_3' , \\
&\max\{\Omega^3|\mathscr{L}_{yyyy}|, \ \Omega^2|\mathscr{L}_{yyyx}|, \ \Omega^2|\mathscr{L}_{yyxy}|, \ \Omega|\mathscr{L}_{yyxx}|, \ \Omega|\mathscr{L}_{yxxy}|, \ |\mathscr{L}_{yxxx}| \,\} \,\Omega \ \leq \ L_4 , \\
&\max\{\Omega^2|\mathscr{L}_{yyyt}|, \ \Omega|\mathscr{L}_{yyxt}|, \ |\mathscr{L}_{yxxt}| \,\} \,\Omega \ \leq \ L_4' , \qquad (\text{where } |\cdot| \equiv |\cdot|_{\mathcal{D}^\xi}) .
\end{aligned}
\tag{3.24}
$$

The various powers of $\Omega \equiv |(\omega, 1)|$ have been introduced for later convenience. It might appear somewhat strange that for a quadratic scheme for an equation involving the first derivative of $\mathscr{L}$, we introduce the fourth derivatives, but, as we shall see, these "extra" derivatives allow us to avoid one more loss in the analyticity domain, which is a very costly operation from the point of view of accurate bounds.

Finally, we will also denote by $s_i(\rho)$, $\rho > 0$, an upper bound on $\sigma_i(\rho)$ (cfr. (3.16)) and for simplicity we assume that $\rho \to s_i(\rho)$ is decreasing:

$$
\sigma_i(\rho) \leq s_i(\rho) \qquad (\rho \to s_i(\rho) \ \text{decreasing}) .
\tag{3.25}
$$

We end up this section noticing the following simple relations among the above parameters:

$$
L\overline{L} \geq 1 , \qquad M \geq 1 , \qquad \overline{M} \geq 1 .
\tag{3.26}
$$

The first inequality is obvious; in fact:

$$
1 = |\mathbb{1}| = |\mathscr{L}_{yy} \mathscr{L}_{yy}^{-1}|_{\mathcal{D}^\xi} \leq L\overline{L} .
\tag{3.27}
$$

Next:

$$
M \equiv |\mathcal{M}|_\xi = |\mathcal{M}^T|_\xi > |\mathcal{M}|_0 = |\mathcal{M}^T|_0 ;
\tag{3.28}
$$

then, if e_1 denotes the d-vector $(1, 0, ..., 0)^T$ $(|e_1| = 1)$,

$$
\mathcal{M}^T e_1 \ = \ e_1 + (\partial_\theta v_1)^T .
\tag{3.29}
$$

Therefore if (θ_0, t_0) is a critical point of the periodic function $v_1(\theta, t)$ then

$$
\mathcal{M}^T(\theta_0, t_0) \, e_1 \ = \ e_1 , \qquad \mathcal{M}^{-T}(\theta, t) \, e_1 = e_1 .
\tag{3.30}
$$

Thus, from (3.28) and (3.30) it follows immediately the second and third inequalities of (3.26).

3.4. Bounds on the Solution of the Linearized Equation

Here we shall provide bounds on $|w|_{\xi-\delta}$ and $|w_\theta|_{\xi-\delta}$, where w is the solution of the linearized equation (2.44) and δ is (at the moment) an arbitrary number such that

$$0 < \delta \leq \xi . \tag{3.31}$$

Since

$$|w|_{\xi-\delta} \equiv |\mathcal{M}z|_{\xi-\delta} \leq |\mathcal{M}|_\xi |z|_\xi \leq M|z|_\xi \tag{3.32}$$

and, by (3.13) applied to $\mathcal{M} : \Delta_\xi \to L(\mathbb{C}^d)$,

$$|w_\theta|_{\xi-\delta} \equiv |\mathcal{M}_\theta z + \mathcal{M}z_\theta|_{\xi-\delta} \leq M(\delta^{-1}|z|_{\xi-\delta} + |z_\theta|_{\xi-\delta}) , \tag{3.33}$$

we see that we have to estimate $|z|_{\xi-\delta}$ and $|z_\theta|_{\xi-\delta}$.

Remark 3.4. Obviously, once a bound on $|z|_{\xi-\delta}$ is established one could immediately estimate $|z_\theta|_{\xi-\delta}$ in, say, $\xi - 2\delta$ by using Cauchy estimates. However, with some more work, it is possible to estimate z_θ *directly* in $\Delta_{\xi-\delta}$. Restricting the domain of analyticity (or, better, the domain where it is possible to estimate the sup-norms) is a *very costly operation* from the point of view of "optimal bounds", and it is, therefore, important to avoid unnecessary analyticity losses.

Let us begin by estimating the constants c_0 and c_1 appearing in the definition of z (cfr. (2.42), (2.43), (2.45)). We need some properties of the twist matrix $\mathcal{T}(\theta, t)$ for (θ, t) real [recall that $|\cdot|_0 \equiv \sup_{\mathbb{T}^{d+1}} |\cdot|$, $\phi_v \equiv (\omega + Dv(\theta, t), \theta + v(\theta, t), t)$]:

Lemma 3.5. *Let* $\mathcal{T} \equiv \mathcal{M}^T \mathcal{L}_{yy} \mathcal{M} \equiv \mathcal{M}^T \mathcal{L}_{yy} \circ \Phi_v \mathcal{M}$ *be the twist matrix of a non degenerate approximate solution* v *and let* $M, \overline{M}, L, \overline{L}$ *denote upper bounds on (respectively)* $|\mathcal{M}|_\xi$, $|\mathcal{M}^{-1}|_\xi$, $|\mathcal{L}_{yy}|_{\mathcal{D}\xi}$, $|\mathcal{L}_{yy}^{-1}|_{\mathcal{D}\xi}$: *then*

$$\begin{aligned}
&(i) \quad \overline{M}^{-2}\overline{L}^{-1} \leq |\mathcal{T}|_0 \leq |\mathcal{T}|_\xi \leq M^2 L \\
&(ii) \quad M^{-2}L^{-1} \leq |\mathcal{T}^{-1}|_0 \leq |\mathcal{T}^{-1}|_\xi \leq \overline{M}^2 \overline{L} \\
&(iii) \quad |(\mathcal{T}^{-1})^{-1}|_0 \leq |\mathcal{T}|_0 \qquad \left(\langle \cdot \rangle \equiv \int_{\mathbb{T}^{d+1}} \cdot \frac{d\theta dt}{(2\pi)^{d+1}} \right) .
\end{aligned}$$

The proof of this simple lemma is given in Appendix 4. From this lemma it follows immediately that

$$|c_0| \leq (M\overline{M})^2 (L\overline{L}) \, |D^{-1}(\mathcal{M}^T \varepsilon)|_0 \tag{3.34}$$

and from (3.15) (used here with $l = 0$ and $\delta = \xi$) it follows

$$|c_0| \leq M(M\overline{M})^2 (L\overline{L}) \, s_0(2\xi)E . \tag{3.35}$$

The estimate of c_1 is analogous. Using twice (3.15) (with $\delta = \xi/2$ each time) one obtains:

$$
\begin{aligned}
|c_1| &\leq M \, |D^{-1} \{(\mathcal{M}^T \mathcal{L}_{yy} \mathcal{M})^{-1} [c_0 - D^{-1}(\mathcal{M}^T \varepsilon)]\}|_0 \\
&\leq M s_0(\xi) \quad |(\mathcal{M}^T \mathcal{L}_{yy} \mathcal{M})^{-1} [c_0 - D^{-1}(\mathcal{M}^T \varepsilon)]|_{\xi/2} \\
&\leq M s_0(\xi) \, |(\mathcal{M}^T \mathcal{L}_{yy} \mathcal{M})^{-1}|_{\xi/2} \, (|c_0| + |D^{-1}(\mathcal{M}^T \varepsilon)|_{\xi/2}) \\
&\leq (M\overline{M})^2 s_0(\xi)^2 \, E\overline{L} \left[1 + (M\overline{M})^2 \, L\overline{L} \, \frac{s_0(2\xi)}{s_0(\xi)} \right] .
\end{aligned}
\tag{3.36}
$$

The estimate of $|z|_{\xi-\delta}$ proceeds along the same lines, using (3.15) twice (with δ replaced here by $\delta/2$ twice):

$$
\begin{aligned}
|z|_{\xi-\delta} &\leq s_0(\delta) \, |(\mathcal{M}^T \mathcal{L}_{yy} \mathcal{M})^{-1} (c_0 - D^{-1}(\mathcal{M}^T \varepsilon))|_{\xi-\delta/2} + |c_1| \\
&\leq s_0(\delta)|(\mathcal{M}^T \mathcal{L}_{yy} \mathcal{M})^{-1}|_\xi \, (|c_0| + |D^{-1}(\mathcal{M}^T \varepsilon)|_{\xi-\delta/2}) + |c_1| \\
&\leq s_0(\delta)|(\mathcal{M}^T \mathcal{L}_{yy} \mathcal{M})^{-1}|_\xi \, (|c_0| + s_0(\delta)|\mathcal{M}^T \varepsilon|_\xi) + |c_1| \\
&\leq E\overline{L} s_0(\delta)^2 (M\overline{M})\overline{M} \left\{ 1 + (M\overline{M})^2 \, L\overline{L} \, \frac{s_0(2\xi)}{s_0(\delta)} \right. \\
&\quad \left. + M \left(\frac{s_0(\xi)}{s_0(\delta)} \right)^2 \left[1 + (M\overline{M})^2 \, L\overline{L} \, \frac{s_0(2\xi)}{s_0(\xi)} \right] \right\} \\
&\equiv E\overline{L} \, \overline{M} \, (M\overline{M}) \, s_0(\delta)^2 \, b \\
&\equiv E\overline{L} a \, M^{-1} ,
\end{aligned}
\tag{3.37}
$$

where last identities define the parameters a and b. Also the estimate of z_θ is similar as long as one uses (3.15) twice but the first term with $l = 1$ (and δ replaced by $\delta/2$):

$$
\begin{aligned}
|z_\theta|_{\xi-\delta} &= |\frac{\partial}{\partial \theta} D^{-1} \{ (\mathcal{M}^T \mathcal{L}_{yy} \mathcal{M})^{-1} \left[c_0 - D^{-1}(\mathcal{M}^T \varepsilon) \right] \}|_{\xi-\delta} \\
&\leq s_1(\delta)|(\mathcal{M}^T \mathcal{L}_{yy} \mathcal{M})^{-1}|_\xi \, (|c_0| + |D^{-1}(\mathcal{M}^T \varepsilon)|_{\xi-\delta/2}) \\
&\leq E\overline{L} M \overline{M}^2 s_0(\delta) s_1(\delta) \left\{ 1 + (M\overline{M})^2 \, L\overline{L} \, \frac{s_0(2\xi)}{s_0(\delta)} \right\} \\
&= E\overline{L} \, M \, (M\overline{M}) \, s_0(\delta) s_1(\delta) \, b_1 ,
\end{aligned}
\tag{3.38}
$$

where the last identity defines the parameter b_1.
Finally, by (3.32) and (3.33) we find:

$$
\begin{aligned}
|w|_{\xi-\delta} &\leq E\overline{L} a , \qquad a \equiv s_0(\delta)^2 \, (M\overline{M})^2 b , \\
b &\equiv 1 + (M\overline{M})^2 L\overline{L} \frac{s_0(2\xi)}{s_0(\delta)} + M \left(\frac{s_0(\xi)}{s_0(\delta)} \right)^2 \left[1 + (M\overline{M})^2 L\overline{L} \frac{s_0(2\xi)}{s_0(\xi)} \right]
\end{aligned}
\tag{3.39}
$$

and

$$
\begin{aligned}
|w_\theta|_{\xi-\delta} &\leq E\overline{L} a \left(\delta^{-1} + \frac{s_1(\delta)}{s_0(\delta)} \frac{b_1}{b} \right) , \\
b_1 &\equiv 1 + (M\overline{M})^2 L\overline{L} \, \frac{s_0(2\xi)}{s_0(\delta)} .
\end{aligned}
\tag{3.40}
$$

3.5. Bounds on the New Error Term

At this point all the machinery is set up and we are ready to estimate the new error term $\varepsilon' \equiv \mathscr{C}(v')$ (cfr. Proposition 2.13) and to check that

$$|\varepsilon'|_{\xi'} \equiv |\varepsilon'|_{\xi-\delta} \leq K\bar{L}E^2 , \qquad (3.41)$$

for a suitable constant K depending on ξ, δ and on the norm-parameters introduced in §3.3. The purpose of this section (and relative appendices where all details are carried out) is to provide an explicit and accurate expression for the constant K.

It is fairly clear and straightforward how to proceed; however the care (apparently quite excessive) we shall put in determining K is justified by *(i)* the need for complete explicitness (expecially in view of concrete applications) and *(ii)* the need of keeping track of the quantitative roles all the different parameters play in the scheme so as, e.g., to avoid dangerous (from the point of view of accuracy) approximations.

Of course in order to get a manageable theorem we shall need to do several semplifications at the expense of accuracy; but one of the points of the present work (and of [CC1], [CC2], [CC3], [CC4], [CFP], [CG]) is that the stringent smallness requirement of such a theorem can be mitigated by a previous iterative application of the set of accurate estimates we are working out here. We will come back on this crucial point in the next section.

After these premises we formulate (without shame) the main result of this section.

Proposition 3.6. *Let* $\varepsilon' = \mathscr{C}(v') \equiv q_1 + q_2 + q_3$ *as in (2.46), (2.47) of Proposition 2.13; let* $0 < \delta < \xi$ *and let* $M, \bar{M}, L, \bar{L}, E, s_i(\rho)$ $(i = 0, 1), L_3, L_4, L_3', L_4'$ *be as in (3.23), (3.24), (3.25) of §3.3. Then (3.41), i.e.*

$$|\varepsilon'|_{\xi'} \leq K\bar{L}E^2 , \qquad \xi' = \xi - \delta , \qquad (3.42)$$

holds with

$$K \equiv a \left\{ \frac{\delta^{-1}}{M} + \frac{a}{2}(c+1)^2 \, \bar{L}\Big[L_3 + L_4(S+M)(1+\delta^{-1}) + \frac{4}{3}E\bar{L}a\delta^{-2}L_4 \right.$$
$$\cdot \left(\frac{\delta}{4} + g + \frac{\delta^2}{4}\frac{s_1(\delta)}{s_0(\delta)}\frac{b_1}{b}\right) + L_4'\Big]$$
$$+ 4L_3' \, a\bar{L}(c+1)\delta^{-2}\left(\frac{\delta}{4} + g + \frac{\delta^2}{4}\frac{s_1(\delta)}{s_0(\delta)}\frac{b_1}{b}\right)$$
$$\left. + \chi_d \cdot 4\delta^{-1}\bar{M}s_0(\delta)\,\frac{a'}{a} \right\} ,$$

where

$$\begin{cases} \chi_d = 0 & d = 1 \\ \chi_d = 1 & d \geq 2, \end{cases}$$

$$a \equiv (M\bar{M})^2 s_0(\delta)^2 b , \qquad b_1 \equiv 1 + (M\bar{M})^2 L\bar{L}\,\frac{s_0(2\xi)}{s_0(\delta)} ,$$

$$b \equiv b_1 + M\left(\frac{s_0(\xi)}{s_0(\delta)}\right)^2 \left[1 + (M\bar{M})^2 L\bar{L}\,\frac{s_0(2\xi)}{s_0(\xi)}\right] , \qquad c \equiv \delta^{-1} + \frac{a'}{\Omega a} ,$$

$$a' \equiv (M\overline{M})^2 s_0(2\delta) b_1' \,, \qquad b_1' \equiv 1 + (M\overline{M})^2 L\overline{L} \, \frac{s_0(2\xi)}{s_0(2\delta)} \,,$$

$$a'' \equiv (M\overline{M})^2 s_0(\delta) b_1 \,, \qquad g \equiv 1 + \frac{\delta}{4} \frac{a'}{a} + \frac{\delta}{4} s_1(\delta) \frac{a''}{a} + \frac{\delta}{2} \frac{a''}{a} \,. \tag{3.43}$$

Proof: The easiest term to treat is q_2: by (3.37) one sees that

$$|q_2|_{\xi'} \equiv |\varepsilon_\theta z|_{\xi'} \leq |\varepsilon_\theta|_{\xi'} |z|_{\xi'} \leq |\varepsilon|_{\xi'} |z|_{\xi'} \, \delta^{-1} \leq E^2 \overline{L} \, \frac{a}{M\delta} \,, \tag{3.44}$$

with

$$a \equiv (M\overline{M})^2 s_0(\delta)^2 b \,, \qquad b_1 \equiv 1 + (M\overline{M})^2 L\overline{L} \, \frac{s_0(2\xi)}{s_0(\delta)}$$

and

$$b \equiv b_1 + M \left(\frac{s_0(\xi)}{s_0(\delta)} \right)^2 \left[1 + (M\overline{M})^2 L\overline{L} \, \frac{s_0(2\xi)}{s_0(\xi)} \right] \,.$$

Next, recalling the definiton of v' (Proposition 2.13) and of ϕ_v, ϕ_v', we see that

$$\begin{aligned}
|q_1|_{\xi'} &\leq \left| D \left[\mathscr{L}_y^T \circ \phi_{v'} - \mathscr{L}_y^T \circ \phi_v - \mathscr{L}_{yy} \circ \phi_v Dw - \mathscr{L}_{yx} \circ \phi_v w \right] \right|_{\xi'} \\
&\quad + \left| \mathscr{L}_x^T \circ \phi_{v'} - \mathscr{L}_x^T \circ \phi_v - \mathscr{L}_{xy} \circ \phi_v Dw - \mathscr{L}_{xx} \circ \phi_v w \right|_{\xi'} \\
&\equiv |Dq_1^{(y)}|_{\xi'} + |q_1^{(x)}|_{\xi'} \,,
\end{aligned} \tag{3.45}$$

where $q_1^{(y)}$ and $q_1^{(x)}$ have been here defined (in the obvious way). Then using the integral formula for the remainder of Taylor's formula (at the second order) and recalling the definition of L_3 one finds

$$\begin{aligned}
|q_1^{(x)}|_{\xi'} &\leq \frac{1}{2} \left[|\mathscr{L}_{xyy}|_{\xi'} |Dw|_{\xi'}^2 + 2|\mathscr{L}_{xxy}|_{\xi'} |w|_{\xi'} |Dw|_{\xi'} + |\mathscr{L}_{xxx}|_{\xi'} |w|_{\xi'}^2 \right] \\
&\leq \frac{L_3}{2} \left[\frac{|Dw|_{\xi'}^2}{\Omega^2} + 2|w|_{\xi'} \frac{|Dw|_{\xi'}}{\Omega} + |w|_{\xi'}^2 \right] \,,
\end{aligned} \tag{3.46}$$

where $|\mathscr{L}.|_{\xi'}$ is short for $|\mathscr{L}. \circ \phi_v|_{\xi'}$. We need therefore a bound on Dw:

$$Dw = D(\mathcal{M}z) = (D\mathcal{M})z + \mathcal{M}Dz \,. \tag{3.47}$$

First of all observe, in general, that if $f : \theta \in \mathbb{R}^m \to \mathbb{R}^d$ and $\alpha \in \mathbb{R}^m$, then

$$(\alpha \cdot \partial_\theta)f = \partial_\theta f \, \alpha \quad \Rightarrow \quad |(\alpha \cdot \partial_\theta)f| \leq |\alpha| \, |\partial_\theta f| \,, \tag{3.48}$$

so that by Lemma 3.2 one finds

$$|D\mathcal{M}|_{\xi'} \leq \Omega \delta^{-1} M \,. \tag{3.49}$$

Recalling (3.37) one obtains

$$
\begin{aligned}
|Dz|_{\xi'} &= |(\mathcal{M}^T \mathcal{L}_{yy} \mathcal{M})^{-1} \left[c_0 - D^{-1}(\mathcal{M}^T \varepsilon) \right]|_{\xi'} \\
&\leq \overline{M}^2 \overline{L} \left[|c_0| + s_0(2\delta)ME \right] \\
&\leq \overline{M}^2 \overline{L} \left[M^3 \overline{M}^2 L \overline{L} s_0(2\xi)E + s_0(2\delta)ME \right] \\
&\equiv E \overline{L} M \overline{M}^2 \, s_0(2\delta) b_1' \\
&\equiv E \overline{L} \, \frac{a'}{M} ,
\end{aligned}
\tag{3.50}
$$

where we have introduced the constants

$$
a' \equiv (M\overline{M})^2 s_0(2\delta) b_1' , \qquad b_1' \equiv 1 + (M\overline{M})^2 L \overline{L} \, \frac{s_0(2\xi)}{s_0(2\delta)} .
\tag{3.51}
$$

Thus, form (3.47), (3.49), (3.50):

$$
\begin{aligned}
|Dw|_{\xi'} &\leq \Omega \delta^{-1} E \overline{L} a + E \overline{L} a' = E \overline{L} a \left(\Omega \delta^{-1} + \frac{a'}{a} \right) \\
&\equiv E \, \overline{L} \, a \, \Omega \, c ,
\end{aligned}
\tag{3.52}
$$

where

$$
c = \delta^{-1} + \frac{a'}{a\Omega} .
$$

And finally,

$$
|q_1^{(x)}|_{\xi'} \leq \frac{L_3}{2} \left[E^2 \overline{L}^2 a^2 c^2 + 2E^2 \overline{L}^2 a^2 c + E^2 \overline{L}^2 a^2 \right] = \frac{L_3}{2} E^2 \overline{L}^2 a^2 \, (c+1)^2 .
\tag{3.53}
$$

To estimate $Dq_1^{(y)}$ we could treat $q_1^{(y)}$ similarly to $q_1^{(x)}$ and then use Lemma 3.2 to estimate $Dq_1^{(y)}$; but this would lead to an extra loss in the domain of analyticity in θ, which we want to avoid. Instead we compute explicitly $Dq_1^{(y)}$ and calling here

$$
f = f(\omega + Dv, \theta + v, t) = \mathcal{L}_y^T(\omega + Dv, \theta + v, t) , \qquad f^+ = f(\omega + Dv + Dw, \theta + v + w, t)
\tag{3.54}
$$

we find

$$
\begin{aligned}
|Dq_1^{(y)}|_{\xi'} &\equiv |D\left[f^+ - f - f_y Dw - f_x w \right]|_{\xi'} \\
&\equiv |\omega \cdot \partial_\theta (f^+ - f - f_y Dw - f_x w) + \partial_t(f^+ - f - f_y Dw - f_x w)|_{\xi'} \\
&\leq |\partial_\theta(f^+ - f - f_y Dw - f_x w) \, \omega|_{\xi'} + |\partial_t(f^+ - f - f_y Dw - f_x w)|_{\xi'} \\
&\equiv A_1 + A_2 .
\end{aligned}
\tag{3.55}
$$

And by the integral formula for the remainder of Taylor's expansion one has:

$$
\begin{aligned}
A_1 &= |\partial_\theta(f^+ - f - f_y Dw - f_x w) \, \omega|_{\xi'} \\
&= |\partial_\theta \left\{ \int_0^1 (1 - \beta) \left[f_{yy} Dw Dw + f_{yx} Dw \, w + f_{xy} w \, Dw + f_{xx} w w \right] d\beta \right\} \omega|_{\xi'}
\end{aligned}
\tag{3.56}
$$

and

$$
A_2 = |\partial_t \left\{ \int_0^1 (1 - \beta) \left[f_{yy} Dw Dw + f_{yx} Dw \, w + f_{xy} w Dw + f_{xx} w w \right] d\beta \right\}|_{\xi'} ,
$$

where the derivatives of f are evaluated at $(\omega + Dv + \beta Dw, \theta + v + \beta w, t)$. Performing the θ and t derivatives one sees that A_i involves, besides derivatives of $\mathcal{L}$ and quantities already controlled, also Dw_θ, w_t, Dw_t. To estimate Dw_θ observe that

$$Dw_\theta = D(\mathcal{M}z_\theta + \mathcal{M}_\theta z) = (D\mathcal{M})z_\theta + \mathcal{M} Dz_\theta + (D\mathcal{M}_\theta)z + \mathcal{M}_\theta Dz \qquad (3.57)$$

and

$$|Dz_\theta|_{\xi'} \leq 2\delta^{-1}|Dz|_{\xi - \frac{\delta}{2}} = 2\delta^{-1}\frac{E\overline{L}a''}{M}, \qquad a'' = (M\overline{M})^2 s_0(\delta)b_1 .$$

Therefore, from the inequality

$$|D\mathcal{M}|_{\xi'} \leq \Omega\delta^{-1}M, \qquad \Omega \equiv \sum_{i=1}^{d} \omega_i + 1 ,$$

one has

$$|Dw_\theta|_{\xi'} \leq \Omega E\overline{L}\delta^{-1} \left[a''s_1(\delta) + 2a'' + 4\delta^{-1}a + a' \right] = 4E\overline{L}\delta^{-2}ag\Omega ,$$

with

$$g \equiv 1 + \frac{\delta}{4}\frac{a'}{a} + \frac{\delta}{4}s_1(\delta)\frac{a''}{a} + \frac{\delta}{2}\frac{a''}{a} .$$

Analogously one obtains

$$|w_t|_{\xi'} \leq E\overline{L}a \left(\delta^{-1} + \frac{s_1(\delta)}{s_0(\delta)}\frac{b_1}{b} \right) ,$$

$$|Dw_t|_{\xi'} \leq 4E\overline{L}\delta^{-2}ag\Omega .$$

With a bit of patience one will now obtain proposition Proposition 3.6 above: see Appendix 5 for complete details. □

4. KAM Algorithm

In the preceeding two sections we saw how, starting from a given approximate solution, one can construct a new approximation leading to a new error term which is quadratically smaller than the original one. If the new approximation is non-degenerate (in the sense of definition Definition 2.6) one can iterate. This procedure will lead to an algorithm that, given a *set of positive numbers* ($\equiv$ the norm-parameters relative to a given approximate solution), produces a *new set of positive numbers* ($\equiv$ the norm-parameters relative to the new approximate solution).

Eventhough the estimates we established look complicated (because we avoided arbitrary approximations), any computer will have little trouble in performing the iteration for us (of course a control of the errors introduced by the machine is needed: see §8).

If the norm of the error term converges to zero (in a suitable way), the solution we are after will be constructed. Of course, in order to establish the convergence of the scheme in a finite number of steps we still need a theorem: such a theorem will be discussed in §5.

4.1. A Self-Contained Description of the KAM Algorithm

Given an initial non-degenerate approximate solution $v \equiv v^{(0)}$, let $\xi_0 \equiv \xi$ and α be as in (3.22). Fix now a sequence δ_j of positive numbers and let

$$\sum_{j=0}^{\infty} \delta_j < \xi_0 , \qquad \xi_{j+1} \equiv \xi_j - \delta_j . \tag{4.1}$$

At this level the choice of the "analyticity-loss-sequence" $\{\delta_j\}$ is rather arbitrary; however it might already be clear that asymptotically it will have to satisfy certain requirements (we shall come back on this point, see §6).

Let $j \geq 0$ and let

$$\mathcal{N}_j \equiv \{M_j, \overline{M}_j, S_j, E_j, \rho_j \equiv (\rho_{1j}, \rho_{2j}), L_j, \overline{L}_j, L_{3j}, L_{4j}, L'_{3j}, L'_{4j}\} \tag{4.2}$$

be the set of positive numbers controlling the norms of $v^{(j)}$ on Δ_{ξ_j} and of $\mathscr{L}$ on $\mathscr{D}_{\rho_j}^{\xi_j}$ [i.e., $M_j \geq |\mathbb{1} + v_{\theta}^{(j)}|_{\xi_j} \equiv |\mathbb{1} + v_{\theta}^{(j)}|_j$, $\overline{M}_j \geq |(\mathbb{1} + v_{\theta}^{(j)})^{-1}|_{\xi_j}$, $S_j \geq |v_t^{(j)}|_{\xi_j}$, $E_j \geq |\mathscr{E}(v^{(j)})|_{\xi_j}$, $L_j \geq |\mathscr{L}_{yy}|_{\mathscr{D}_{\rho_j}^{\xi_j}} \equiv |\mathscr{L}_{yy}|_j$, $\overline{L}_j \geq |\mathscr{L}_{yy}^{-1}|_{\mathscr{D}_{\rho_j}^{\xi_j}}$, the remaining parameters being as in (3.24) with $|\cdot|$ replaced by $|\cdot|_{\mathscr{D}_{\rho_j}^{\xi_j}}$]. The estimates of §3 can be encoded in the following rule defining $\mathcal{N}_{j+1}$ in terms of $\mathcal{N}_j$:

$$M_{j+1} \equiv M_j + E_j \overline{L} a_j \left(\delta_j^{-1} + \frac{s_1(\delta_j)}{s_0(\delta_j)} \frac{b_{1j}}{b_j} \right)$$

$$\overline{M}_{j+1} \equiv \begin{cases} \overline{M}_j \left(1 - \overline{M}_j E_j \overline{L} a_j \left(\delta_j^{-1} + \frac{s_1(\delta_j)}{s_0(\delta_j)} \frac{b_{1j}}{b_j} \right) \right)^{-1} & \text{if } |w_{\theta}^{(j)}|_j < 1 \\ \infty & \text{if } |w_{\theta}^{(j)}|_j \geq 1 \end{cases}$$

$$S_{j+1} \equiv S_j + E_j \overline{L} a_j \left(\delta_j^{-1} + \frac{s_1(\delta_j)}{s_0(\delta_j)} \frac{b_{1j}}{b_j} \right)$$

$$E_{j+1} \equiv K_j E_j^2 \overline{L} , \qquad \text{with}$$

$$K_j \equiv a_j \left\{ \frac{\delta_j^{-1}}{M_j} + \frac{a_j}{2}(c_j + 1)^2 \, \overline{L} \Big[L_3 + L_4(S_j + M_j)(1 + \delta_j^{-1}) + \frac{4}{3} E_j \overline{L} a_j \delta_j^{-2} L_4 \right.$$
$$\cdot \left(g_j + \frac{\delta_j}{4} + \frac{\delta_j^2}{4} \frac{s_1(\delta_j)}{s_0(\delta_j)} \frac{b_{1j}}{b_j} \right) + L'_4 \Big]$$
$$+ 4L'_3 \, \overline{L} a_j (c_j + 1) \delta_j^{-2} \left(g_j + \frac{\delta_j}{4} + \frac{\delta_j^2}{4} \frac{s_1(\delta_j)}{s_0(\delta_j)} \frac{b_{1j}}{b_j} \right)$$
$$+ \chi_d \cdot 4 \delta_j^{-1} \overline{M}_j s_0(\delta_j) \frac{a'_j}{a_j} \Big\} ,$$

$$\rho_{1,j+1} \equiv \rho_{1j} + \Omega E_j \overline{L} a_j \left(\delta_j^{-1} + \frac{s_1(\delta_j)}{s_0(\delta_j)} \frac{b_{1j}}{b_j} \right) ,$$

$$\rho_{2,j+1} \equiv \rho_{2j} + E_j \overline{L} a_j , \tag{4.3}$$

where $s_\lambda(\delta)$ are upper bounds on $\sigma_l(\delta)$ (cfr. (3.16), (3.25)) and $\chi_d, a_j, b_j, c_j, g_j$ are as in (3.43) with $\delta, M, \overline{M},...$ replaced by $\delta_j, M_j, \overline{M}_j,...$.

The definitions of M_{j+1}, S_{j+1} and E_{j+1} come immediately from from $v^{(j+1)} = v^{(j)} + w^{(j)}$ and from (3.42) (obviously here $v^{(j)}$ plays the role of v, $\varepsilon^{(j)}$ of ε, etc.). One easily obtains $\overline{M}_{j+1}$ observing that

$$(\mathcal{M}^{(j+1)})^{-1} = (\mathcal{M}^{(j)} + w_\theta^{(j)})^{-1} = (\mathcal{M}^{(j)})^{-1}(\mathbb{1} + (\mathcal{M}^{(j)})^{-1} w_\theta^{(j)})$$
$$\leq |(\mathcal{M}^{(j)})^{-1}|\,(1 - |(\mathcal{M}^{(j)})^{-1}|\,|w_\theta^{(j)}|)^{-1}\,.$$

Finally, one has $\rho_{1,j+1} = \rho_{1j} + \Omega |w_\theta^{(j)}|_{\xi_j}$ and $\rho_{2,j+1} = \rho_{2j} + |w^{(j)}|_{\xi_j}$. Now, if $\overline{M}_{j+1} < \infty$, the function $v^{(j+1)} = v^{(j)} + w^{(j)}$ (again $w^{(j)}$ here plays the role of w of §2–§3) is a non-degenerate approximate solution and iteration is possible.

Clearly the problem is now to give criteria for the convergence of

$$v^{(j)} \equiv v^{(0)} + \sum_{i=0}^{j-1} w^{(i)} \tag{4.4}$$

to the solution $u(\theta, t)$ we are after. Here, we establish a criterium which however is not practical (involving the check of infinitely many conditions) and we postpone to the next sections a complete discussion.

Proposition 4.1. *Let* $p \geq 2$, $\xi_\infty \equiv \xi_0 - \sum_{i=0}^{\infty} \delta_i > 0$ *and recall the definition of* α *(see (3.22)). Then, if*

$$\overline{M}_j < \infty \qquad (\forall j \geq 0)\,,$$
$$\sum_{j=1}^{\infty} \rho_{1j} \leq \Omega\alpha\,, \qquad \sum_{j=1}^{\infty} \rho_{2j} \delta_j^{-p} < \infty\,,$$

then the KAM algorithm converges, i.e., $v^{(j)} = v^{(0)} + \sum_{i=0}^{j-1} w^{(i)}$ *converges in the* C^p-*norm on* Δ_{ξ_∞} *to a solution* u *of* (2.4).

After realizing that ρ_{1j} is a bound on $|w_\theta^{(j)}|_j$, that $\rho_{2j} \leq \rho_{1j}$ and that (by Lemma 3.2)

$$\sum_{i=0}^{j-1} |\partial_\theta^p w^{(i)}|_{\xi_\infty} \leq \sum_{i=0}^{j-1} \frac{1}{(\xi_i - \xi_\infty)^p} |w^{(i)}|_{\xi_i} \leq \sum_{i=0}^{j-1} \delta_i^{-p} |w^{(i)}|_{\xi_i}$$

the above statement becomes obvious.

5. A KAM Theorem

Here we prove a theorem which provides a "simple" quantitative criterion for the existence (and local uniqueness) of a solution of (2.4) close to an approximate solution v. The theorem is formulated in a way that makes it possible to apply it to approximate solutions obtained via the KAM algorithm of §4 (in which case v below would correspond to v_N if the KAM algorithm has been applied N times, M to M_N, etc.).

Theorem 5.1. *Let ω satisfy Assumption 2.11 and let v be a real-analytic approximate solution of (2.4). Let $r \equiv 1/67^4$ and let $M, \overline{M}, L, \overline{L}, S, L_3, L_3', L_4, L_4', E$, be as in §3.3, (3.23), (3.24): the norms relative to $\mathscr{L}$ in (3.24) are $|\cdot| \equiv |\cdot|_{\mathscr{D}_{(r,r)}^{\xi}}$; let K_l be as in (3.17), assume (for simplicity) $0 < \xi < 1$; finally define*

$$K \equiv 1664^2 \cdot M^2 (M\overline{M})^8 (L\overline{L})^2 (S+M) K_1 K_0^3 \lambda \gamma^4 \; \xi^{-4\tau-3} \, 2^{12\tau} , \tag{5.1}$$

where $\lambda \equiv \max (\overline{L}L_3, \overline{L}L_3', \overline{L}L_4, \overline{L}L_4', 1)$ and γ, τ are the diophantine constants of ω (see (2.33)). If

$$K E \overline{L} \leq 1 , \tag{5.2}$$

then (2.4) has a unique real-analytic solution u with $\langle u \rangle = \langle v \rangle$ admitting an analytic extension to $\Delta_{\xi/2}$, such that $\mathbb{1} + u_\theta$ is invertible on $\Delta_{\frac{\xi}{2}}$ and

$$|u - v|_{\frac{\xi}{2}} < K E \overline{L} \frac{\xi}{32 \cdot 12} , \tag{5.3}$$

$$|u_\theta - v_\theta|_{\frac{\xi}{2}} < \frac{K E \overline{L}}{125^2} . \tag{5.4}$$

Local uniqueness holds in the following sense. If u and u' are two non-degenerate solutions in Δ_ξ with vanishing mean-value ($\langle u \rangle = \langle u' \rangle = 0$) and if the parameters $M, \overline{M},...$ in the definition of the above constant K are, here, defined replacing v with u, then

$$\sqrt{K} \, |u - u'|_\xi \leq 1 \quad \Longrightarrow \quad u \equiv u' \; (\text{in } \Delta_\xi) . \tag{5.5}$$

If $\mathscr{L}$ depends analytically on parameters $\mu \in \mathscr{P}$ ($\mathscr{P}$ being a compact subset of $\mathbb{C}^m$) and the above norms ($M, \overline{M}$, etc.) are defined replacing Δ_ξ by $\Delta_\xi \times \mathscr{P}$, then the solution u is analytic also in $\mu \in \mathscr{P}$ and (5.3), (5.4) hold on $\Delta_\xi \times \mathscr{P}$.

Proof: Let $v^{(0)} \equiv v$, $\varepsilon^{(0)} \equiv \varepsilon$ and, for $j \geq 0$, let

$$\xi_j \equiv \frac{\xi}{2} + \frac{\xi}{2^{j+1}} , \qquad \delta_j \equiv \xi_j - \xi_{j+1} = \frac{\xi}{2^{j+2}} , \tag{5.6}$$

and (cfr. Lemma 3.3):

$$\begin{aligned} s_0(\delta) &= K_0 \gamma \delta^{-\tau} , & K_0 &= 2^{d+2-\tau} \sqrt{\Gamma(2\tau + 1)} , \\ s_1(\delta) &= K_1 \gamma \delta^{-(\tau+1)} , & K_1 &= 2^{d+1-\tau} \sqrt{\Gamma(2\tau + 3)} . \end{aligned} \tag{5.7}$$

We claim that if (5.2) holds, then we can construct via Proposition 2.13 a sequence of non-degenerate approximate solutions $v^{(j)} \equiv v^{(j-1)} + w^{(j-1)}$ for all $j \geq 1$ (this means that we can apply iteratively for $j \geq 1$ Proposition 2.13 with $v = v^{(j-1)}$, $w = w^{(j-1)}$, $\varepsilon = \varepsilon^{(j-1)}$ and $v' = v^{(j)}$, $\varepsilon' = \varepsilon^{(j)}$, $\mathcal{M} = \mathcal{M}^{(j)} \equiv (\mathbb{1} + v_\theta^{(j)})$, etc.). Moreover if $V_j, V_{1j}, W_j, W_{1j}, W_{tj}, M_j, \overline{M}_j, E_j$ denote bounds on the corresponding norms [i.e. $V_j \geq |v^{(j)}|_{\xi_j}$, $V_{1,j} \geq |v_\theta^{(j)}|_{\xi_j}$, $W_j \geq |w^{(j)}|_{\xi_j}$, $W_{1,j} \geq |w_\theta^{(j)}|_{\xi_j}$, $W_{t,j} \geq |w_t^{(j)}|_{\xi_j}$, $M_j \geq |\mathcal{M}^{(j)}|_{\xi_j}$, $\overline{M}_j \geq |(\mathcal{M}^{(j)})^{-1}|_{\xi_j}$], then $\overline{M}_j < \infty$ for all j and the following estimates are true for every j:

$$E_j \overline{L} \leq (K E \overline{L})^{2^j} \tag{5.8}_j$$

$$|\sum_{k=0}^{j} D w^{(k)}|_{\xi_j} \leq r \Omega , \qquad r \equiv \frac{1}{67^4} \tag{5.9}_j$$

$$\sum_{k=0}^{j} W_k \le r \tag{5.10$_j$}$$

$$M_j \le 2M \tag{5.11$_j$}$$

$$\overline{M}_j \le 2\overline{M} \tag{5.12$_j$}$$

$$S_j + M_j \le 2S + 2M . \tag{5.13$_j$}$$

Observe that (5.9) and (5.10) show the consistency of the choice of the domain $\mathscr{D}_{(r,r)}^{\xi}$, which can be kept fixed during the iteration.
To check the claim, observe first, that from the recursive definition of V_j, V_{1j}, M_j, $\overline{M}_j$, S_j, it is:

$$V_{j+1} \equiv V + \sum_{i=0}^{j} W_i$$

$$V_{1,j+1} \equiv V_1 + \sum_{i=0}^{j} W_{1i}$$

$$M_{j+1} \equiv M + \sum_{i=0}^{j} W_{1i}$$

$$\overline{M}_{j+1} \equiv \begin{cases} \overline{M} \cdot (1 - \overline{M} \sum_{i=0}^{j} W_{1i})^{-1} & \text{if } \sum_{i=0}^{j} W_{1i} < 1 \\ \infty & \text{if } \sum_{i=0}^{j} W_{1i} \ge 1 \end{cases}$$

$$S_{j+1} \equiv S + \sum_{i=0}^{j} W_{ti} .$$

We now want to prove (5.8),...,(5.13) by induction on j: $j = 0$ is obvious. Assume the claim true for $0,...,j$; we want to prove it for $j+1$. Let $\lambda \equiv \max(\overline{L}L_3, \overline{L}L'_3, \overline{L}L_4, \overline{L}L'_4, 1)$; it is not difficult (just a bit tedious) to check that for $i \le j$:

$$\begin{aligned} E_{i+1}\overline{L} &\le (E_i\overline{L})^2 \beta_0 \gamma_0{}^i \\ W_i &\le E_i\overline{L}\beta_1\gamma_1{}^i \\ W_{1i} &\le E_i\overline{L}\beta_2\gamma_2{}^i \\ |Dw^{(i)}| &\le E_i\overline{L}\beta_3\gamma_3{}^i \Omega \end{aligned} \tag{5.14}$$

with

$$\begin{aligned} \beta_0 &\equiv 8 \cdot 208^2 M^2 (M\overline{M})^8 (L\overline{L})^2 (S+M) K_1 K_0^3 \lambda \gamma^4 2^{8\tau} \xi^{-4\tau-3} , & \gamma_0 &\equiv 2^{4\tau+3} \\ \beta_1 &\equiv 13 \cdot M(M\overline{M})^4 L\overline{L} \, K_0^2 \gamma^2 2^{4\tau+1} \xi^{-2\tau} , & \gamma_1 &\equiv 2^{2\tau} \\ \beta_2 &\equiv 165 \cdot M(M\overline{M})^4 L\overline{L} \, K_1 K_0 \gamma^2 2^{4\tau} \xi^{-2\tau-1} , & \gamma_2 &\equiv 2^{2\tau+1} \\ \beta_3 &\equiv 130 \cdot M(M\overline{M})^4 L\overline{L} \, K_0^2 \gamma^2 2^{4\tau} \xi^{-2\tau-1} , & \gamma_3 &\equiv 2^{2\tau+1} \end{aligned}$$

(see Appendix 6 for details).

Next: $(5.12)_j$ is equivalent to

$$2\overline{M} \sum_{i=0}^{j-1} W_{1i} \leq 1, \qquad (5.15)_j$$

and this condition implies $(5.11)_j$. In view of these comments, to prove the claim we have to show that

$$E_j \overline{L} \leq (K E \overline{L})^{2^j}, \qquad K \equiv \beta_0 \gamma_0, \qquad (5.16)$$

$$|\sum_{i=0}^{j} D w^{(i)}|_{\xi_j} \leq r\Omega, \qquad r \equiv \frac{1}{67^4}, \qquad (5.17)$$

$$\sum_{k=0}^{j} W_k \leq r, \qquad (5.18)$$

$$2\overline{M} \sum_{k=0}^{j-1} W_{1k} \leq 1, \qquad (5.19)$$

$$|\sum_{k=0}^{j} w_t^{(k)}| \leq S. \qquad (5.20)$$

Proof of (5.16):

$$E_{j+1}\overline{L} \leq (E_j \overline{L})^2 \beta_0 \gamma_0{}^j \leq (E\overline{L})^{2^{j+1}} \prod_{i=0}^{j} (\beta_0 \gamma_0{}^{j-i})^{2^i}$$

$$= \left[E\overline{L} \, \beta_0^{\sum_{i=1}^{j+1} \frac{1}{2^i}} \gamma_0^{\sum_{i=1}^{j+1} \frac{i-1}{2^i}} \right]^{2^{j+1}} < (E\overline{L}\beta_0\gamma_0)^{2^{j+1}},$$

and, since

$$\sum_{i=1}^{j+1} \frac{1}{2^i} = 1 - \frac{1}{2^{j+1}}, \qquad \sum_{i=1}^{j+1} \frac{i-1}{2^i} = 1 - \frac{j+2}{2^{j+1}},$$

one obtains:

$$E_{j+1}\overline{L} \leq \left[E\overline{L}\beta_0^{1-\frac{1}{2^{j+1}}} \gamma_0^{1-\frac{j+2}{2^{j+1}}} \right]^{2^{j+1}} = \frac{(E\overline{L}\beta_0\gamma_0)^{2^{j+1}}}{\beta_0\gamma_0{}^{j+2}}.$$

Proof of (5.17):

$$|\sum_{i=0}^{j} Dw^{(i)}|_{\xi_j} \leq \sum_{i=0}^{j} |Dw^{(i)}|_{\xi_j} \leq \beta_3 \Omega \sum_{i=0}^{j} E_i \overline{L} \gamma_3^i$$

$$< \frac{\beta_3}{\beta_0 \gamma_0} \Omega \sum_{i=0}^{j} (E\overline{L}\beta_0\gamma_0)^{2^i} (\frac{\gamma_3}{\gamma_0})^i < \frac{\beta_3}{\beta_0 \gamma_0} \Omega \sum_{i=0}^{j} (\frac{1}{2^4})^i$$

$$< \frac{16\beta_3}{15\beta_0\gamma_0} \Omega \leq r\Omega , \qquad r \equiv \frac{1}{67^4} ,$$

where (5.2) and $\frac{16\beta_3}{15\beta_0\gamma_0} \leq r$ have been used.

Proof of (5.18):

$$\sum_{k=0}^{j} W_k \leq \beta_1 \sum_{i=0}^{j} E_i \overline{L} \gamma_1^i$$

$$< \frac{\beta_1}{\beta_0\gamma_0} \sum_{i=0}^{j} (E\overline{L}\beta_0\gamma_0)^{2^i} (\frac{\gamma_1}{\gamma_0})^i < \frac{\beta_1}{\beta_0\gamma_0} \sum_{i=0}^{j} (\frac{1}{2^5})^i$$

$$< \frac{32\beta_1}{31\beta_0\gamma_0} \leq r .$$

Proof of (5.19):

$$2\overline{M} \sum_{i=0}^{j-1} W_{1i} \leq 2\overline{M} \beta_2 \sum_{i=0}^{j-1} E_i \overline{L} \gamma_2^i$$

$$< \frac{2\overline{M}\beta_2}{\beta_0\gamma_0} \sum_{i=0}^{j-1} (E\overline{L}\beta_0\gamma_0)^{2^i} (\frac{\gamma_2}{\gamma_0})^i < \frac{2\overline{M}\beta_2}{\beta_0\gamma_0} \sum_{i=0}^{j-1} (\frac{1}{2^4})^i$$

$$< \frac{32\overline{M}\beta_2}{15\beta_0\gamma_0} \leq 1 .$$

Proof of (5.20): since

$$S_j + M_j \leq S + \sum_{i=0}^{j-1} |w_t^{(i)}| + M + \sum_{i=0}^{j-1} |w_\theta^{(i)}| ;$$

we have to show that

$$\sum_{i=0}^{j-1} |w_t^{(i)}| + \sum_{i=0}^{j-1} |w_\theta^{(i)}| \leq S + M .$$

From $w^{(i)} = \mathcal{M}^{(i)} z^{(i)}$, one has $w_t^{(i)} = \mathcal{M}_t^{(i)} z + z_t^{(i)} \mathcal{M}$, and

$$|w_t^{(i)}| \leq \delta_i^{-1} S_i \frac{E_i \overline{L} a_i}{M_i} + E_i \overline{L} a_i \frac{s_1(\delta_i)}{s_0(\delta_i)} \frac{b_{1i}}{b_i} ;$$

hence

$$\sum_{i=0}^{j-1} |w_t^{(i)}| + \sum_{i=0}^{j-1} |w_\theta^{(i)}| \le$$

$$\le \sum_{i=0}^{j-1} \frac{E_i \overline{L} a_i \delta_i^{-1} S_i}{M_i} + \sum_{i=0}^{j-1} E_i \overline{L} a_i \delta_i^{-1} \left(1 + 2\delta_i \frac{s_1(\delta_i)}{s_0(\delta_i)} \frac{b_{1i}}{b_i} \right)$$

$$\le 2 \sum_{i=0}^{j-1} E_i \overline{L} a_i \delta_i^{-1} \left[\frac{S+M}{M_i} + \delta_i \frac{s_1(\delta_i)}{s_0(\delta_i)} \frac{b_{1i}}{b_i} \right]$$

$$\le S + M .$$

The inductive argument is complete and (5.8) ÷ (5.13) hold for all j.
From the above estimates it now follows that

$$|v^{(j)} - v|_{\xi_j} \le |\sum_{i=0}^{j-1} w^{(i)}|_{\xi_j} < \frac{32\beta_1}{31\beta_0\gamma_0} K E\overline{L} < K E\overline{L} \frac{\xi}{321^2} \tag{5.21}$$

and

$$|v_\theta^{(j)} - v_\theta| \le |\sum_{i=0}^{j-1} w_\theta^{(i)}|_{\xi_j} < \frac{16\beta_2}{15\beta_0\gamma_0} K E\overline{L} < \frac{K E\overline{L}}{125^2} . \tag{5.22}$$

This shows that the series $v + \sum_{i \ge 0} w^{(i)}$ converges uniformly in the complex compact domain $\Delta_{\xi_\infty} \equiv \Delta_{\xi/2}$ in the C^1-norm to a (real-)analytic function $u \equiv v + \sum_{i \ge 0} w^{(i)}$, which by (5.21) and (5.22) satisfies (5.3) and (5.4). Obviously, by construction, u will satisfy (2.4) and by (5.12)

$$|(1 + u_\theta)^{-1}|_{\xi/2} = \lim_{j \to \infty} |(1 + v_\theta^{(j)})^{-1}|_{\xi/2} \le \overline{M}_j \le 2\overline{M} , \tag{5.23}$$

showing that u is a non-degenerate solution of (2.4).
The condition

$$E\overline{L} \beta_0 \gamma_0 < 1$$

is equivalent to $K E\overline{L} \le 1$ if one sets:

$$K \equiv \beta_0\gamma_0 \equiv 1664^2 \cdot M^2 (M\overline{M})^8 (L\overline{L})^2 (S+M) K_1 K_0^3 \lambda \gamma^4 \xi^{-4\tau - 3} 2^{12\tau} .$$

We sketch now the proof of *local uniqueness*. Let u, u' be two non-degenerate solutions of (2.4) with $\langle u \rangle = \langle u' \rangle = 0$, real-analytic in some Δ_ξ ($\xi < 1$) and let $w \equiv u' - u$ and set $\eta \equiv |w|_\xi$.
Let $M, \overline{M},...$be bounds on $|(1 + u_\theta)|_\xi$, $|(1 + u_\theta)^{-1}|_\xi,...$and let ξ_j, δ_j be as in (5.6). Since u and u' are both solutions we see that

$$0 \equiv \mathcal{E}(u') = \mathcal{E}(u) + \mathcal{E}'(u)w + q_1 , \qquad \text{i.e.} \quad \mathcal{E}'(u)w + q_1 = 0 , \tag{5.24}$$

where q_1 is as in (2.24) with v, v' replaced by, respectively, $u, u + w \equiv u'$. Now, q_1 can be estimated on $\Delta_{\xi_0 - \delta_0/2}$ by (cfr. (3.46), (3.48)) [$30 \cdot L_3 \Omega \eta^2 \xi^{-3}$]; here we are disregarding that many terms in (2.24) actually cancel. On the other hand one can check that setting $z \equiv \mathcal{M}^{-1} w$, one has (cfr. (2.32) and note that in the present case $q_2 = 0$,

$\mathcal{A} = 0$):

$$\mathcal{E}'(u)w = \mathcal{M}^{-T}[D(\mathcal{J}Dz)] , \qquad (5.25)$$

where, of course, the twist-matrix $\mathcal{J}$ is defined as in (2.18) with u replacing v. Thus:

$$w = -\mathcal{M} D^{-1}\left\{\mathcal{J}^{-1}[D^{-1}(\mathcal{M}^T q_1) + c_1]\right\} + \mathcal{M}c_2 , \qquad (5.26)$$

with c_1 defined so that the term in curly brackets has vanishing mean-value and c_2 so that $\langle w \rangle = 0$ (recall that u, u' have vanishing mean-value and that $\langle \mathcal{M} \rangle = 1$):

$$c_1 = (\mathcal{J}^{-1})^{-1}\langle \mathcal{J}^{-1}D^{-1}[\mathcal{M}^T q_1]\rangle , \qquad c_2 = \langle \mathcal{M}D^{-1}\left\{\mathcal{J}^{-1}[D^{-1}(\mathcal{M}^T q_1) + c_1]\right\}\rangle . \qquad (5.27)$$

But then one can check that w can be estimated on Δ_{ξ_1} by (compare with (3.39)):

$$|w|_{\xi_1} \le |q_1|_{\xi_0}\overline{L}a , \qquad (5.28)$$

where a is defined in (3.39) with $\xi = \xi_0$ and $\delta = \delta_0$. Thus we see that the size of w in Δ_{ξ_1} is *quadratically* smaller than its size in Δ_{ξ_0}. Now, iterating such argument, mimicking the estimates in this section, one can easily check that

$$|w|_{\xi_j} \le \left(K\eta^2\overline{L}\right)^{2^j} , \qquad (5.29)$$

which shows that, if (5.2) holds, then $|w|_{\xi_\infty} = 0$, so that, by analyticity we can conclude that $w \equiv (u' - u) \equiv 0$ in Δ_ξ.

Finally, the last statement in Theorem 5.1 on the dependence on parameters μ is obvious by the uniformity of the bounds. $\square$

6. Application of the KAM Algorithm to Problems with Parameters

In this section we shall describe how one may apply the machinery of §2–§5 in an efficient way to problems with parameters.

6.1. Convergent Power-Series (Lindstedt-Poincaré-Moser Series)

In many physical problems one deals with a one-parameter family of Lagrangians $\mathcal{L}(\dot{y}, x, t; \mu)$ such that for $\mu = 0$ an explicit solution $u_0(\theta, t)$ of (2.4) is known. For example if

$$\mathcal{L} \equiv \mathcal{L}_0(\dot{y}) + \mu\mathcal{L}_1(\dot{y}, x, t)$$

is a nearly-integrable Lagrangian, $u_0 \equiv 0$ is the trivial solution of (2.4) with $\mu = 0$ and any ω. If we assume that $\mathcal{L}$ depends analytically on $\mu \in \mathcal{P}$, where $\mathcal{P}$ is some compact domain in $\mathbb{C}$ containing $\mu = 0$ and if all the sup-norms $|\cdot|.$ of §2–§5 are replaced by $\sup_{\mathcal{P}}|\cdot|.$, if $E = \sup_{\mathcal{P}}\sup_{\Delta_\xi}|\mathcal{E}(\theta;\mu)|$ satisfies (5.2) then by the uniformity of all the limits, we can conclude that $u = u(\theta, t; \mu)$ will be *real-analytic in $\Delta_\xi \times \mathcal{P}$*; therefore the solution u admits a convergent power series expansion in the *complex parameter μ*. For example in the case of nearly-integrable Lagrangians, as already noted, we can take $v = 0$ (which clearly is non-degenerate) so that $\varepsilon = \mu\mathcal{L}_1 \equiv \mathcal{L}_1(\omega, \theta, t)$; now if we take

$\mathcal{P} = \{\mu \in \mathbb{C}, \ |\mu| \le \mu_0\}$ one sees that

$$E = \mu_0 \, |\mathcal{L}_1|_\xi \, .$$

Therefore if

$$\mu_0 \ < \ \left(K \, |\mathcal{L}_1|_\xi \, \overline{L} \right)^{-1}, \tag{6.1}$$

we can conclude by Theorem 5.1 the existence of $u(\theta, t; \mu)$ real-analytic on $\Delta_{\xi/2} \times \{\mu \in \mathbb{C} : \ |\mu| \le \mu_0\}$ and μ_0 close to 0.

The problem of the convergence of such series in μ is as old as Celestial Mechanics and was considered by Lindstedt (see also [A2] for more informations). Poincaré longly studied this problem, which he considered as one of the central problems in Celestial Mechanics, but did not arrive to any conclusion (actually he thought quite unlikely the convergence of such series but did not exclude it). It was only with the use of KAM techniques that it was possible to answer positively the question. In particular J. Moser devoted a beautiful paper setting completely such a problem ([Mo3]; for different "direct" proofs see [EkI] and [ChF2]).

As one can see from (6.1) the smallness requirement dictated by a *tout-court* application of the KAM theorem gives a radius of convergence absurdely small and certainly of little practical interest. Instead one can proceed as follows.

6.2. Improving the Lower Bound on the Radius of Convergence

First step. Compute as many as you can (and if you are not Gauss or Delaunay you may want to use a computer) Taylor-Fourier coefficients of the expansion

$$u(\theta, t; \mu) \equiv \sum_{k \ge 1} u_k(\theta, t) \, \mu^k \equiv \sum_{k \ge 1} \sum_{(n,m) \in \mathbb{Z}^{d+1}} u_{kn} \, e^{i(n \cdot \theta + mt)} \mu^k \, .$$

Notice that expanding in μ the tori equation and equating the kth-coefficient in μ will yield a linear equation for u_k of the form:

$$D^2 u_k(\theta, t) \ = \ \Phi_k(\theta, t) \, ,$$

where Φ_k is a function (with vanishing mean value on $\mathbb{T}^{d+1}$) depending upon $u_1, ..., u_{k-1}$ and of the derivatives of $\partial_{y_i} \partial_{x_j} \mathcal{L}|_{\mu=0}$ (explicit formulae and a self-contained proof of $\langle \Phi_k \rangle = 0$ is given in §7.1 below).

Second step. Let

$$v \equiv v^{(0)} \equiv \sum_{k \le k_0} \sum_{|(n,m)| \le N_k} u_{kn} \, e^{i(n \cdot \theta + mt)} \mu^k \tag{6.2}$$

(k_0 and N_k depending on your computational ability). Make a guess μ_0 for the true radius of convergence (eventually using numerical methods to get a hint). Fix a ξ so that $\Phi_v(\Delta_\xi)$ is contained in the holomorphic domain of $\mathcal{L}$ and evaluate the starting parameters for the KAM algorithm, i.e. evaluate

$$M \ = \ \sup_{|\mu| \le \mu_0} \sup_{\Delta_\xi} |\mathbb{1} + v_\theta| \, , \qquad \overline{M} \ = \ \sup_{|\mu| \le \mu_0} \sup_{\Delta_\xi} |(\mathbb{1} + v_\theta)^{-1}| \, , \qquad \text{etc.} \, ,$$

making sure that $\overline{M} < \infty$ (otherwise reduce the value of μ and/or ξ). This step may also involve computer-aided calculations.

Third step. Fix N (a good starting choice may be $10 \leq N \leq 20$), fix $\delta_0,...,\delta_N$ so that $\xi - \sum\limits_{i=0}^{N} \delta_i > 0$ and apply recursively the KAM algorithm, i.e. compute $\mathcal{N}_j$, $0 \leq j \leq N$ provided, of course, $\overline{M}_j < \infty$ $\forall j$. In spite of the complications of the formulae involved, the application of the KAM algorithm is trivial from the computational point of view, *apart* from the evaluation of $s_i(\delta_j)$ ($\equiv$ upper bound on $\sigma_i(\delta_j)$, see (3.17)). Clearly it is possible by using truncations of $\sigma_i(\delta)$ to give an upper bound estimate $s_i(\delta)$ as close as one wishes to the actual value; however such an operation is in general quite time-consuming.

Fourth step. If $\overline{M}_j < \infty$ $\forall 0 \leq j \leq N$, plug the values M_N, $\overline{M}_N$, E_N, $\xi_N = \xi_*$, L, $\overline{L}$, etc., in the formula for K (see (5.1)) of the KAM theorem. If $KE_N \, \overline{L} \leq 1$ we can apply the theorem and conclude the existence of a true solution u, $(KE_N \, \overline{L})$-close to the approximant $v^{(N)}$. Otherwise go back to the second step and vary the parameters μ_0, ξ, N, $\delta_0,...,\delta_N$ (typically one would reduce μ_0, etc.).

Let us be a bit more formal. Fixed the starting approximation v (e.g. as in (6.2)) the above procedure can be viewed as a finite algorithm

$$\Lambda_N(\mu_0; \xi, \delta_0, ..., \delta_N) \, ,$$

where $\Lambda_N = 1$ if $KE_N \, \overline{L} \leq 1$, $\Lambda_N = 0$ otherwise. What we are after is

$$\mu_N \equiv \sup \{\mu_0 : \, \Lambda_N(\mu_0; \xi, \delta_0, ..., \delta_N) = 1 \quad \text{for some } \xi, \delta_0, ..., \delta_N\} \tag{6.3}$$

and since it is fairly clear that μ_N is increasing with N, one wants to get a good approximation of

$$\mu_\infty \equiv \mu_{KAM} \equiv \sup_N \mu_N \, .$$

Experience teaches us that $N = 20$ is usually a good approximation of ∞; however approximating μ_N is a difficult nonlinear programing problem which we believe is interesting by itself ([ChF]). Notice that in principle (i.e. if we can compute v for any k_0 and N_k, if we can estimate efficiently the norms relative to v, etc.), $\sup\limits_{k_0, N_k} \mu_\infty = \rho_a$, where

$$\rho_a \equiv \inf_\theta \{\text{radius of convergence of } \sum_{k \geq 1} u_k \mu^k\} \, . \tag{6.4}$$

It is therefore a pure (and highly non-trivial) computational problem to give accurate lower bounds on ρ_a.

There is however a much deeper theoretical problem beyond this approach. Namely, let

$$\rho_c \equiv \sup\{\rho : \exists \text{ a solution } u(\theta, t, \mu) \text{ of (2.4) which is } C(\mathbb{T}^{d+1} \times [0, \rho)) \cap C^2(\mathbb{T}^{d+1})\} \, . \tag{6.5}$$

What is the relation between ρ_a and ρ_c? There is experimental evidence that in simple models (e.g. the so-called standard map) it is $\rho_a = \rho_c$ (see [BC]); however in general this will not be the case ([BCCF]). An obvious weaker and more realistic approach would be to replace ρ_a above with

$$\rho_r \equiv \sup \{\rho > 0 : \exists \; \text{real} - \text{analytic extension of} \; u(\theta, t, \mu) \; \text{to} \; \mathbb{T}^{d+1} \times [0, \rho)\} . \quad (6.6)$$

Of course, in this case, one would have to replace the above computation of v with other methods taking into account the possibility of μ-analyticity domains different from circles.

Note that, obviously $\rho_a \leq \rho_r \leq \rho_c$.

One may also think of solving the equation at *given fixed* μ, simplifying, therefore, significantly the first step above, which is by far the most time-consuming; this approach (which has been pursued in [LR]) however yields no informations on the μ-dependence of the solution and in particular cannot be used to give lower estimates on the above critical parameters $\rho_.$.

7. Power Series Expansions and Estimate of the Error Term

As already noted if one considers a one-parameter family of analytic Lagrangians, such that for $\mu = 0$ a solution u_0 is known, it follows that there exists for μ sufficiently small an analytic solution $u(\theta, t; \mu)$ of the Euler-Lagrange equation. This solution can be expanded in power series of the perturbing parameter μ. An approximate solution of the Euler-Lagrange equation can be obtained truncating the power series. The approximate solution will satisfy the Euler-Lagrange equation up to an error term. An indication of estimating this error is provided in the last part of this paragraph.

7.1. Power Series Expansions

Let us consider, for simplicity, a special class of nearly integrable Lagrangians (cfr. [CC2]) given by

$$\mathscr{L}(y, x, t) \equiv \frac{1}{2} \sum_{i=1}^{d} \frac{y_i^2}{2} + \mu V(x, t) , \quad (7.1)$$

where $y \in \mathbb{R}^d$ and $(x, t) \in \mathbb{T}^{d+1}$. For this Lagrangian, equation (2.4) takes the form

$$D^2 u = \mu \, V_x(\theta + u, t) , \quad (7.2)$$

where $V_x \equiv (\frac{\partial V}{\partial x_1}, ..., \frac{\partial V}{\partial x_d})$. As mentioned in §6, if ω is Diophantine a non-degenerate solution u exists and is analytic in the parameter μ in a small neighbourhood of the origin; therefore one can expand u in power series as

$$u(\theta, t) \equiv \sum_{k=1}^{\infty} u_k(\theta, t)\mu^k . \quad (7.3)$$

We proceed now to describe a method for finding a recursive relation among the coefficients u_k. Let $f(\theta, t) \equiv V_x(\theta, t)$ and expand $f(\theta, t)$ in Fourier series as

$$f(\theta, t) \equiv \sum_{(n,m) \in \mathbb{Z}^{d+1} \setminus \{0\}} \hat{f}_{(n,m)} \, e^{i(n \cdot \theta + mt)} . \quad (7.4)$$

Following an idea that we learned in [Her3] (cfr. also [Go]), we define for any $(n, m) \in \mathbb{Z}^{d+1}$ with $(n, m) \neq 0$ the complex-analytic functions $b_k^{(n,m)}(\theta, t)$ as the coefficients of the series expansion in powers of μ of $e^{i(n \cdot (\theta + u) + mt)}$:

$$e^{i(n \cdot (\theta + u) + mt)} \equiv \sum_{k=0}^{\infty} b_k^{(n,m)}(\theta, t) \mu^k . \tag{7.5}$$

Differentiating (7.5) with respect to μ one has:

$$in \cdot u' \, e^{i(n \cdot (\theta + u) + mt)} = \sum_{k=1}^{\infty} k b_k^{(n,m)} \mu^{k-1} ,$$

namely

$$i \, n \cdot \sum_{h=1}^{\infty} \sum_{k=0}^{\infty} h u_h b_k^{(n,m)} \mu^{h-1} \mu^k = \sum_{k=1}^{\infty} k b_k^{(n,m)} \mu^{k-1}$$

or

$$i \sum_{k=1}^{\infty} \left(n \cdot \sum_{h=1}^{\infty} h u_h b_{k-h}^{(n,m)} \right) \mu^{k-1} = \sum_{k=1}^{\infty} k b_k^{(n,m)} \mu^{k-1} . \tag{7.6}$$

A comparison between terms of the same order in μ in the equality (7.6) shows that

$$b_0^{(n,m)} = e^{i(n \cdot \theta + mt)}$$
$$b_k^{(n,m)} = \frac{i}{k} \, n \cdot \sum_{h=1}^{k} h u_h b_{k-h}^{(n,m)} , \quad k \geq 1 . \tag{7.7}$$

Therefore, by (7.2)

$$D^2 u = \mu \sum_{(n,m) \in \mathbb{Z}^{d+1} \setminus \{0\}} \hat{f}_{(n,m)} \, e^{i(n \cdot (\theta + u) + mt)} ,$$

and by (7.5)

$$D^2 u_k = \sum_{(n,m) \in \mathbb{Z}^{d+1} \setminus \{0\}} \hat{f}_{(n,m)} \, b_{k-1}^{(n,m)} . \tag{7.8}$$

This equation makes sense provided that the right hand side has mean average zero over $\mathbb{T}^{d+1}$. Actually, we already know that this fact is true, since we proved in §6 that if μ_0 is small enough then $\sum_{k \geq 1} u_k \mu^k$ is an absolutely convergent series for $|\mu| \leq \mu_0$. However, for completeness we shall give now a purely algebraic check of the vanishing of the r.h.s. of (7.8). Let us denote by $[\cdot]_k$ the k-th coefficient of the μ-power series expansion:

$$\text{if} \quad g \equiv \sum_{k=1}^{\infty} g_k \mu^k \quad \text{then} \quad [g]_k = g_k .$$

From equation (7.2) we can rewrite (7.8) as

$$D^2 u_k = [\mu V_x(\theta + u, t)]_k .$$

Proposition 7.1. *Let $u_0 \equiv$ constant and let $k \geq 1$. Assume that there exist $u_0, ..., u_{k-1}$ such that for every $0 \leq l \leq k - 1$ one has*

$$D^2 u_l = [\mu \, V_x(\theta + \sum_{i=1}^{l-1} u_i \mu^i, t)]_l . \tag{7.9}$$

Then

$$\int_{\mathbb{T}^{d+1}} [\mu \, V_x(\theta + \sum_{i=1}^{k-1} u_i \mu^i, t)]_k \, d\theta dt = 0 . \tag{7.10}$$

Proof: Notice that for $k = 1$, $u_0 \equiv$ const. does satisfy (7.9).
Now, for any function $G = G(x, t)$,

$$\int_{\mathbb{T}^d} \partial_\theta [G(\theta + u, t)] d\theta = 0 = \int_{\mathbb{T}^d} (\mathbb{1} + u_\theta) \, G_x(\theta + u, t) d\theta .$$

Let $G(x, t) \equiv \mu V_x(x, t)$; then

$$\int_{\mathbb{T}^d} (\mathbb{1} + u_\theta) \, \mu V_x(\theta + u, t) d\theta = 0$$

and since $[\cdot]$ is a linear operator, for any $l \geq 0$

$$[\int_{\mathbb{T}^d} (\mathbb{1} + u_\theta) \, \mu V_x(\theta + u, t) d\theta]_l = 0$$
$$= \int_{\mathbb{T}^d} [\mu V_x]_l d\theta + \int_{\mathbb{T}^d} [\mu u_\theta V_x]_l d\theta . \tag{7.11}$$

Now by (7.9) and recalling that u_0 is independent of θ, one sees that

$$\int_{\mathbb{T}^{d+1}} [\mu u_\theta V_x]_k \, d\theta dt = \sum_{\substack{j+i=k \\ 1 \leq j \leq k, \; 0 \leq i \leq k-1}} \int_{\mathbb{T}^{d+1}} (u_j)_\theta \, [\mu V_x]_i \, d\theta dt$$

$$= \sum_{j=1}^{k} \int_{\mathbb{T}^{d+1}} (u_j)_\theta \, D^2 u_{k-j} = \sum_{j=1}^{k-1} \int_{\mathbb{T}^{d+1}} (u_j)_\theta \, D^2 u_{k-j}$$

$$= \frac{1}{2} \sum_{j=1}^{k-1} \int_{\mathbb{T}^{d+1}} \left\{ (u_j)_\theta \, D^2 u_{k-j} + (u_{k-j})_\theta \, D^2 u_j \right\} d\theta dt .$$

Finally, integrating by parts three times one finds

$$\int_{\mathbb{T}^{d+1}} (u_j)_\theta \, D^2 u_{k-j} \, d\theta dt = - \int_{\mathbb{T}^{d+1}} D^2 u_j \, (u_{k-j})_\theta \, d\theta dt$$

(notice that in this identity we have to integrate over t too). Therefore

$$\int_{\mathbb{T}^{d+1}} [\mu u_\theta V_x]_k \, d\theta dt = 0$$

and integrating (7.11) over t, one obtains (7.10). $\square$

Thus one can invert the operator D^2 in (7.8) to get

$$u_k = D^{-2} \sum_{(n,m)\in\mathbb{Z}^{d+1}\setminus\{0\}} \hat{f}_{(n,m)} \; b_{k-1}^{(n,m)} , \tag{7.12}$$

which defines u_k ($\langle u_k \rangle = 0$), in terms of the preceding functions $u_0,..., u_{k-1}$. Notice that it is not legitimate to interchange the order of the summation and of D^{-2} as the functions $b_k^{(n,m)}$ *may not* have vanishing mean-value.

7.2. Truncated Series as Initial Approximations and the Majorant Method

We choose now, as initial approximate solution of (7.2) a truncation of the μ-expansion of u and discuss the estimates on the associated error function.

Thus, if $\{u_l\}$, $l \geq 1$ ($u_0 = 0$ as $\langle u_i \rangle = 0$), are the functions defined in the previous paragraph, we set

$$v^{(0)}(\theta, t) \equiv \sum_{l=1}^{l_0} u_l(\theta, t)\mu^l , \tag{7.13}$$

for a suitable $l_0 \in \mathbb{Z}_+$. Notice that if V is a trigonometric polynomial, then so are the u_l and the computation of $v^{(0)}$ reduces to a finite number of steps. In general, one can introduce truncations in Fourier space according to the desidered accuracy.

Recalling from §3.3 the definition of the norm-parameters, we see that in the present case [(7.1)], $L = \bar{L} = 1$, $L_3' = L_4 = L_4' = 0$; the vector ρ can be replaced by ρ_2 (as no geometry in the y-variables comes really in). Thus, the only parameters we have to evaluate are S_0, M_0, $\overline{M}_0$, E_0, which are upper bounds on the norms of $v_t^{(0)}$, $\mathbb{1} + v_\theta^{(0)}$, $(\mathbb{1} + v_\theta^{(0)})^{-1}$, $\varepsilon^{(0)}$ (where $\varepsilon^{(0)} \equiv \mathscr{E}(v^{(0)})$) in the domain $\Delta_\xi \times \mathscr{P} \equiv \Delta_\xi \times \{\mu \in \mathbb{C} : |\mu| \leq \mu_0\}$; we also need to evaluate the parameter $L_3 \geq \mu_0 |V_{xxx}(\theta + v^{(0)}, t)|_{\Delta_{\rho_2}^\xi}$ so that $\lambda \equiv$ max$\{L_3, 1\}$. The estimate of $v_t^{(0)}$ can be obtained using

$$|\partial_t v^{(0)}|_{\xi,\mu_0} \leq \sum_{l=1}^{l_0} |\partial_t u_l|_{\xi,\mu_0} \; \mu_0^l , \qquad |\cdot|_{\xi,\mu_0} \equiv \sup_{\Delta_\xi \times \mathscr{P}} |\cdot| ,$$

and analogously for $v_\theta^{(0)}$. Then M_0 and $\overline{M}_0$ can be estimated respectively by $1+|v_\theta^{(0)}|_{\xi,\mu_0}$ and $(1 - |v_\theta^{(0)}|_{\xi,\mu_0})^{-1}$, *provided*

$$|v_\theta^{(0)}|_{\xi,\mu_0} < 1 .$$

We also set

$$V_0 \geq \sum_{l=1}^{l_0} |u_l|_\xi \; \mu_0^l \geq |v^{(0)}|_{\xi,\mu_0}$$

and

$$\rho \equiv \rho_2 \equiv \xi + V + r \qquad (r \equiv \frac{1}{674})$$

and we take $L_3 \geq \mu_0 |V_{xxx}|_{\Delta_\rho}$.

It remains to estimate the error term $\varepsilon^{(0)}$. Let $f(\theta, t) \equiv V_x(\theta, t)$; by (7.9)

$$
\begin{aligned}
\varepsilon^{(0)} &= D^2 v^{(0)} - \mu f(\theta + v^{(0)}, t) \\
&= \sum_{l=1}^{l_0} D^2 u_l \, \mu^l - \mu f(\theta + v^{(0)}, t) \\
&= \sum_{l=1}^{l_0} D^2 u_l \, \mu^l - \mu \sum_{(n,m)\neq 0} \hat{f}_{(n,m)} \, e^{i(n\cdot(\theta+v^{(0)})+mt)} \\
&= \sum_{l=1}^{l_0} D^2 u_l \, \mu^l - \sum_{(n,m)\neq 0} \hat{f}_{(n,m)} \sum_{h=1}^{\infty} d_{h-1}^{(n,m)}(\theta, t)\mu^h ,
\end{aligned}
\tag{7.14}
$$

where the functions $d_h^{(n,m)}(\theta, t)$ are defined as the coefficients of the power series expansion

$$
e^{i(n\cdot(\theta+v^{(0)})+mt)} \equiv \sum_{h\geq 0} d_h^{(n,m)}(\theta, t)\mu^h .
\tag{7.15}
$$

Therefore, one has:

$$
\begin{aligned}
\varepsilon^{(0)}(\theta, t) &= \left[\sum_{l=1}^{l_0} \mu^l \left(D^2 u_l - \sum_{(n,m)\neq 0} \hat{f}_{(n,m)} d_{l-1}^{(n,m)} \right) \right] \\
&\quad - \left[\mu \sum_{(n,m)\neq 0} \hat{f}_{(n,m)} \sum_{l=l_0}^{\infty} \mu^l d_l^{(n,m)} \right] \\
&\equiv F_{l_0} + R_{l_0} = R_{l_0} ,
\end{aligned}
\tag{7.16}
$$

since $F_{l_0} = 0$ because of the definition (7.13) of $v^{(0)}$. (If V is not a trigonometric polynomial one can replace $|F_{l_0}|$ with an arbitrarily small positive number.)

To estimate $|R_{l_0}|$ we shall make use of an old technique, which we shall refer to as the *majorant method*. Such a technique, used e.g. by C. L. Siegel in [S], consists, roughly speaking, in comparing the supremum of an analytic function with the *value* of another analytic function with positive coefficients. More precisely:

Lemma 7.2. *Let $v^{(0)}(\theta, t)$ and $d_h^{(n,m)}(\theta, t)$ be as in (7.13) and (7.15) respectively. For any $\xi > 0$ define the sequence $\{a_l^{(n,m)}(\xi)\}$, $0 \leq l \leq l_0$, by*

$$
\begin{aligned}
a_0^{(n,m)} &= (|n| + |m|)\xi \\
a_l^{(n,m)} &\geq |n \cdot u_l(\theta, t)|_\xi , \quad 1 \leq l \leq l_0
\end{aligned}
\tag{7.17}
$$

and for $l \geq 0$ let $\delta_l^{(n,m)}$ be defined by the identity

$$
\exp\left(\sum_{l=0}^{l_0} a_l^{(n,m)} \mu^l \right) = \sum_{l=0}^{\infty} \delta_l^{(n,m)} \mu^l .
\tag{7.18}
$$

Then one has

$$
|d_l^{(n,m)}(\theta, t)|_\xi \leq \delta_l^{(n,m)}
\tag{7.19}
$$

and for any $\mu_0 > 0$,

$$|\sum_{l=l_0}^{\infty} d_l^{(n,m)}(\theta,t)\mu^l|_{\xi,\mu_0} \leq \exp\left(\sum_{l=0}^{l_0} a_l^{(n,m)}\mu_0^l\right) - \sum_{l=0}^{l_0-1} \delta_l^{(n,m)}\mu_0^l, \qquad (7.20)$$

where, as above, $|\cdot|_{\xi,\mu_0} \equiv \sup_{\Delta_\xi \times \{|\mu| \leq \mu_0\}}$.

Proof: As in the discussion on the b's (cfr. (7.5) ÷ (7.7)), it is easy to check that $d_k^{(n,m)}$ and $\delta_k^{(n,m)}$ verify the recursive relations

$$d_0^{(n,m)}(\theta,t) = e^{i(n\cdot\theta+mt)}$$

$$d_l^{(n,m)}(\theta,t) = \frac{i}{l}n \cdot \sum_{h=1}^{\min(l,l_0)} hu_h(\theta,t)d_{l-h}^{(n,m)}, \quad l \geq 1,$$

$$\delta_0^{(n,m)} = e^{(|n|+|m|)\xi}$$

$$\delta_l^{(n,m)} = \frac{1}{l}\sum_{h=1}^{\min(l,l_0)} ha_h^{(n,m)}\delta_{l-h}^{(n,m)}, \quad l \geq 1. \qquad (7.21)$$

We prove (7.19) by induction on l. For $l = 0$,

$$|d_0^{(n,m)}(\theta,t)|_\xi \leq e^{(|n|+|m|)\xi} \equiv \delta_0^{(n,m)}.$$

Let now $l \geq 1$ and *assume* (7.19) for $0 \leq h \leq l-1$; by (7.17) and the inductive hypotheses one has:

$$|d_l^{(n,m)}(\theta,t)|_{\xi,\mu_0} \leq \frac{1}{l}\sum_{h=1}^{\min(l,l_0)} h|n \cdot u_h(\theta,t)|_\xi|d_{l-h}^{(n,m)}(\theta,t)|_\xi$$

$$\leq \frac{1}{l}\sum_{h=1}^{\min(l,l_0)} ha_h^{(n,m)}\delta_{l-h}^{(n,m)} \equiv \delta_l^{(n,m)}.$$

Inequality (7.20) now follows from (7.19) and the definition of the a_l's in (7.18). □

Therefore the estimate of the error term (7.16) on the domain $\Delta_\xi \times \{\mu \in \mathbb{C} : |\mu| \leq \mu_0\}$ is given by

$$|\varepsilon^{(0)}(\theta,t)|_{\xi,\mu_0} \leq \mu_0 \sum_{(n,m)\neq 0} |\hat{f}_{(n,m)}| \left\{\exp\left(\sum_{l=0}^{l_0} a_l^{(n,m)}\mu_0^l\right) - \sum_{l=0}^{l_0-1} \delta_l^{(n,m)}\mu_0^l\right\}, \quad (7.22)$$

where $a_l^{(n,m)}$ and $\delta_l^{(n,m)}$ are given in (7.17), (7.21).

The strategy outlined in this section was carried out in [CC2] and led to existence estimates (for the standard map and a forced pendulum) that are away from the experimentally observed "break-down values" by (respectively) a factor ~ 0.55 and 0.67.

However a serious computational hindrance is hidden in this approach. In fact, in order to carry out the above steps one needs to *rigorously* control the computational errors introduced by mechanical calculations. To do this one may use the so-called *interval arithmetic* (see §8 below for a more detailed discussion), which consists, basically, in trapping the result of a computation performed by a computer in an interval whose end-points are representable by the machine and which is sure to contain the actual result of the given computation. Now, the computation of the u_l's contains a lot of division by "small divisors" ($\omega \cdot n + m$) which have the effect of spreading very quickly the size of the intervals controlling the u_l's.

To avoid such a problem one would have to turn to "arbitrary" accuracy computations, which are, obviously, very time-consuming (see [CC3] for more informations on this phenomenon).

A different approach is the following.

7.3. Numerical Initial Approximations

A quite different approach is to *compute numerically* (i.e. without caring about errors) the functions u_l by means of (7.12) and (7.7) and *define* the initial approximation as

$$v^{(0)}(\theta, t) \equiv \sum_{l=1}^{l_0} \bar{u}_l(\theta, t)\mu^l , \quad l_0 \in \mathbb{Z}_+ ,$$

where the $\bar{u}_l$'s are the result, given by a computer, of the implementation of (7.12) and (7.7). With this choice of $v^{(0)}$, one has

$$F_{l_0} \equiv \sum_{l=1}^{l_0} \mu^l \left(D^2 \bar{u}_l - \sum_{(n,m)\neq 0} \hat{f}_{(n,m)} \, d_{l-1}^{(n,m)} \right) ,$$

which, *eventhough will not vanish anymore*, can be estimated with a finite number of operations. The estimate of R_{l_0} is obtained, instead, applying directly Lemma 7.2, with u_l replaced by $\bar{u}_l$.

The advantage of this approach is that interval arithmetic is not used *directly* to control small divisors.

This strategy has been implemented in [CC4] on various models and yields indeed sensibly better results; for example, the existence of the "golden-mean" invariant curve for the standard map is established for values of the parameter away by a factor 1.16 from optimal (see also §10).

8. Computer Assisted Methods

In order to apply the method outlined in the preceding sections, one may have to perform lengthy but straightforward calculations (e.g. to calculate a "good" initial approximation together with the associated norms), in which case the use of computers may be helpful. In this section we briefly discuss the so-called *interval arithmetic*, the implementation of which allows to take care of rounding-off and propagation errors introduced by computers.

8.1. Representable Numbers and Intervals

A computer can represent exactly a finite set of numbers, which we shall call here the set of *representable numbers* $\mathfrak{R}$ (of course $\mathfrak{R}$ depends on the particular computer we are considering). Such numbers are encoded by strings of "bits", i.e., 0's or 1's. For example, if $x = \sum_{j=1}^{N} \epsilon_j 2^{-j}$, $(\epsilon_j = 0$ or 1) is the binary expansion of the rational number $x \in [0, 1)$, one can identify x with $\{\epsilon_1, ..., \epsilon_N\}$. To represent other rational numbers in $\mathbb{R}$, the computer uses extra bits in a symbolic way (see next section for a discussion of how VAXes handle that). Now, to deal with *real* numbers without making approximations one can try to trap them within the "smallest" possible interval whose end-points are in $\mathfrak{R}$. In this way, operations among numbers are replaced (in a quite straightforward way) by operations among intervals.

Before going into more details, let us discuss how computers perform "elementary" operations (i.e., additions, subtractions, multiplications and divisions).

In general, the result of an elementary operation between representable numbers is *not* a representable number and therefore the computer will, in general, approximate such a result. For some computers (like VAXes) the approximation rule is the following. Let us call *rounding bit* the first bit lost in the truncation of the theoretical result. Then:

(i) if the rounding bit is 0, the rounded result is equal to the chopped number;

(ii) if the rounding bit is 1, the chopped result is increased by one bit.

It is therefore clear that by modifying suitably certain bits one can find intervals of representable numbers which contain the theoretical result.

8.2. Intervals on VAXes

As an example, let us consider a VAX. The procedure to create *upper* and *lower* bounds on the result of an elementary operation depends on the structure of data one is working with. Therefore we start by illustrating the different kind of precisions available on a VAX (see [Vax]).

Real numbers are represented in "floating point" notation by a sign, an exponent and a fraction. The size of the floating point data may be of 32, 64, 128 bits. Correspondingly one distinguishes between F-floating (i.e. *simple* precision), D or G-floating (i.e. *double* precision) and H-floating (i.e. *quadruple* precision). The difference between D and G-floating is that G-floating reserves more space to the exponent (allowing numbers in the range of $\sim 0.56 \cdot 10^{-308}$ to $\sim 0.9 \cdot 10^{308}$) with consequent loss of precision of the fraction. In our computations we use G-floating data which we are going to describe in full detail. A G-floating datum is composed by 64 bits (i.e. a set of 0's and 1's), labelled from 0 to 63. The first bit denotes the sign of the number; bits 1 to 11 correspond to the exponent (one bit is reserved for the sign of the exponent), while the remaining 52 bits individuate the fraction. The precision of a G-floating number is approximately of 15 decimal digits. Moreover, there are two extra hidden *guard* bits which guarantee the result of an elementary operation "up to 1/2 of the last significant bit" (see [Vax], appendix H; the quoted sentence is related to the approximation rule discussed above and roughly speaking it means that the bit before the last one is always correct, while the last one is used to get "the closest guess" to the true result and therefore may not coincide with the corresponding bit of the theoretical outcome).

With this representation of real numbers, it can be shown that the upper and lower bounds on the result of an elementary operation are obtained, respectively, *by increasing or decreasing by one bit the last bit of the mantissa, taking eventually care of the propagation of the carry.* (The Fortran procedures to create upper and lower bounds of a real number are described in Appendix 7).

8.3. Interval Operations

We end this section by discussing in more detail interval operations.
Let $\odot$ denote one of the four elementary operations and, for $a, b \in \mathfrak{R}$, let $\langle a \odot b \rangle$ be the *result produced* by a VAX; let $\text{Up}(a)$, $\text{Down}(a)$ be the representable number obtained by, respectively, increasing or decreasing the last significant bit (taking care of the possible carry). Then it is clear that:

$$a \odot b, \ \langle a \odot b \rangle \in (\text{Down}(\langle a \odot b \rangle), \text{Up}(\langle a \odot b \rangle)), \qquad a, b \in \mathfrak{R} .$$

Now, consider first additions. If $a \in (a_-, a_+)$ and $b \in (b_-, b_+)$ where $a, b \in \mathbb{R}$ and $a_\pm, b_\pm \in \mathfrak{R}$, then, as above,

$$a + b \in (\text{Down}(\langle a_- + b_- \rangle), \text{Up}(\langle a_+ + b_+ \rangle)) \equiv (a_-, a_+) + (b_-, b_+) ,$$

which serves as definition of addition (and subtraction) between intervals.
The multiplication is slightly more complicated and several subcases must be considered in order to properly define $(a_-, a_+) * (b_-, b_+) \equiv (c_-, c_+)$, $(a_\pm, b_\pm, c_\pm \in \mathfrak{R})$. In computer-like language:

(i): Let $a_- \geq 0$.
If $b_- \geq 0$ then $(c_-, c_+) \equiv (\text{Down}\langle a_- * b_- \rangle, \text{Up}\langle a_+ * b_+ \rangle)$;
if $b_+ \leq 0$ then $(c_-, c_+) = (\text{Down}\langle a_+ * b_- \rangle, \text{Up}\langle a_- * b_+ \rangle)$.
Finally, if $b_- < 0$ and $b_+ > 0$, then $(c_-, c_+) = (\text{Down}\langle a_+ * b_- \rangle, \text{Up}\langle a_+ * b_+ \rangle)$.

(ii): Let $a_+ \leq 0$.
If $b_- \geq 0$ then $(c_-, c_+) = (\text{Down}\langle a_- * b_+ \rangle, \text{Up}\langle a_+ * b_- \rangle)$;
if $b_+ \leq 0$ then $(c_-, c_+) = (\text{Down}\langle a_+ * b_+ \rangle, \text{Up}\langle a_- * b_- \rangle)$.
Finally, if $b_- < 0$ and $b_+ > 0$, then $(c_-, c_+) = (\text{Down}\langle a_- * b_+ \rangle, \text{Up}\langle a_- * b_- \rangle)$.

(iii): Let $a_- < 0$ and $a_+ > 0$.
If $b_- \geq 0$ then $(c_-, c_+) = (\text{Down}\langle a_- * b_+ \rangle, \text{Up}\langle a_+ * b_+ \rangle)$;
if $b_+ \leq 0$ then $(c_-, c_+) = (\text{Down}\langle a_+ * b_- \rangle, \text{Up}\langle a_- * b_- \rangle)$.
Finally, if $b_- < 0$ and $b_+ > 0$, then
$(c_-, c_+) = (\min\{\text{Down}\langle a_- * b_+ \rangle, \text{Down}\langle a_+ * b_- \rangle, \}, \ \max\{\text{Up}\langle a_- * b_- \rangle, \text{Up}\langle a_+ * b_+ \rangle) .$

The division is treated in an analogous way.
Elementary transcendental functions (exponentials, logarithms, trigonometric functions, etc.) may be approximated by a finite sequence of elementary operations using Taylor expansions and simple inequalities to truncate the expansion at a certain (arbitrarly defined) order; for example:

$$\sum_{n=0}^{N-1} \frac{x^n}{n!} < e^x < \sum_{n=0}^{N-1} \frac{x^n}{n!} + \frac{x^N}{N!} \frac{N}{N-1}, \qquad \forall N \geq 2, \quad \forall 0 < x \leq 1 .$$

9. Applications: Three-Dimensional Phase Space Systems

Here and in the following section 10, we briefly illustrate the computer-assisted application of the above theory to a few concrete models. Some of the results presented here are new (compare end of §9.1), while most of them were obtained in preceding works of the authors: see [CC1], [CC2], [CC3], [CC4], [C1], [C2], to which we refer also for complete details.

9.1. A Forced Pendulum

One of the simplest non-integrable conservative system is a periodically forced classical pendulum described by the one-dimensional, time-dependent Lagrangian

$$\mathcal{L}(y, x, t) \equiv \frac{y^2}{2} - \mu\left[\cos x + \cos(x - t)\right], \qquad (9.1)$$

where $y \in \mathbb{R}$ and $(x, t) \in \mathbb{T}^2$.

This model, which is the central object of the renormalization theory of Escande and Doveil ([ED], [Es]) can also be viewed as describing the motion of a particle with charge μ, in the field of a potential of two longitudinal (electrostatic) waves.

Here, we apply the method presented in the previous sections to (9.1) in order to construct the "*golden-mean* torus" $\mathcal{T}(\omega_g) \equiv \mathcal{T}(\frac{\sqrt{5}-1}{2})$, for values of the non-linearity parameter μ of the same order of magnitude of the expected "break-down" threshold (see below for definitions and for an experimental recipe to compute such a threshold).

The interest in the stability properties of this particular torus comes from various considerations: the main being that the golden-mean ω_g, which satisfies (2.33) with $\gamma = (\sqrt{5} + 3)/2$ and $\tau = 1$ (cfr. Appendix 8), is, in a suitable sense, the "most irrational" number in $(0, 1)$ and one expects this fact to show up in the μ-power series, which involves the small divisors $(\omega_g n + m)$. Of course, this is a rather naive observation that might lead to the belief that the ω_g-torus is always the most stable, while (cfr. §10) there are examples pointing in different directions.

Numerical methods for determining the critical value ρ_c (see §6.2) have been developed in [Gr], [C], [ED]. The remarkable method developed by J. Greene in [Gr] is based on the following idea. Let $\{p_j/q_j\}$ be the sequence of rational approximants to the irrational number ω (see Appendix 8 for more informations) and let $\mathcal{P}(p_j/q_j)$ denote a periodic orbit with period q_j and rotation number p_j/q_j. Greene conjectures that the disappearance of the torus $\mathcal{T}(\omega)$ is related to a sudden change, from stability to instability, of the periodic orbits $\mathcal{P}(p_j/q_j)$, which, as $j \to \infty$, approximates the torus $\mathcal{T}(\omega)$. This criterion applied to the forced pendulum (9.1) indicates that $\mu_c(\omega_g)$ is $\simeq 0.027$.

Now a brief history. Our first attempt ([CC1],[CFP]) to obtain stability estimates for $\mathcal{T}_\mu(\omega_g)$ in the above model was based on refining Arnold's version of the KAM theorem ([A1], see also [G1]). This strategy allowed us to establish existence for $0 \le \mu \le 6.75 \cdot 10^{-4}$, a value which was later increased up to $1.42 \cdot 10^{-3}$ in [CG]. However, this approach, which, as is well known, is based on a sequence of canonical (symplectic) transformations, presents an intrinsic difficulty related to the geometry of the domain where the the canonical transformations are defined. In fact, in order to control the resonances (i.e., phase points where the frequencies are rationally dependent), such domains have to be taken smaller and smaller as the iteration is carried out and to obtain sharp quantitative results one is led to the difficult analysis of the domain of holomorphy of each canonical transformation.

This problem is bypassed by considering directly the parametric equation for the tori as discussed in the present work. The first implementation of this new strategy was carried out in [CC2] where the Euler-Lagrange equation

$$D^2 u = \mu \left[\sin(\theta + u) + \sin(\theta + u - t) \right] \qquad (9.2)$$

is solved using as initial approximation the finite power series

$$v^{(0)}(\theta, t) \equiv \sum_{l=1}^{l_0} u_l(\theta, t) \mu^l , \qquad (9.3)$$

the u_l's being the Taylor coefficients of the (convergent) expansion around $\mu = 0$ of the solution (see §7.2).

Since, for $|\mu| < \rho_a$ (cfr. §6.2), $v^{(0)} \to u$ as $l_0 \to \infty$, we shall get better initial approximation by taking l_0 large. However the number of Fourier coefficients of u_l grows rapidly with the order l: u_1 has 4 Fourier coefficients, u_{10} has 120 coefficients and u_{60} has 3720 coefficients. Therefore, computer-time limitations (if nothing else) forces to stop at relatively small orders. In [CC2] we computed the functions u_l using the general formula

$$D^2 u_{l+1} = \sum_{h \in \mathcal{H}_l} \partial_\theta^{|h|} f \prod_{i=1}^{l} \frac{u_i^{h_i}}{h_i!} , \qquad (9.4)$$

where $\mathcal{H}_l \equiv \{ h \in \mathbb{N}^l : h_1 + 2h_2 + ... + lh_l = l \}$, with $f \equiv \sin \theta + \sin(\theta - t)$. Such a general formula presents, however, serious combinatorial problems as l gets large. Using (9.4) (and about two hours of CPU on a VAX 8600) we computed, *using interval arithmetic*, (9.3) for $l_0 = 24$ and proved the existence of $\mathcal{T}(\frac{\sqrt{5}-1}{2})$ for $|\mu| < 0.015$.

In [CC3], using formulae (7.7) and (7.12), which reduce considerably the combinatorics problems, we could compute (with about the same computer time) (9.3) with $l_0 = 40$. At this order the existence of the golden-mean torus can be established (as above, via a KAM algorithm very similar to the one presented in the present work) for $|\mu| \leq 0.018$. Finally, using the strategy described in §7.3, we computed numerically (i.e. without interval arithmetic) v_0 up to order $l_0 = 60$ and establish the existence of the golden mean torus for $|\mu| \leq 0.019$. This result, which is new and to our present knowledge is the best rigorous result in a hamiltonian setting, is in agreement of the 70% with the numerical guess provided by Greene's method.

To obtain this existence result we solve the Euler-Lagrange equation (9.2) using the initial approximate solution $v^{(0)} = \sum_1^{l_0} \tilde{u}_l \mu^l$ where the functions $\tilde{u}_l(\theta, t)$ are *numerically computed* using the recursive formulae of §7, namely

$$b_0^{(n,m)} = e^{i(n\theta + mt)}$$

$$b_l^{(n,m)} = \frac{i}{l} n \cdot \sum_{h=1}^{l} h \tilde{u}_h b_{l-h}^{(n,m)} , \qquad l \geq 1$$

and

$$\tilde{u}_l = \frac{1}{2i} D^{-2} \left[b_{l-1}^{(1,0)} - b_{l-1}^{(-1,0)} + b_{l-1}^{(1,-1)} - b_{l-1}^{(-1,1)} \right]$$

(which are easily recognized as (7.12) with $f(x, t) = \sin(x) + \sin(x - t)$).

The computation of the functions $\tilde{u}_l$ has been performed on a VAX 6000; the computer time necessary to evaluate the function u_{60} was about 24 minutes.

Then, following the (quite straightforward) steps of §6.2, we obtained the following

Theorem 9.1. *Let $\omega = \frac{\sqrt{5}-1}{2}$ and let $\xi = 0.08$, $\rho = 0.019$. Then equation (9.2) admits a locally unique real-analytic solution $u(\theta, t; \mu)$ with $\langle u \rangle = 0$ on $\mathbb{T}^2$, analytic in $\Delta_\xi \times \{\mu \in \mathbb{C} : |\mu| \le \rho\}$. Moreover, one can construct a polynomial approximation $v(\theta, t; \mu) \equiv \sum_{l=1}^{60} \tilde{u}_l(\theta, t)\mu^l$, where $\tilde{u}_l$ are trigonometric polynomials, satisfying*

$$|u - v|_{\xi,\rho} < 0.2526 , \qquad |u_\theta - v_\theta|_{\xi,\rho} < 0.3824$$

where $| \cdot |_{\xi,\rho} \equiv \sup_{\Delta_\xi, |\mu| \le \rho} | \cdot |$.

9.2. Spin-Orbit Coupling in Celestial Mechanics

We discuss now an example drown from Celestial Mechanics, which has been investigated in [C1], [C2], [CF] using the methods developped in [CC2].

One of the most astonishing phenomena in the mechanics of our solar system is that all the *evolved* satellites of the solar system always point the same face toward the host planet, as in the Moon-Earth case. The only exception to this rule is the Mercury-Sun system, as radar observations have shown that the period of revolution of Mercury around the Sun is $\frac{3}{2}$ of the period of rotation about its spin-axis.

Exact commensurabilities between the period of rotation and the period of revolution go under the name of *spin-orbit resonances* ([GP], [He], [W]). More precisely, for an oblate satellite S orbiting around a central body P one has a $p : q$ resonance (for any $p, q \in \mathbb{Z}_+$) when the motion of S is periodic and the ratio between the periods of revolution and rotation of the satellite is $\frac{p}{q}$; we shall denote by $\mathcal{P}(p/q)$ the set of all such orbits.

A natural question is whether the motion of satellites observed in spin-orbit resonance is stable or not.

Following [C1], [C2], [CF], we shall use the theory of invariant surfaces to give a positive answer to such a question under suitable simplifying assumptions.

Let us start by introducing the model we want to study.

Let S be a triaxial homogeneous ellipsoidal satellite with principal moments of inertia $A < B < C$ subject to the gravitational attraction of a (fixed) central planet P and assume that:

i) the orbit of the center of mass of S around the central body P is a fixed Keplerian ellipse;

ii) the spin-axis of S coincides with its shortest physical axis and is perpendicular to the orbit plane;

iii) all the dissipative forces as well as perturbations due to other bodies are negligible (and therefore ignored).

Then, the equations of motion can be derived from the standard Euler's equations for a
rigid body ([D]). Normalizing the period of revolution to 2π, one obtains

$$\ddot{x} + \frac{3}{2}\frac{B-A}{C}\left(\frac{a}{r}\right)^3 \sin(2x-2f) = 0, \tag{9.5}$$

where a is the semimajor axis of the Keplerian ellipse, r the orbital "radius" (i.e. the
distance between the centers of mass of S and P), f the true anomaly (i.e. the angle
between the planet-satellite direction and the periapsis line) and x is the angle between
the longest axis of the ellipsoid and the periapsis line.

From assumption i) and the theory of the two-body problem, it follows that the quantities
r and f are periodic functions of the time, $r(t+2\pi) = r(t)$, $f(t+2\pi) = f(t)$, and that
they are analytic functions of the orbital eccentricity e. Therefore expanding the second
term of (9.5) in series, one obtains:

$$\ddot{x} + \mu \sum_{m\neq 0, m=-\infty}^{\infty} V(\frac{m}{2}, e)\ \sin(2x-mt) = 0, \tag{9.6}$$

where $\mu \equiv \frac{3}{2}\frac{B-A}{C}$ and the coefficients $V(\frac{m}{2}, e)$ are analytic functions of the eccentricity
e:

$$V(\frac{m}{2}, e) \equiv e^{|m-2|}\sum_{k\geq 0}a_k e^{2k},$$

for suitable $a_k \in \mathbb{R}$.

In writing (9.5) (or equivalently (9.6)) we have ignored the dissipative forces acting on
the system. The major dissipative contribution is originated by the internal non-rigidity
of the satellite and goes under the name of "tidal torque".

Having ignored such a force allow us to ignore *all* the quantities which are of comparable
size. This leads us to consider equations of the form:

$$\ddot{x} + \mu \sum_{m\neq 0, m=N_1}^{N_2} W(\frac{m}{2}, e)\ \sin(2x-mt) = 0, \quad N_1, N_2 \in \mathbb{Z}, \tag{9.7}$$

where $W(\frac{m}{2}, e)$ are truncations, to a suitable order in the eccentricity, of the coefficients
$V(\frac{m}{2}, e)$: of course, such truncations will depend upon the specific model at hand, on the
physical values of μ, e and on the size of the observed tidal torque.

Under the above simplifications, we can investigate the stability of the system using the
following argument.

The phase space $\mathcal{S}$ associated to (9.7)

$$\mathcal{S} = \{(y, x, t): \ y = \dot{x} \in \mathbb{R}, \ (x, t) \in \mathbb{T}^2\}$$

has dimension three. Therefore bidimensional invariant surfaces $\mathcal{T}(\omega)$ divide the phase
space into invariant compartments, with the property that any orbit starting in one of these
regions would remain forever in it. Now, one can show that, in the parameter regions we
shall consider below, the Poincarè map $(y, x) \rightarrow \phi^{2\pi}(y, x)$ [$\phi^t(y, x)$ $\equiv$ solution at time
t starting at $x(0) = x$, $\dot{x}(0) = y$] is a smooth "monotone twist map": $(\partial x'/\partial y) > 0$ (see
[Mo5], [MK] for general informations). Invariant tori for (9.7) correspond to *invariant
circles* for $\phi^{2\pi}$ and the Poincarè rotation number for such a circle coincide with the
frequency ω associated to the invariant torus. It is not difficult to show (see, e.g., [Her1])

that if $\{z_i\}_{i=1,...,q}$, $z_i \equiv (y_i, x_i)$, is a periodic orbit with rotation number p/q and if Γ_ω is an invariant circle with rotation number $\omega >$ (resp. $<$) p/q, then Γ_ω lies above (resp. below) $\{z_i\}$, i.e., $\Gamma_\omega \cap \{x = x_i\} >$ (resp. $<$) y_i.

We make use of this property to trap a periodic orbit $\mathcal{P}(\frac{p}{q})$, associated to the $p : q$ resonance, between invariant surfaces $\mathcal{T}(\omega_1)$ and $\mathcal{T}(\omega_2)$, with $\omega_1 < \frac{p}{q} < \omega_2$. Obviously one is interested in taking ω_i as close as possible to p/q.

Therefore we select the two sequences of irrational frequencies

$$\Gamma_k^{(p/q)} \equiv \frac{p}{q} - \frac{1}{k+\omega_g}, \qquad \Delta_k^{(p/q)} \equiv \frac{p}{q} + \frac{1}{k+\omega_g}, \qquad k \in \mathbb{Z}, \ k \geq 2$$

($\omega_g \equiv \frac{\sqrt{5}-1}{2}$), which approach $\frac{p}{q}$ from below and, respectively, above and satisfy the diophantine condition (Assumption 2.11) with the constant $\gamma \equiv \gamma_k = q^2(k+\omega_g)$.

Let us consider now the Moon-Earth system (Moon $\equiv S$, Earth $\equiv P$) and let us look at the synchronous resonance $\mathcal{P}(1/1)$, which human kind has been observing for quite a while.

According to our simplifications, we are led to study the Lagrangian

$$\mathcal{L}(y, x, t) \equiv \frac{y^2}{2} + \mu \left[(-\frac{e}{4} + \frac{e^3}{32}) \cos(2x - t) \right.$$
$$+ (\frac{1}{2} - \frac{5}{4}e^2 + \frac{13}{32}e^4) \cos(2x - 2t) + (\frac{7}{4}e - \frac{123}{32}e^3) \cos(2x - 3t)$$
$$+ (\frac{17}{4}e^2 - \frac{115}{12}e^4) \cos(2x - 4t) + (\frac{845}{96}e^3 - \frac{32525}{1536}e^5) \cos(2x - 5t)$$
$$\left. + \frac{533}{32}e^4 \cos(2x - 6t) + \frac{228347}{7680}e^5 \cos(2x - 7t) \right],$$

$$\tag{9.8}$$

where the physical value of the perturbing parameter $\mu \equiv \frac{3}{2}\frac{B-A}{C}$ is $3.45 \cdot 10^{-4}$ and that of the eccentricity e is 0.0549. Using the techiques of §1 ÷ §9 (with the choice in §7.2) one can construct the surfaces $\mathcal{T}(\Gamma_k^{(1)})$ and $\mathcal{T}(\Delta_k^{(1)})$ for $k = 2, 3, ..., 35$ and from the above discussion it then follows that the motion of the Moon, *as ruled by the approximate Lagrangian (9.8)*, will be forever trapped in the region enclosed by $\mathcal{T}(\Gamma_{35}^{(1)})$ and $\mathcal{T}(\Delta_{35}^{(1)})$, which, in turns, is shown to be a subset of $\{(y, x, t) : (x, t) \in \mathbb{T}^2, 0.97 \leq y \leq 1.03\}$.

Let us now consider the system Mercury-Sun, which is complicated by the fact that its eccentricity is relatively large: $e = 0.2056$. Therefore we have to retain a larger number of terms in (9.7); in particular we consider the lagrangian function

$$\mathcal{L}(y, x, t) \equiv \frac{y^2}{2} + \frac{\mu}{2} \sum_{m \neq 0, m=-11}^{3} W(\frac{m}{2}, e) \cos(2x - mt) . \tag{9.9}$$

We are still able to conclude the stability of the 3:2 resonance, in which Mercury is actually observed for the astronomical value of the perturbing parameter, i.e. $\mu = 1.5 \cdot 10^{-4}$. The tori closest to the periodic orbits in $\mathcal{P}(\frac{3}{2})$ are those with rotation numbers $\Gamma_{70}^{(3/2)}$ and $\Delta_{70}^{(3/2)}$; the outcoming trapping region is contained in $\{(y, x, t) : (x, t) \in \mathbb{T}^2, 1.48 \leq y \leq 1.52\}$.

10. Applications: Symplectic Maps

In this section we briefly discuss how the KAM techniques of §1 ÷ §8 can be adapted to deal with symplectic diffeomorphisms of plane regions.

Eventhough there is a tight connection (via Poincarè maps) between symplectic diffeomorphisms and Hamiltonian flows (cfr. [Do], [Mo6], [SZ]), it is interesting and often useful, to have *direct* formalisms and methods. The direct method discussed below was introduced in [CC2]; see also [CC4].

10.1. Formalism

Here we shall consider a special class of symplectic (area-preserving) twist diffeomorphisms of the cylinder $\mathscr{C} \equiv \mathbb{R} \times S^1$, $(S^1 \equiv \mathbb{T}^1 \equiv \mathbb{R}/2\pi\mathbb{Z})$, namely:

$$F : (y, x) \in \mathscr{C} \longmapsto (y', x') \equiv (y + f(x), x + y + f(x)) \in \mathscr{C} , \qquad \langle f \rangle = 0 , \qquad (10.1)$$

where f is a real-analytic function on S^1 (i.e., a real-analytic function on $\mathbb{R}$ with period 2π) with vanishing mean-value. As above, we shall also consider one-parameter families obtained by replacing f with μf.

The word "twist" in the definition of the present model refers to the following property. If we look at the universal covering, $\mathbb{R}^2$, of the cylinder and consider a lift $\tilde{F}$ of F (for example, replace, in (10.1), $\mathscr{C}$ with $\mathbb{R}^2$), then $\tilde{F}$ maps vertical lines $\{x = x_0\}$ into graphs of increasing functions of the x-variable; analytically: $(\partial x'/\partial y) > 0$. In our case, $(\partial x'/\partial y) = 1$.

The problem is to study the behaviour of the orbits $(y_n, x_n) \equiv F^n(y_0, x_0)$, where F^n denotes F composed with itself n times.

The observation that, for (10.1), $y' = x' - x$, allows to eliminate the y-variable: (y_n, x_n) is an F-orbit if and only if the sequence $\{x_n\}$ satisfies

$$x_{n+1} - 2x_n + x_{n-1} = f(x_n) ; \qquad (10.2)$$

obviously, given a solution $\{x_n\}$ of (10.2) the associated F-orbit is simply $(x_n - x_{n-1}, x_n)$.

Analogously to Definition 2.1 we give the following

Definition 10.1. *A solution $\{x_n\}$ of (10.2) is called quasi-periodic with frequency $\omega \in \mathbb{R}$, if $\omega/2\pi$ is irrational and if there exists a continuous periodic function $u: \theta \in S^1 \to u(\theta) \in \mathbb{R}$, such that*

$$x_n \equiv \omega n + u(\omega n) , \qquad (\text{mod } 2\pi) . \qquad (10.3)$$

and, analogously to Definition 2.5:

Definition 10.2. *We shall say that a quasi-periodic solution is non-degenerate if $\forall \theta \in S^1$*

$$(1 + u_\theta) \neq 0 . \qquad (10.4)$$

As for flows, non-degenerate quasi-periodic solutions correspond to invariant surfaces, which, in the present case, are *invariant circles*: the map

$$\theta \in S^1 \longmapsto \left(\omega + u(\theta) - u(\theta - \omega), \theta + u(\theta) \right) \qquad (10.5)$$

yields a non-contractible embedding of S^1 into $\mathscr{C}$.

To require that $u(\theta)$ is a non-degenerate quasi-periodic solution of (10.2) with frequency ω is equivalent to require that u satisfies the following non-linear finite-difference equation:

$$D^2 u = f(\theta + u(\theta)) , \qquad (1 + u_\theta \neq 0) , \tag{10.6}$$

where D, here, denotes the symmetrized finite-difference operator with step ω

$$Du \equiv u(\theta + \frac{\omega}{2}) - u(\theta - \frac{\omega}{2}) \equiv u^+(\theta) - u^-(\theta) . \tag{10.7}$$

The abuse of language in denoting with the same symbol different objects will be forgiven in view of the complete analogy of the present situation with the Lagrangian case.

10.2. The Newton Scheme, the Linearized Equation, etc.

The strategy of §2 can be carried out in this context and it is actually simpler because of the dimension and of the pecularity of the maps we are considering.

First notice that, since for any periodic function g the S^1-average of Dg vanishes, in dealing with the equation

$$Dg = h \tag{10.8}$$

one has to require that $\langle h \rangle = 0$, and in such a case, the unique solution with vanishing mean-value of (10.8) is given by:

$$g = D^{-1}h \equiv \sum_{n \in \mathbb{Z}, n \neq 0} \frac{\hat{h}_n}{2i \sin\left(\frac{n\omega}{2}\right)} e^{in\theta} , \qquad \left(h \equiv \sum_{n \in \mathbb{Z}, n \neq 0} \hat{h}_n e^{in\theta} \right) , \tag{10.9}$$

and we see the reason for the irrationality of $\omega/2\pi$ in Definition 10.1.

However, $n\omega/2$ will come arbitrarily close to 0, π (mod 2π) and we need a Diophantine asumption; from now on we shall assume that ω satisfies, for some $\gamma, \tau \geq 1$,

$$| \frac{\omega}{2\pi} n + m | \geq \frac{1}{\gamma |n|^\tau} , \qquad \forall n \in \mathbb{Z} \setminus \{0\} , \forall m \in \mathbb{Z} . \tag{10.10}$$

Let us now introduce the function spaces. Let $\Delta^1_\xi \equiv \{\theta \in \mathbb{C} : |\text{Im } \theta| \leq \xi\}$ and denote by $| \cdot |_\xi$ the sup norm over Δ^1_ξ. Besides Lemma 3.2, which will be used with $d = 1$, we need the analogous of Lemma 3.3. For $l \geq 0$ we let $s_l(\delta)$ be an upper bound on the small-divisor series

$$[\sum_{n=1}^{\infty} (\frac{n^l}{\sin \frac{n\omega}{2}})^2 e^{-\delta n}]^{1/2} \leq s_l(\delta) . \tag{10.11}$$

Then:

Lemma 10.3. *Let h be a real-analytic function on $\Delta^1_\xi \times \mathscr{P}$, ($\mathscr{P}$ being a compact set of $\mathbb{C}$) and let $l \geq 1$. Then ($| \cdot |_{\xi,\mathscr{P}} \equiv \sup_{\Delta^1_\xi \times \mathscr{P}} | \cdot |$)*

$$|D^{-1} \partial^l_\theta h|_{\xi - \delta, \mathscr{P}} \leq s_l(2\delta) |h|_{\xi, \mathscr{P}} . \tag{10.12}$$

The same estimate holds for $l = 0$ provided h has vanishing mean value over S^1.

An explicit estimate of $s_l(\rho)$ for ω's satisfying (10.10) is given in Appendix 9.

Let us now discuss the *Newton scheme*. As above, we shall call v a *non-degenerate approximate solution* of (10.6) any real-analytic periodic function such that $1 + v_\theta \neq 0$, and we shall associate to it its *error function* $\mathcal{E}(v) \equiv \varepsilon(\theta)$, given by

$$\mathcal{E}(v) \equiv \varepsilon(\theta) \equiv D^2 v - f(\theta + v) . \tag{10.13}$$

With the proficiency acquired in the more complicate case of §2, §3, the reader will have no trouble in checking the following Proposition (cfr. Proposition 2.13):

Proposition 10.4. *Let ω satisfy (10.10), let v be a (non-degenerate) approximate solution of equation (10.6) and let $\varepsilon(\theta)$ be the associated error-function. Let*

$$\mathcal{M} \equiv 1 + v_\theta . \tag{10.14}$$

Then $\langle \mathcal{M}\varepsilon \rangle = 0$ and if we set:

$$w \equiv \mathcal{M} \Big\{ D^{-1}[(\mathcal{M}^+\mathcal{M}^-)^{-1}(-D^{-1}(\mathcal{M}\varepsilon) + c_1)] + c_2 \Big\} , \tag{10.15}$$

with

$$
\begin{aligned}
c_1 &\equiv \frac{\langle (\mathcal{M}^+\mathcal{M}^-)^{-1}[-D^{-1}(\mathcal{M}\varepsilon)] \rangle}{\langle (\mathcal{M}^+\mathcal{M}^-)^{-1} \rangle} , \\
c_2 &\equiv \langle \mathcal{M}D^{-1}[(\mathcal{M}^+\mathcal{M}^-)^{-1}(-D^{-1}(\mathcal{M}\varepsilon) + c_1)] \rangle ,
\end{aligned}
\tag{10.16}
$$

and $v' \equiv v + w$, then $\langle w \rangle = 0$ and

$$\mathcal{E}(v') \equiv \varepsilon' = \varepsilon_\theta \mathcal{M}^{-1} w - \{ f(\theta + v') - f(\theta + v) - f_x(\theta + v)w \} \tag{10.17}$$

The estimates leading to the KAM algorithm and to the existence KAM theorem for the present situation, can be obtained, at this point, in a completely straightforward way; however, for completeness and convenience of the reader we shall collect the main estimates in Appendix 10.

10.3. Results

The above methods, together with the strategy outlined in §7.3, have been used in [CC4], to study the stability of various invariant circles for the following one-parameter families of twist maps:

$$F : (y, x) \in \mathcal{C} \longmapsto (y', x') \equiv (y + \mu f(x), x + y + \mu f(x)) \in \mathcal{C} , \tag{10.18}$$

with

$$
\begin{aligned}
(M) \qquad & f = \sin x , \qquad \text{(standard map)} , \\
(M') \qquad & f = \sin x + \frac{1}{50}\sin(5x) .
\end{aligned}
\tag{10.19}
$$

The rotation numbers considered are:

$$\frac{\omega_1}{2\pi} \equiv \frac{\sqrt{5}-1}{2} , \qquad \frac{\omega_2}{2\pi} \equiv \frac{\sqrt{5}+5}{10} , \qquad \frac{\omega_3}{2\pi} \equiv \frac{\sqrt{2}}{2} . \tag{10.20}$$

Now, let $\Gamma_\mu(\omega_k)$, respectively $\Gamma'_\mu(\omega_k)$, be the invariant circles for (M), resp. (M'). The stability results are summarized in the following table:

Curve	l_0	ρ	N	ξ
$\Gamma_\mu(\omega_1)$	190	0.838	6	$5.07 \cdot 10^{-3}$
$\Gamma_\mu(\omega_2)$	190	0.77	5	$5.15 \cdot 10^{-3}$
$\Gamma_\mu(\omega_3)$	160	0.76	5	$5.15 \cdot 10^{-3}$
$\Gamma'_\mu(\omega_1)$	60	0.4	7	$5.03 \cdot 10^{-3}$
$\Gamma'_\mu(\omega_2)$	60	0.39	7	$5.03 \cdot 10^{-3}$

where l_0 is the order of the polynomial initial guess (cfr. §7.3) $v = \sum_{l=1}^{l_0} \bar{u}_l(\theta)\mu^l$, ρ and ξ measure the size of the analyticity domain of the solution u: $u(\theta; \mu)$ is real-analytic on $\Delta_\xi^1 \times \{\mu \in \mathbb{C} : |\mu| \le \rho\}$, finally N denotes the number of times the KAM algorithm has been used before applying the KAM theorem.

These results should be compared with the experimental prediction given by Greene's method discussed above;

Curve	Greene's threshold
$\Gamma_\mu(\omega_1)$	0.9716
$\Gamma_\mu(\omega_2)$	$0.9044 - 0.9045$
$\Gamma_\mu(\omega_3)$	$0.908 - 0.909$
$\Gamma'_\mu(\omega_1)$	$0.6013 - 0.6014$
$\Gamma'_\mu(\omega_2)$	$0.7213 - 0.7214$

Hence, our theoretical results are in an agreement ranging within $86\% \div 54\%$. The reason for the more sensible discrepancy for the map (M') seems to be related to the distribution of the μ singularities (θ almost everywhere in S^1) of the solution u. In particular there are numerical evidences that μ-domain of analyticity of the solutions u for the maps (M) and (M') (for the given rotation numbers) has (for almost every θ) a natural boundary.

For the map (M'), it seems that $d_r > d_i$, if $d_{r/i}$ denotes the distance from the origin with the first real/purely imaginary singularity: see [BC], [BCCF].

Appendix 1: Proof of (2.37)

We prove in this appendix the formula (2.37) of §2, i.e.

$$\mathscr{A} = D^{-1}(\mathscr{M}^T \varepsilon_\theta)^A , \tag{10.21}$$

where $\mathscr{A}$ is defined in (2.31) as

$$\mathscr{A} \equiv (\mathscr{M}^T \partial_\theta \mathscr{L}_y^T)^A$$

and the superscript A denotes the antisymmetric part of a matrix, namely

$$\begin{aligned} D^{-1}(\mathscr{M}\varepsilon_\theta)^A &\equiv D^{-1}(\mathscr{M}^T \varepsilon_\theta) - D^{-1}(\varepsilon_\theta^T \mathscr{M}) \\ &= D^{-1}\left[\mathscr{M}^T \varepsilon_\theta - \varepsilon_\theta^T \mathscr{M}\right] . \end{aligned} \tag{10.22}$$

Taking the gradient with respect to θ of the definition (2.12) of the error function

$$\varepsilon(\theta, t) = D\mathscr{L}_y^T(\omega + Dv, \theta + v, t) - \mathscr{L}_x^T(\omega + Dv, \theta + v, t) , \tag{10.23}$$

one has

$$\varepsilon_\theta = D[\mathcal{L}_{yy}D\mathcal{M} + \mathcal{L}_{yx}\mathcal{M}] - \mathcal{L}_{xy}D\mathcal{M} - \mathcal{L}_{xx}\mathcal{M} \ . \tag{10.24}$$

Multiplying (10.24) by $\mathcal{M}^T$ and taking the antisymmetric part, one has:

$$
\begin{aligned}
\mathcal{M}^T\varepsilon_\theta - \varepsilon_\theta^T\mathcal{M} &= \mathcal{M}^T\, D[\mathcal{L}_{yy}D\mathcal{M} + \mathcal{L}_{yx}\mathcal{M}] - \mathcal{M}^T\mathcal{L}_{xy}D\mathcal{M} \\
&\quad - D[D\mathcal{M}^T\mathcal{L}_{yy} + \mathcal{M}^T\mathcal{L}_{xy}]\mathcal{M} + D\mathcal{M}^T\mathcal{L}_{yx}\mathcal{M} \\
&= D[\mathcal{M}^T\mathcal{L}_{yy}D\mathcal{M} + \mathcal{M}^T\mathcal{L}_{yx}\mathcal{M}] - D(D\mathcal{M}^T\mathcal{L}_{yy}\mathcal{M} + \mathcal{M}^T\mathcal{L}_{xy}\mathcal{M}) \ .
\end{aligned}
$$

Finally, recalling the definition of $\mathcal{A}$

$$
\begin{aligned}
\mathcal{A} &\equiv (\mathcal{M}^T\partial_\theta\mathcal{L}_y^T)^A \equiv \mathcal{M}^T\partial_\theta\mathcal{L}_y^T - (\partial_\theta\mathcal{L}_y^T)^T\mathcal{M} \\
&= \mathcal{M}^T\mathcal{L}_{yy}D\mathcal{M} + \mathcal{M}^T\mathcal{L}_{yx}\mathcal{M} - D\mathcal{M}^T\mathcal{L}_{yy}\mathcal{M} - \mathcal{M}^T\mathcal{L}_{xy}\mathcal{M} \ ,
\end{aligned}
\tag{10.25}
$$

one obtains

$$\mathcal{M}^T\varepsilon_\theta - \varepsilon_\theta^T\mathcal{M} = D\mathcal{A} \ . \quad \square$$

Appendix 2: Proof of Lemma 3.2

In this appendix we prove the Lemma 3.2 of §3.2.

Lemma: *Let h be an analytic map from $\Omega \times \mathcal{P} \to \mathbb{C}$, where Ω is a (smooth) domain in $\mathbb{C}^d$ and $\mathcal{P} \subset \mathbb{C}^k$ a space of parameters; then for any subdomain $\Omega' \subset \Omega$ with $\mathrm{dist}(\Omega', \partial\Omega) \equiv \delta > 0$ and for any multi index $m = (m_1, ..., m_d) \in \mathbb{N}^d$ one has*

$$\sup_{\Omega'\times\mathcal{P}} |\partial_z^m h| \equiv \sup_{\Omega'\times\mathcal{P}} |\frac{\partial^{|m|}h}{\partial z_1^{m_1}...\partial z_d^{m_d}}| \le m!\, \delta^{-|m|} \sup_{\Omega\times\mathcal{P}} |h| \tag{10.26}$$

($|m| = m_1 + ... + m_d$). Moreover, if h is an analytic map, $h : \Omega \times \mathcal{P} \to L^p(\mathbb{C}^d)$ for some $p \in \mathbb{N}$ ($L^0(\mathbb{C}^d) \equiv \mathbb{C}^d$), then $\forall l \in \mathbb{Z}_+$, $\partial_z^l h \in L^{p+l}(\mathbb{C}^d)$ and

$$\sup_{\Omega'\times\mathcal{P}} |\partial_z^l h| \le l!\, \delta^{-l} \sup_{\Omega\times\mathcal{P}} |h| \ . \tag{10.27}$$

Proof: Consider first a holomorphic function $h_0 : \Omega \times \mathcal{P} \to \mathbb{C}$. Then, Cauchy's integral formula implies

$$
\begin{aligned}
\sup_{\Omega'\times\mathcal{P}} |\frac{\partial^{|m|}h_0}{\partial z_1^{m_1}...\partial z_d^{m_d}}| &= \\
&= \sup_{\Omega'\times\mathcal{P}} |\frac{m!}{(2\pi i)^d} \oint_{|\zeta_1-z_1|=\delta,...,|\zeta_d-z_d|=\delta} \frac{h_0(\zeta_1, ..., \zeta_d)}{\prod\limits_{k=1}^{d}(\zeta_k - z_k)^{m_k+1}} \, d\zeta_1...d\zeta_d | \\
&\le m!\, \delta^{-|m|} \sup_{\Omega\times\mathcal{P}} |h_0| \ .
\end{aligned}
\tag{10.28}
$$

Now, if $h : \Omega \times \mathcal{P} \to L^p(\mathbb{C}^d)$ for some $p \in \mathbb{N}$, then (10.28) implies

$$
\begin{aligned}
\sup_{\Omega' \times \mathcal{P}} |\partial_z^l h| &\equiv \sup_{|c_1|=...=|c_l|=1} \sup_{\Omega' \times \mathcal{P}} |\partial_z^l h \ c_1...c_l| \\
&\leq \sup_{|c_1|=...=|c_l|=1} \sup_{\Omega' \times \mathcal{P}} \left(|\partial_z^l h| \ |c_1|...|c_l| \right) \\
&\leq \sup_{|c_1|=...=|c_l|=1} \left(l! \ \delta^{-l} \sup_{\Omega \times \mathcal{P}} |h| \ |c_1|...|c_l| \right) \\
&= l! \ \delta^{-l} \sup_{\Omega \times \mathcal{P}} |h| \ . \quad \square
\end{aligned}
$$

Appendix 3: Proof of Lemma 3.3

This appendix is devoted to the proof of the Lemma 3.3 of §3.2.

Lemma: *Let $h = h(\theta, t; \mu)$ be a real-analytic map of $\Delta_\xi \times \mathcal{P}$ into $\mathcal{H}$, where $\mathcal{H}$ is either $\mathbb{C}$, or $\mathbb{C}^d$ or $L^p(\mathbb{C}^d)$ and let $l \geq 1$. Then*

$$|D^{-1}\partial_\theta^l h|_{\xi-\delta,\mathcal{P}} \leq \sigma_l(2\delta) \ |h|_{\xi,\mathcal{P}} \ , \tag{10.29}$$

where

$$\sigma_l(\rho) = \left[2^{d+1} \sum_{(n,m)\in\mathbb{Z}^{d+1}\backslash(0,0)} \left(\frac{\|n\|^l}{\omega \cdot n + m}\right)^2 e^{-\rho(|n|+|m|)} \right]^{\frac{1}{2}} \ , \tag{10.30}$$

$(\|n\| \equiv (\sum_{i=1}^d |n_i|^2)^{1/2}, |n| \equiv \sum_{i=1}^d |n_i|)$. *Moreover the same estimate holds for $l = 0$ provided h has vanishing mean value over $\mathbb{T}^{d+1}$. If $(\omega, 1)$ verifies Assumption 2.11 then*

$$\sigma_l(\rho) < K_l \gamma \delta^{-(\tau+l)} \ , \qquad K_l \equiv 2^{d+2-(\tau+l)} \sqrt{\Gamma(2(\tau+l)+1)} \ ,$$

Γ *being Euler's gamma function.*

Proof: We prove first (10.29) for a holomorphic function $h_0 : \Delta_\xi \times \mathcal{P} \to \mathbb{C}$ with vanishing mean value. Denote by $\| \cdot \|_{\xi,\mathcal{P}}$ the L^2-norm

$$\|h_0\|_{\xi,\mathcal{P}}^2 \equiv \sup_{\mathcal{P}} \sup_{\substack{|a_1|,...,|a_d|\leq\xi \\ |b|\leq\xi}} \int_{\mathbb{T}^{d+1}} |h_0(\theta + ia, t + ib)|^2 \ \frac{d\theta dt}{(2\pi)^{d+1}} \ .$$

Then, for any $\nu = (\nu_1, ..., \nu_d) \in \{-1, 1\}^d$, $\lambda \in \{-1, 1\}$, one has

$$\sup_{\mathcal{P}} \sum_{(n,m)} e^{2(n\cdot\nu+m\lambda)\xi} |\hat{h}_{0_{(n,m)}}|^2 \leq \|h_0\|_{\xi,\mathcal{P}}^2 \ . \tag{10.31}$$

To prove (10.31), let $\xi' < \xi$ and consider the function

$$h_0' \equiv h_0(\theta - i\nu\xi', t - i\lambda\xi') \ .$$

By Cauchy's theorem we have:

$$\hat{h}_{0'_{(n,m)}} = e^{\xi'(n\cdot\nu+m\lambda)} \hat{h}_{0_{(n,m)}} \ .$$

Then, Parseval's identity yields

$$\sum |\hat{h}_{0_{(n,m)}}|^2 \, e^{2\xi'(n\cdot\nu+m\lambda)} = \int_{\mathbb{T}^{d+1}} |h'_0|^2 \, \frac{d\theta dt}{(2\pi)^{d+1}} \leq \|h_0\|^2_{\xi,\mathcal{P}} \, .$$

Taking the supremum over $\xi' < \xi$ one obtains (10.31). From the maximum principle, Schwarz inequality, Assumption 2.11 and (10.31), it follows (dropping the index 0)

$$| \, \partial^l_{\theta_j} \, D^{-1} h \, |_{\xi-\delta,\mathcal{P}} = | \sum_{(n,m)\neq 0} \hat{h}_{(n,m)} \, \frac{n^l_j}{(\omega\cdot n+m)} \, e^{i(n\cdot\theta+mt)}|_{\xi-\delta,\mathcal{P}}$$

$$= \sup_{\mathcal{P}} \sup_{(\nu,\lambda)\in\{-1,1\}^{d+1}} | \sum_{(n,m)\neq 0} \hat{h}_{(n,m)} \, \frac{n^l_j}{(\omega\cdot n+m)} \, e^{(n\cdot\nu+m\lambda)(\xi-\delta)}|$$

$$\leq \sup_{\mathcal{P}} \sum_{(n,m)\neq 0} |\hat{h}_{(n,m)}| \left(\sum_{\lambda,\nu} e^{2(n\cdot\nu+m\lambda)\xi} \right)^{1/2} e^{-\delta(|n|+|m|)} \frac{|n_j|^l}{|\omega\cdot n+m|} \qquad (10.32)$$

$$\leq \sigma_l(2\delta) \sup_{\mathcal{P}} \left(\frac{1}{2^{d+1}} \sum_{(n,m)} |\hat{h}_{(n,m)}|^2 \sum_{\lambda,\nu} e^{2(n\cdot\nu+m\lambda)\xi} \right)^{1/2}$$

$$\leq \sigma_l(2\delta) \, \|h\|_{\xi,\mathcal{P}} \leq \sigma_l(2\delta) \, |h|_{\xi,\mathcal{P}} \, .$$

Now, let $h : \Delta_\xi \times \mathcal{P} \to \mathbb{C}^d$ and $l = 0$. Then, by (10.32)

$$|D^{-1} h|_{\xi-\delta,\mathcal{P}} \equiv \sum_i |D^{-1} h_i|_{\xi-\delta,\mathcal{P}}$$

$$\leq \sigma_0(2\delta) \sum_i |h_i|_{\xi,\mathcal{P}} \equiv \sigma_0(2\delta) \, |h|_{\xi,\mathcal{P}} \, .$$

If $l \geq 1$, then for $c_1 \in \mathbb{C}^d,...,c_l \in \mathbb{C}^d$,

$$|\partial^l_\theta \, D^{-1} h|_{\xi-\delta,\mathcal{P}} \equiv \sup_{|c_1|=1,...,|c_l|=1} |D^{-1} \, \partial^l_\theta h \, c_1...c_l|_{\xi-\delta,\mathcal{P}}$$

$$\leq \sup_{|c_1|=1,...,|c_l|=1} |\partial^l_\theta \, D^{-1} h|_{\xi-\delta,\mathcal{P}} \, |c_1|...|c_l|$$

$$\leq \sup_{|c_1|=1,...,|c_l|=1} \sigma_l(2\delta) \, |h|_{\xi,\mathcal{P}} \, |c_1|...|c_l|$$

$$= \sigma_l(2\delta) \, |h|_{\xi,\mathcal{P}} \, .$$

Finally, if $h : \Delta_\xi \times \mathcal{P} \to L^p(\mathbb{C}^d)$, applying again (10.32) one has

$$|\partial^l_\theta \, D^{-1} h|_{\xi-\delta,\mathcal{P}} = \sup_{|c_1|=1,...,|c_l|=1} |D^{-1} \, \partial^l_\theta h \, c_1...c_l|_{\xi-\delta,\mathcal{P}}$$

$$\leq \sigma_l(2\delta) \, |h|_{\xi,\mathcal{P}} \, .$$

Now we want to show that if $(\omega, 1)$ verifies Assumption 2.11 then

$$\sigma_l(\rho) < K_l \gamma \delta^{-(\tau+l)} \, , \qquad K_l \equiv 2^{d+2-(\tau+l)} \sqrt{\Gamma(2(\tau+l)+1)} \, , \qquad (10.33)$$

where Γ is the Euler's gamma function. Assume for ω the diophantine condition

$$|\omega\cdot n+m|^{-1} < \gamma|n|^\tau \, .$$

The solution g of the equation $Dg = h$ is given by

$$g = D^{-1}h = \sum_{(n,m)\in\mathbb{Z}^{d+1}\setminus(0,0)} \frac{\hat{h}_{(n,m)}\, e^{i(n\cdot\theta+mt)}}{i(n\cdot\omega+m)} \; ;$$

therefore

$$\partial_{\theta_i}^l g(\theta,t) = \sum_{(n,m)\in\mathbb{Z}^{d+1}\setminus(0,0)} \frac{n_i^l}{n\cdot\omega+m}\, \hat{h}_{(n,m)}\, e^{i(n\cdot\theta+mt)} .$$

From the inequality (see [R3] p.180, formula (9.4)),

$$\sum_{n\in\mathbb{Z}^d,m\in\mathbb{Z}} |\hat{h}_{(n,m)}|^2\, e^{2\xi(|n|+|m|)} < 2^{d+1}|h|^2_{\xi,\mathscr{P}} , \tag{10.34}$$

the term $\partial_{\theta_i}^l g(\theta,t)$ can be estimated using Schwarz's inequality as

$$|\partial_{\theta_i}^l g(\theta,t)| \leq \sum_{(n,m)} |\frac{n_i^l}{n\cdot\omega+m}\hat{h}_{(n,m)}\, e^{i(n\cdot\theta+mt)}|$$

$$\leq \sum_{(n,m)} |\frac{n_i^l}{n\cdot\omega+m}|e^{-\delta(|n|+|m|)}\, |\hat{h}_{(n,m)}|\, e^{\xi(|n|+|m|)}$$

$$\leq \sqrt{\sum_{(n,m)} |\hat{h}_{(n,m)}|^2\, e^{2\xi(|n|+|m|)}} \sqrt{\sum_{(n,m)} |\frac{n_i^l}{n\cdot\omega+m}|^2\, e^{-2\delta(|n|+|m|)}}$$

$$\leq 2^{\frac{d+1}{2}}\|h\|\sqrt{\Psi(\delta)} < 2^{\frac{d+1}{2}}|h|\sqrt{\Psi(\delta)} ,$$

where $\Psi(\delta) \equiv \sum_{(n,m)} |\frac{n_i^l}{n\cdot\omega+m}|^2\, e^{-2\delta(|n|+|m|)}$. Let us estimate $\Psi(\delta)$ as follows:

$$\Psi(\delta) \leq \sum_{k=1}^{\infty}\sum_{|n|+|m|=k} |\frac{n_i^l}{n\cdot\omega+m}|^2\, e^{-2\delta k} \leq \sum_{k=1}^{\infty}\left(\sum_{|n|+|m|=k} |\frac{1}{n\cdot\omega+m}|^2\right) k^{2l}e^{-2\delta k} .$$

Finally defining $b_0 = 0$, $b_k = \sum_{0<|n|+|m|\leq k} |\frac{1}{n\cdot\omega+m}|^2$, one has:

$$\Psi(\delta) = \sum_{k=1}^{\infty}(b_k - b_{k-1})k^{2l}\, e^{-2\delta k} = \sum_{k=1}^{\infty} k^{2l}b_k e^{-2\delta k} - \sum_{k=1}^{\infty}(k+1)^{2l}b_k e^{-2\delta(k+1)}$$

$$\leq (1-e^{-2\delta})\cdot 2^{d+3}\sum_{k=1}^{\infty}\frac{k^{2l}e^{-2\delta k}}{D_k^2} ,$$

where in the last inequality we used $b_k = \sum_{0<|n|+|m|\leq k} |\frac{1}{n\cdot\omega+m}|^2 \leq \frac{2^{d+3}}{D_k^2}$, where

$$D_k = \min_{0<|n|+|m|\leq k} |n\cdot\omega+m|$$

(see the proof in [R2], [R4] suitably adapted to be valid with the actual choice of the norms). Since

$$D_k \geq \min_{0<|n|+|m|\leq k}\left(\frac{1}{\gamma|n|^\tau}\right) = \frac{1}{\gamma k^\tau}$$

or $\frac{1}{D_k} \leq \gamma k^\tau$, it follows

$$\Psi(\delta) \leq (1 - e^{-2\delta}) \, 2^{d+3} \gamma^2 \sum_{k=1}^{\infty} k^{2l+2\tau} \, e^{-2\delta k} \ .$$

Let us estimate the sum as follows. Let $s = 2\tau$ and $\beta = 2\delta$:

$$\sum_{k=1}^{\infty} k^s e^{-\beta k} = \sum_{k=1}^{\infty} \int_k^{k+1} k^s \beta e^{-\beta x} \, \frac{dx}{1 - e^{-\beta}}$$

$$= \frac{\beta}{1 - e^{-\beta}} \sum_{k=1}^{\infty} \int_k^{k+1} k^s \, e^{-\beta x} dx \leq \frac{\beta}{1 - e^{-\beta}} \sum_{k=1}^{\infty} \int_k^{k+1} x^s e^{-\beta x} dx$$

$$\leq \frac{\beta^{-s}}{1 - e^{-\beta}} \int_0^{\infty} y^s e^{-y} dy = \frac{\beta^{-s}}{1 - e^{-\beta}} \, \Gamma(s+1) \ .$$

Therefore,

$$\Psi(\delta) \leq 2^{d+3-(2\tau+2l)} \gamma^2 \delta^{-(2\tau+2l)} \, \Gamma(2(\tau + l) + 1)$$

and finally

$$|\partial^l_{\theta_i} g(\theta, t)| \leq 2^{d+2-(\tau+l)} \gamma \, |h|_{\xi, \mathcal{P}} \, \delta^{-(\tau+l)} \, \sqrt{\Gamma(2(\tau + l) + 1)} \ . \quad \square$$

Appendix 4: Proof of Lemma 3.5

This appendix is devoted to the proof of the Lemma 3.5 of §3.4.

Lemma: *Let $\mathcal{T} \equiv \mathcal{M}^T \mathcal{L}_{yy} \mathcal{M}$ be the twist matrix of a non degenerate approximate solution v and let M, $\overline{M}$, L, $\overline{L}$ denote upper bounds on (respectively) $|\mathcal{M}|_\xi$, $|\mathcal{M}^{-1}|_\xi$, $|\mathcal{L}_{yy}|_{\mathcal{D}\xi}$, $|\mathcal{L}_{yy}^{-1}|_{\mathcal{D}\xi}$. Then*

$$(i) \quad \overline{M}^{-2} \overline{L}^{-1} \leq |\mathcal{T}|_0 \leq |\mathcal{T}|_\xi \leq M^2 L$$

$$(ii) \quad M^{-2} L^{-1} \leq |\mathcal{T}^{-1}|_0 \leq |\mathcal{T}^{-1}|_\xi \leq \overline{M}^2 \overline{L}$$

$$(iii) \quad |\langle \mathcal{T}^{-1} \rangle^{-1}|_0 \leq |\mathcal{T}|_0 \qquad \left(\langle \cdot \rangle \equiv \int_{\mathbb{T}^{d+1}} \cdot \, \frac{d\theta dt}{(2\pi)^{d+1}} \right) \ .$$

Proof:

(i) By $|\mathcal{T}| = |\mathcal{T}^T|$, $|\mathcal{T}^{-1}| \geq |\mathcal{T}|^{-1}$ (since $1 = |\mathcal{T}\mathcal{T}^{-1}| \leq |\mathcal{T}||\mathcal{T}^{-1}|$) and the positivity of the matrix $\mathcal{T}$ one has:

$$|\mathcal{T}| \leq M^2 L$$

and

$$|\mathcal{T}|^{-1} \leq |\mathcal{T}^{-1}| \leq |(\mathcal{T}^T)^{-1}| |(\mathcal{L}_{yy})^{-1}| |\mathcal{T}^{-1}| \leq \overline{M}^2 \overline{L}$$

or

$$|\mathcal{T}| \geq \overline{M}^{-2} \overline{L}^{-1} \ .$$

(ii) The inequality $|\mathcal{T}^{-1}| \le \overline{M}^2 \underline{L}$ has been proven in *(i)*. Moreover, $|\mathcal{T}| \le M^2 L$ implies

$$|\mathcal{T}^{-1}| \ge |\mathcal{T}|^{-1} \ge M^{-2}L^{-1} .$$

(iii) $\mathcal{T} > 0$ (i.e. $\mathcal{T} \ge 0$ and $\mathcal{T}$ invertible) implies $|(\mathcal{T}^{-1})^{-1}| \le |\mathcal{T}|$. $\mathcal{T} > 0$ follows from the following four general facts:
(a) $A = A^*$ (with $A^* = \overline{A}^T$) $\Rightarrow$ $A \le |A|$;
(b) $a \in \mathbb{R}_+$, $A \ge 0$, $A \le a$ $\Rightarrow$ $|A| \le a$;
(c) $a \in \mathbb{R}_+$, $A \ge 0$, $A \ge a$ $\Rightarrow$ $A^{-1} \le a^{-1}$;
(d) $A, B \ge 0$, $[A, B] \equiv AB - BA = 0$ $\Rightarrow$ $AB \ge 0$.
   Proof of *(a)*, *(b)*, *(c)*, *(d)*:
(a) $\langle Ax, x \rangle \le |Ax||x| \le |A||x|^2 = \langle |A|x, x \rangle$
(b) $A^* = A$ $\Rightarrow$ $|A| = \sup_{|x|=1} \langle Ax, x \rangle \le a$
(c) $\langle Ax, x \rangle \ge a\langle x, x \rangle$ $\Leftrightarrow$ $\langle A^{1/2}x, A^{1/2}x \rangle \ge a\langle x, x \rangle$; setting $A^{1/2}x = y$ one has

$$a^{-1}\langle y, y \rangle \ge \langle A^{-1/2}y, A^{-1/2}y \rangle = \langle A^{-1}y, y \rangle$$

(d) $[A, B] = 0$ $\Rightarrow$ $[A, f(B)] = 0$ $\forall f$ continuous; thus,

$$\langle ABx, x \rangle = \langle AB^{1/2}B^{1/2}x, x \rangle = \langle B^{1/2}AB^{1/2}x, x \rangle = \langle AB^{1/2}x, B^{1/2}x \rangle \ge 0 .$$

By *(a)* $|\mathcal{T}| - \mathcal{T} \ge 0$ and since $|\mathcal{T}| - \mathcal{T}$ commutes with $\mathcal{T}^{-1}$, while by *(d)* $(|\mathcal{T}| - \mathcal{T})\mathcal{T}^{-1} \ge 0$ i.e. $|\mathcal{T}|\mathcal{T}^{-1} \ge \mathbb{1}$ and by averaging $\mathbb{1} \le |\mathcal{T}|\langle \mathcal{T}^{-1} \rangle$ or, by *(c)*, $(\mathcal{T}^{-1})^{-1} \le |\mathcal{T}|$; finally *(b)* yields $\mathcal{T} > 0$. $\square$

Appendix 5: Proof of Proposition 3.6

In this appendix we provide the details of the proof of Proposition 3.6 of §3.5. In particular we want to bound the error function

$$\varepsilon' \equiv q_1 + q_2 + q_3 ,$$

where q_1, q_2, q_3 have been defined in (2.47).
From (3.45) we have

$$|q_1|_{\xi'} \le |Dq_1^{(y)}|_{\xi'} + |q_1^{(x)}|_{\xi'} .$$

The term $|q_1^{(x)}|_{\xi'}$ can be bounded as in (3.53) by

$$|q_1^{(x)}|_{\xi'} \le \frac{L_3}{2} E^2 \overline{L}^2 a^2 (c + 1)^2 .$$

Now let

$$f = f(\omega + Dv, \theta + v, t) = \mathcal{L}_y^T(\omega + Dv, \theta + v, t) , \qquad f^+ = f(\omega + Dv + Dw, \theta + v + w, t) ,$$

as in (3.54); then from (3.55)

$$
\begin{aligned}
|Dq_1^{(y)}|_{\xi'} &\equiv |D\left[f^+ - f - f_y Dw - f_x w\right]|_{\xi'}\\
&\equiv |\omega \cdot \partial_\theta(f^+ - f - f_y Dw - f_x w) + \partial_t(f^+ - f - f_y Dw - f_x w)|_{\xi'}\\
&\le |\partial_\theta(f^+ - f - f_y Dw - f_x w)\ \omega|_{\xi'} + |\partial_t(f^+ - f - f_y Dw - f_x w)|_{\xi'}\\
&\equiv A_1 + A_2,
\end{aligned}
\tag{10.35}
$$

where

$$
\begin{aligned}
A_1 &= |\partial_\theta(f^+ - f - f_y Dw - f_x w)\ \omega|_{\xi'}\\
&= |\partial_\theta\left\{\int_0^1 (1-\beta)\left[f_{yy} Dw Dw + f_{yx} Dw\ w + f_{xy} w Dw + f_{xx} ww\right] d\beta\right\}\ \omega|_{\xi'}
\end{aligned}
\tag{10.36}
$$

and

$$
A_2 = |\partial_t\left\{\int_0^1 (1-\beta)\left[f_{yy} Dw Dw + f_{yx} Dw\ w + f_{xy} w Dw + f_{xx} ww\right] d\beta\right\}|_{\xi'}
$$

(notice that the derivatives of f are evaluated at $(\omega + Dv + \beta Dw, \theta + v + \beta w, t)$).

Let us start with the estimate of A_1. By (10.36) we have:

$$
\begin{aligned}
A_1 &\le \left\{\int_0^1 (1-\beta)\left\{\left[|f_{yyy}|\ (|D\mathcal{M}| + \beta|Dw_\theta|) + |f_{yyx}|\ (|\mathcal{M}| + \beta|w_\theta|)\right]|Dw|^2\right.\right.\\
&\quad + 2|f_{yy}||Dw_\theta||Dw| + 2\left[|f_{yxy}|\ (|D\mathcal{M}| + \beta|Dw_\theta|)\right.\\
&\quad + |f_{yxx}|\ (|\mathcal{M}| + \beta|w_\theta|)\Big]|Dw||w|\\
&\quad + 2|f_{yx}|\ (|Dw_\theta||w| + |Dw||w_\theta|) + \left[|f_{xxy}|\ (|D\mathcal{M}| + \beta|Dw_\theta|)\right.\\
&\quad + |f_{xxx}|\ (|\mathcal{M}| + \beta|w_\theta|)\Big]|w|^2 + 2|f_{xx}|\ |w_\theta||w|\bigg\} d\beta\bigg\}\ |\omega|\\
&\le \left\{\frac{1}{2}\left[|f_{yyy}||D\mathcal{M}| + M|f_{yyx}|\right]|Dw|^2 + |f_{yy}||Dw||Dw_\theta| + \left[|f_{yxy}||D\mathcal{M}|\right.\right.\\
&\quad + |f_{yxx}|M\Big]|w||Dw| + |f_{yx}|\left[|w||Dw_\theta| + |w_\theta||Dw|\right]\\
&\quad + \frac{1}{2}\left[|f_{xxy}||D\mathcal{M}| + |f_{xxx}|M\right]|w|^2 + |f_{xx}||w||w_\theta|\\
&\quad + \frac{1}{6}\left[|f_{yyy}||Dw_\theta| + |f_{yyx}||w_\theta|\right]|Dw|^2 + \frac{1}{3}\left[|f_{yxy}||Dw_\theta| + |f_{yxx}||w_\theta|\right]|w||Dw|\\
&\quad + \frac{1}{6}\left[|f_{xxy}||Dw_\theta| + |f_{xxx}||w_\theta|\right]|w|^2\bigg\}\ |\omega|\ .
\end{aligned}
$$

Finally, denoting by

$$
\begin{aligned}
\lambda_3' &\equiv \max\ \{\Omega^2|\mathcal{L}_{yyy}|,\ \Omega|\mathcal{L}_{yyx}|,\ |\mathcal{L}_{yxx}|\}\ ,\\
\lambda_4 &\equiv \max\ \{\Omega^3|\mathcal{L}_{yyyy}|,\ \Omega^2|\mathcal{L}_{yyyx}|,\ \Omega^2|\mathcal{L}_{yyxy}|,\ \Omega|\mathcal{L}_{yyxx}|,\ \Omega|\mathcal{L}_{yxxy}|,\ |\mathcal{L}_{yxxx}|\}\ ,
\end{aligned}
$$

one has:

$$A_1 \leq \left\{ \frac{\lambda_4}{2} E^2 \overline{L}^2 a^2 (c+1)^2 \left[(1+\delta^{-1})M + \frac{4}{3} E\overline{L}a\delta^{-2}(g + \frac{\delta}{4} + \frac{\delta^2}{4}\frac{s_1(\delta)}{s_0(\delta)}\frac{b_1}{b}) \right] \right.$$
$$\left. + 4\lambda_3' E^2 \overline{L}^2 a^2 \delta^{-2}\left(g + \frac{\delta}{4} + \frac{\delta^2}{4}\frac{s_1(\delta)}{s_0(\delta)}\frac{b_1}{b} \right)(c+1) \right\} |\omega| .$$

Analogously let us estimate the second term A_2 as follows:

$$A_2 \leq \int_0^1 (1-\beta)\left\{ \left[|f_{yyy}|\,(|Dv_t| + \beta|Dw_t|) + |f_{yyx}|\,(|v_t| + \beta|w_t|) \right]|Dw|^2 \right.$$
$$+ 2|f_{yy}|\,|Dw_t||Dw| + 2\Big[|f_{yxy}|\,(|Dv_t| + \beta|Dw_t|)$$
$$+ |f_{yxx}|\,(|v_t| + \beta|w_t|)\Big]|Dw||w|$$
$$+ 2|f_{yx}|\left[|Dw_t||w| + |Dw||w_t| \right] + \Big[|f_{xxy}|\,(|Dv_t| + \beta|Dw_t|)$$
$$+ |f_{xxx}|\,(|v_t| + \beta|w_t|)\Big]|w|^2 + 2|f_{xx}|\,|w_t||w|$$
$$+ |f_{yyt}|\,|Dw|^2 + 2|f_{yxt}|\,|w||Dw| + |f_{xxt}|\,|w|^2 \Big\} d\beta$$
$$\leq \frac{1}{2}\Big[|f_{yyy}||Dv_t| + |f_{yyx}||v_t| \Big]|Dw|^2 + |f_{yy}||Dw||Dw_t|$$
$$+ \Big[|f_{yxy}||Dv_t| + |f_{yxx}||v_t| \Big]|w||Dw| + |f_{yx}|\Big[|w||Dw_t| + |w_t||Dw| \Big]$$
$$+ \frac{1}{2}\Big[|f_{xxy}||Dv_t| + |f_{xxx}||v_t| \Big]|w|^2 + |f_{xx}||w||w_t|$$
$$+ \frac{1}{6}\Big[|f_{yyy}||Dw_t| + |f_{yyx}||w_t| \Big]|Dw|^2$$
$$+ \frac{1}{3}\Big[|f_{yxy}||Dw_t| + |f_{yxx}||w_t| \Big]|w||Dw|$$
$$+ \frac{1}{6}\Big[|f_{xxy}||Dw_t| + |f_{xxx}||w_t| \Big]|w|^2$$
$$+ \frac{1}{2}|f_{yyt}||Dw|^2 + |f_{yxt}||w||Dw| + \frac{1}{2}|f_{xxt}||w|^2 .$$

We recall that if S is an upper bound on $|v_t|_{\xi,\rho}$, then

$$|w_t|_{\xi'} \leq E\overline{L}a \left(\delta^{-1} + \frac{s_1(\delta)}{s_0(\delta)}\frac{b_1}{b} \right) ,$$
$$|Dw|_{\xi'} \leq E\overline{L}ac\Omega ,$$
$$|w|_{\xi'} \leq E\overline{L}a ,$$
$$|Dv_t|_{\xi'} \leq S\delta^{-1}\Omega ,$$
$$|Dw_t|_{\xi'} \leq 4E\overline{L}\delta^{-2}ag\Omega .$$

Denoting by

$$L_4' \equiv \max\{\Omega^2|\mathscr{L}_{yyyt}|, \ \Omega|\mathscr{L}_{yyxt}|, \ |\mathscr{L}_{yxxt}| \} ,$$

one has

$$A_2 \leq \frac{\lambda_4}{2} E^2 \overline{L}^2 a^2 (c+1)^2 \left[(1+\delta^{-1})S + \frac{4}{3} E\overline{L}a\delta^{-2}(g + \frac{\delta}{4} + \frac{\delta^2}{4}\frac{s_1(\delta)}{s_0(\delta)}\frac{b_1}{b}) \right]$$
$$+ 4\lambda_3' E^2 \overline{L}^2 a^2 \delta^{-2} \left(g + \frac{\delta}{4} + \frac{\delta^2}{4}\frac{s_1(\delta)}{s_0(\delta)}\frac{b_1}{b} \right)(c+1)$$
$$+ \frac{L_4'}{2} E^2 \overline{L}^2 a^2 (c+1)^2 .$$

Therefore

$$A_1 + A_2 \leq \frac{E^2 \overline{L}^2 a^2}{2}(c+1)^2 \left\{ \lambda_4 (1+\delta^{-1})(S+M\Omega) + \frac{4}{3}\lambda_4 \Omega E\overline{L}a\delta^{-2} \cdot \right.$$
$$\cdot (g + \frac{\delta}{4} + \frac{\delta^2}{4}\frac{s_1(\delta)}{s_0(\delta)}\frac{b_1}{b}) + L_4' \right\} + 4\lambda_3' \Omega E^2 \overline{L}^2 a^2 \delta^{-2}(c+1) \cdot$$
$$\cdot \left(g + \frac{\delta}{4} + \frac{\delta^2}{4}\frac{s_1(\delta)}{s_0(\delta)}\frac{b_1}{b} \right) .$$

Let now

$$L_3' \equiv \max(\Omega\lambda_3') , \qquad L_4 \equiv \max(\Omega\lambda_4) ;$$

then one has

$$|q_1|_{\xi'} \leq \frac{E^2 \overline{L}^2 a^2}{2}(c+1)^2 \left[L_3 + L_4(1+\delta^{-1})(S+M) + \frac{4}{3}L_4 \, E\overline{L}a\delta^{-2} \left(g + \frac{\delta}{4} \right. \right.$$
$$\left. + \frac{\delta^2}{4}\frac{s_1(\delta)}{s_0(\delta)}\frac{b_1}{b} \right) + L_4' \right] + 4L_3' E^2 \overline{L}^2 a^2 \delta^{-2}(c+1) \left(g + \frac{\delta}{4} + \frac{\delta^2}{4}\frac{s_1(\delta)}{s_0(\delta)}\frac{b_1}{b} \right) .$$

The term $|q_2|_{\xi'}$ can be bounded as in (3.44) by

$$|q_2|_{\xi'} \leq E^2 \overline{L} \frac{a}{M\delta} .$$

Finally, from

$$|Dz|_{\xi'} \leq \frac{E\overline{L}a'}{M} ,$$

one finds

$$|q_3|_{\xi'} \leq |(\mathcal{M}^T)^{-1}|_{\xi'} \cdot |D^{-1}(\mathcal{M}^T \varepsilon_\theta - \varepsilon_\theta^T \mathcal{M})|_{\xi'} \cdot |Dz|_{\xi'}$$
$$\leq 4E^2 \overline{L} \, a' \delta^{-1} \overline{M} s_0(\delta) .$$

Collecting the above estimates, one has:

$$|\varepsilon|_{\xi-\delta} \equiv E^2 \overline{L} a \left\{ \frac{\delta^{-1}}{M} + \frac{a}{2}(c+1)^2 \overline{L} \left[L_3 + L_4(S+M)(1+\delta^{-1}) + \frac{4}{3}E\overline{L}a\delta^{-2}L_4 \cdot \right. \right.$$
$$\cdot \left(g + \frac{\delta}{4} + \frac{\delta^2}{4}\frac{s_1(\delta)}{s_0(\delta)}\frac{b_1}{b} \right) + L_4' \right] + 4L_3'\overline{L} \, a(c+1)\delta^{-2} \left(g + \frac{\delta}{4} + \frac{\delta^2}{4}\frac{s_1(\delta)}{s_0(\delta)}\frac{b_1}{b} \right)$$
$$+ \chi_d \cdot 4\delta^{-1}\overline{M}s_0(\delta) \frac{a'}{a} \right\} ,$$

where

$$\begin{cases} \chi_d = 0 & d = 1 \\ \chi_d = 1 & d \geq 2, \end{cases}$$

$$a \equiv (M\overline{M})^2 s_0(\delta)^2 b \, , \qquad b_1 \equiv 1 + (M\overline{M})^2 L\overline{L} \, \frac{s_0(2\xi)}{s_0(\delta)} \, ,$$

$$b \equiv b_1 + M \left(\frac{s_0(\xi)}{s_0(\delta)}\right)^2 \left[1 + (M\overline{M})^2 L\overline{L} \, \frac{s_0(2\xi)}{s_0(\xi)}\right] \, , \qquad c \equiv \delta^{-1} + \frac{a'}{\Omega \, a} \, ,$$

$$a' \equiv (M\overline{M})^2 s_0(2\delta) b_1' \, , \qquad b_1' \equiv 1 + (M\overline{M})^2 L\overline{L} \, \frac{s_0(2\xi)}{s_0(2\delta)} \, ,$$

$$a'' \equiv (M\overline{M})^2 s_0(\delta) b_1 \, , \qquad g \equiv 1 + \frac{\delta}{4} \frac{a'}{a} + \frac{\delta}{4} s_1(\delta) \frac{a''}{a} + \frac{\delta}{2} \frac{a''}{a} \, . \ \square$$

Appendix 6: Proof of (5.14)

In this appendix we want to prove the inequalities (5.14) of §5, i.e.

$$\begin{aligned} E_{i+1}\overline{L} &\leq (E_i\overline{L})^2 \beta_0 \gamma_0^{\,i} \\ W_i &\leq E_i\overline{L}\beta_1\gamma_1^{\,i} \\ W_{1i} &\leq E_i\overline{L}\beta_2\gamma_2^{\,i} \\ |Dw^{(i)}| &\leq E_i\overline{L}\beta_3\gamma_3^{\,i} \, \Omega \end{aligned}$$ (10.37)

with

$$\begin{aligned} \beta_0 &\equiv 8 \cdot 208^2 M^2 (M\overline{M})^8 (L\overline{L})^2 (S + M) K_1 K_0^3 \lambda \gamma^4 2^{8\tau} \xi^{-4\tau - 3} \, , & \gamma_0 &\equiv 2^{4\tau + 3} \\ \beta_1 &\equiv 13 \cdot M(M\overline{M})^4 L\overline{L} \, K_0^2 \gamma^2 2^{4\tau + 1} \xi^{-2\tau} \, , & \gamma_1 &\equiv 2^{2\tau} \\ \beta_2 &\equiv 165 \cdot M(M\overline{M})^4 L\overline{L} \, K_1 K_0 \gamma^2 2^{4\tau} \xi^{-2\tau - 1} \, , & \gamma_2 &\equiv 2^{2\tau + 1} \\ \beta_3 &\equiv 130 \cdot M(M\overline{M})^4 L\overline{L} \, K_0^2 \gamma^2 2^{4\tau} \xi^{-2\tau - 1} \, , & \gamma_3 &\equiv 2^{2\tau + 1} \, , \end{aligned}$$

where $\lambda \equiv \max(\overline{L}L_3, \overline{L}L_3', \overline{L}L_4, \overline{L}L_4', 1)$.

Let us start to estimate $w^{(i)}$. To this end we need an upper bound on the two quantities a_i and b_i, where

$$a_i \equiv (M_i\overline{M}_i)^2 s_0(\delta_i)^2 b_i \, ,$$

and

$$b_i \equiv b_{1i} + M_i \left(\frac{s_0(\xi_i)}{s_0(\delta_i)}\right)^2 \left[1 + (M_i\overline{M}_i)^2 L\overline{L}\frac{s_0(2\xi_i)}{s_0(\xi_i)}\right] \, .$$

$$b_{1i} \equiv 1 + (M_i\overline{M}_i)^2 L\overline{L} \, \frac{s_0(2\xi_i)}{s_0(\delta_i)} \, ,$$

From the definition of M_i, $\overline{M}_i$ we easily obtain

$$M_i \, \overline{M}_i \leq 3M\overline{M} \, ;$$

moreover, from the estimates:

$$\frac{s_0(2\xi_i)}{s_0(\delta_i)} = (\frac{\delta_i}{2\xi_i})^\tau < (\frac{1}{8})^\tau < \frac{1}{8}$$

$$\left(\frac{s_0(\xi_i)}{s_0(\delta_i)}\right)^2 = (\frac{1}{4^\tau})^2 < \frac{1}{16}$$

$$\frac{s_0(\xi_i)s_0(2\xi_i)}{s_0(\delta_i)^2} = \frac{1}{2^\tau} (\frac{\delta_i}{\xi_i})^{2\tau} < (\frac{1}{32})^\tau ,$$

one has (notice that here we need to indicate the subscript i):

$$b_i = 1 + (M_i\overline{M}_i)^2 L\overline{L}\left(\frac{s_0(2\xi_i)}{s_0(\delta_i)}\right) + M_i \left(\frac{s_0(\xi_i)}{s_0(\delta_i)}\right)^2 \left[1 + (M_i\overline{M}_i)^2 L\overline{L}\frac{s_0(2\xi_i)}{s_0(\xi_i)}\right]$$

$$< 1 + 9(M\overline{M})^2 L\overline{L} (\frac{1}{8})^\tau + 2M (\frac{1}{4})^{2\tau} \left[1 + 9(M\overline{M})^2 L\overline{L}\frac{1}{2^\tau}\right]$$

and therefore

$$a_i = (M_i\overline{M}_i)^2 s_0(\delta_i)^2 b_i$$

$$< 9 (M\overline{M})^2 K_0^2\gamma^2 \left(\frac{2^{i+2}}{\xi}\right)^{2\tau} \left\{1 + 9(M\overline{M})^2 L\overline{L} (\frac{1}{8})^\tau + 2M(\frac{1}{4})^{2\tau}\right.$$

$$\left.\cdot\left[1 + \frac{9}{2^\tau} (M\overline{M})^2 L\overline{L}\right]\right\}$$

$$< 9 (M\overline{M})^4 K_0^2\gamma^2 2^{4\tau+2\tau i}\xi^{-2\tau} \left\{1 + \frac{9}{8}L\overline{L} + \frac{2M}{16}\left(1 + \frac{9}{2}L\overline{L}\right)\right\}$$

$$< 26 M(M\overline{M})^4 (L\overline{L})K_0^2\gamma^2 2^{4\tau}2^{2\tau i}\xi^{-2\tau} .$$

Finally, we obtain

$$W_i \equiv E_i\overline{L}a_i \le E_i\overline{L}\beta_1\gamma_1^i ,$$

with

$$\beta_1 \equiv 13 M(M\overline{M})^4 L\overline{L}K_0^2\gamma^2 2^{4\tau+1}\xi^{-2\tau} , \qquad \gamma_1 = 2^{2\tau} .$$

Next we estimate $w_\theta^{(i)}$. Since $\frac{b_{1i}}{b_i} \le 1$ and $\frac{K_0}{K_1} \le \frac{1}{\sqrt{3}}$, one has:

$$W_{1i} \le E_i\overline{L}a_i \left(\delta_i^{-1} + \frac{s_1(\delta_i)}{s_0(\delta_i)}\right)$$

$$\le E_i\overline{L} \cdot 26M(M\overline{M})^4 K_0^2\gamma^2 2^{4\tau}2^{2\tau i}\xi^{-2\tau}(L\overline{L})\left(\frac{2^{i+2}}{\xi} + \frac{K_1}{K_0}\frac{2^{i+2}}{\xi}\right)$$

$$\le E_i\overline{L} \cdot 165(M\overline{M})^4 MK_1K_0\gamma^2 2^{4\tau} L\overline{L} \, 2^{(2\tau+1)i}\xi^{-2\tau-1} ,$$

namely,

$$W_{i1} \le E_i\overline{L} \, \beta_2\gamma_2^i ,$$

with

$$\beta_2 \equiv 165 \cdot M(M\overline{M})^4 K_1K_0\gamma^2 2^{4\tau}\xi^{-2\tau-1} L\overline{L} , \qquad \gamma_2 \equiv 2^{2\tau+1} .$$

From $\frac{a_i'}{a_i} < \frac{1}{4}$, we have:

$$
\begin{aligned}
|Dw^{(i)}| &\leq E_i \overline{L} a_i \Omega c_i \equiv E_i \overline{L} a_i \Omega \left(\delta_i^{-1} + \frac{a_i'}{a_i \Omega} \right) \\
&\leq E_i \overline{L} \cdot 130 M (M\overline{M})^4 (L\overline{L}) K_0^2 \gamma^2 \, 2^{4\tau} 2^i 2^{2\tau i} \xi^{-2\tau-1} \Omega \\
&= E_i \overline{L} \, \Omega \, \beta_3 \gamma_3^{\,i} \ ,
\end{aligned}
$$

with

$$
\beta_3 \equiv 130 \, M (M\overline{M})^4 (L\overline{L}) K_0^2 \gamma^2 \, 2^{4\tau} \xi^{-2\tau-1} \ , \qquad \gamma_3 \equiv 2^{2\tau+1} \ .
$$

We finally come to the estimate of E_{i+1}. First we can bound a_i from below as

$$
\begin{aligned}
|a_i| &= 9 \, (M\overline{M})^2 K_0^2 \gamma^2 \left(\frac{2^{i+2}}{\xi} \right)^{2\tau} \left\{ 1 + \frac{9}{8^\tau} (M\overline{M})^2 L\overline{L} \right. \\
&\left. + \frac{2M}{4^{2\tau}} \left(1 + \frac{9}{2^\tau} (M\overline{M})^2 L\overline{L} \right) \right\} \\
&> 132^2 \cdot 2^{2\tau i} \xi^{-2\tau} (M\overline{M})^2 \ .
\end{aligned}
$$

Therefore a_i is bounded from above and below as

$$
132^2 \cdot 2^{2\tau i} \xi^{-2\tau} (M\overline{M})^2 \ < |a_i| < \ 26 \cdot 2^{2\tau i} 2^{4\tau} \xi^{-2\tau} K_0^2 \gamma^2 (L\overline{L}) M (M\overline{M})^4 \ .
$$

Before proceding we need also the following estimates, which can be easily obtained from the definition of the various quantities:

$$
\frac{b_1'}{b} < 2^\tau \ , \qquad \frac{a'}{a} = \frac{s_0(2\delta)}{s_0(\delta)^2} \frac{b_1'}{b} < 1 \ , \qquad \frac{a''}{a} < \frac{\delta^\tau}{K_0 \gamma} \ , \qquad g < \frac{K_1}{K_0} \ .
$$

and

$$
\Omega\gamma \geq 1, \quad \frac{b_1}{b} \leq 1, \quad \delta_i \equiv \frac{\xi}{2^{i+2}} < \frac{1}{4}, \quad \delta_i^\tau < \frac{1}{4}
$$

$$
\gamma > 2, \qquad \tau \geq 1, \qquad K_0 \geq 11 \cdot 2^{-\tau} \ .
$$

Moreover from the definition of K_0 and K_1 (see (3.17)) we find

$$
K_1 = \frac{K_0}{2} \sqrt{\frac{\Gamma(2\tau+3)}{\Gamma(2\tau+1)}} = \frac{K_0}{2} \sqrt{(2\tau+2)(2\tau+1)} \ , \qquad K_1 \geq \sqrt{3} K_0 \ .
$$

Therefore, denoting by A_i an upper bound on the norm of a_i and by $\lambda \equiv \max(\overline{L}L_3, \overline{L}L_3', \overline{L}L_4, \overline{L}L_4', 1)$, we obtain

$$E_{i+1} < E_i^2 \overline{L} A_i^2 \lambda \left\{ \frac{\delta_i^{-1}}{A_i M_i} + \frac{4\overline{M}_i \delta_i^{-1}}{A_i} + \frac{1}{2}\left(\delta_i^{-1} + \frac{s_0(2\delta_i)}{s_0(\delta_i)^2} \frac{b_{1i}'}{b_i \Omega} + 1\right)^2 \right.$$

$$\left[1 + (S_i + M_i)(1 + \delta_i^{-1}) + \frac{4}{3}E_i \overline{L} A_i \delta_i^{-2}\left(\frac{\delta_i}{4} + g_i + \frac{\delta_i^2}{4}\frac{s_1(\delta_i)}{s_0(\delta_i)}\frac{b_{1i}}{b_i}\right) + 1\right]$$

$$\left. + 4\,\delta_i^{-2}\left(\delta_i^{-1} + \frac{s_0(2\delta_i)}{s_0(\delta_i)^2}\frac{b_{1i}'}{b_i \Omega} + 1\right)\left(\frac{\delta_i}{4} + \frac{\delta_i^2}{4}\frac{s_1(\delta_i)}{s_0(\delta_i)}\frac{b_{1i}}{b_i} + g_i\right)\right\}$$

$$< E_i^2 \overline{L} A_i^2 \lambda \left\{ \frac{2^{i+2}}{132^2\,2^{2\tau i}M_i(M_i\overline{M}_i)^2} + \frac{4\,\overline{M}_i 2^{i+2}}{132^2\,2^{2\tau i}(M_i\overline{M}_i)^2} \right.$$

$$+ \frac{1}{2}\delta_i^{-3}\left(1 + \frac{\xi}{K_0 \gamma \Omega\,2^{i+2}} + \frac{\xi}{2^{i+2}}\right)^2$$

$$\left[\frac{\xi}{2^i} + (S_i + M_i)(1 + \frac{\xi}{2^{i+2}}) + \frac{4}{3}\delta_i^{-1}E_i \overline{L} A_i\left(\frac{\delta_i}{4} + \frac{K_1}{K_0} + \frac{\delta_i}{4}\frac{K_1}{K_0}\right) + 1\right]$$

$$\left. + 4\delta_i^{-3}\left(1 + \frac{\xi}{K_0 \gamma \Omega 2^{i+2}} + \frac{\xi}{2^{i+2}}\right)\left(\frac{\delta_i}{4} + \frac{K_1}{K_0} + \frac{\delta_i}{4}\frac{K_1}{K_0}\right)\right\}$$

$$< E_i^2 \overline{L} A_i^2 \lambda \frac{K_1}{K_0}\delta_i^{-3}(S + M)\left\{ \frac{4}{2^6 132^2\sqrt{3}} + \frac{16}{2^6 132^2\sqrt{3}} + \frac{1}{2\sqrt{3}}(1 + \frac{1}{20} + \frac{1}{4})^2 \cdot \right.$$

$$\cdot \left[2 + 2(1 + \frac{1}{4}) + \frac{4}{3}E_i \overline{L} A_i \delta_i^{-1}(\frac{1}{16\sqrt{3}} + 1 + \frac{1}{16})\right]$$

$$\left. + 4(1 + \frac{1}{20} + \frac{1}{4})(\frac{1}{2^4\sqrt{3}} + 1 + \frac{1}{2^4})\right\}$$

$$< E_i^2 \overline{L} A_i^2 \lambda \frac{K_1}{K_0}\delta_i^{-3}(S + M)\left[\frac{198}{25} + \frac{8}{11}E_i \overline{L} A_i \delta_i^{-1}\right]$$

$$< E_i^2 \overline{L} \cdot 26^2 M^2 (M\overline{M})^8 (L\overline{L})^2 (S + M)K_1 K_0^3 \lambda \gamma^4\,2^{8\tau}2^{4\tau i}\xi^{-4\tau-3}2^{3i}2^6$$

$$\cdot \left[\frac{198}{25} + \frac{8}{11}E_i \overline{L} A_i \delta_i^{-1}\right].$$

Defining

$$\beta_0 \equiv 8 \cdot 208^2 M^2 (M\overline{M})^8 (L\overline{L})^2 (S + M)K_1 K_0^3 \lambda \gamma^4\,2^{8\tau}\xi^{-4\tau-3}\,, \qquad \gamma_0 \equiv 2^{4\tau+3}\,,$$

one has:

$$E_i \overline{L} < \frac{(E\overline{L}\beta_0 \gamma_0)^{2^i}}{\beta_0 \gamma_0^{i+1}}$$

and by the hypothesis

$$E\overline{L}\beta_0 \gamma_0 < 1\,,$$

one has

$$E_i \overline{L} < \frac{1}{\beta_0 \gamma_0^{i+1}}\,.$$

Therefore,

$$E_{i+1} < 208^2 \cdot 8M^2 (M\overline{M})^8 (L\overline{L})^2 (S + M)K_1 K_0^3 \lambda \gamma^4\,2^{8\tau}2^{(4\tau+3)i}\xi^{-4\tau-3}\,(E_i^2 \overline{L})$$
$$< E_i^2 \overline{L}\,\beta_0 \gamma_0^i\,. \quad \square$$

Appendix 7: Up and Down of Real Numbers

Upper and lower bounds on the result of elementary operations can be obtained increasing or decreasing by one bit the last bit of the mantissa, with an eventual propagation of the carry.

In the Fortran function listed below it is shown how to obtain upper bounds on real numbers in G-floating representation.

The real number r, represented by 64-bits, is initially decomposed in 4 bytes (each one of 8 bits) labelled $kp(1), ..., kp(4)$ by the Fortran "EQUIVALENCE" statement. Degenerate cases (i.e. bytes of all 0's or 1's) are treated properly.

```fortran
Double precision function Up(r)
integer*2 kp(4)
real*8 r,x
equivalence (x,kp(1))
x=r
if (x.gt.0.) then
if (kp(4).eq.32767) then
kp(4)=-32768
Up=x
return
endif
kp(4)=kp(4)+1
if (kp(4).ne.0) then
Up=x
return
endif
if (kp(3).eq.32767) then
kp(3)=-32768
Up=x
return
endif
kp(3)=kp(3)+1
if (kp(3).ne.0) then
Up=x
return
endif
if (kp(2).eq.32767) then
kp(2)=-32768
Up=x
return
endif
kp(2)=kp(2)+1
if (kp(2).ne.0) then
Up=x
return
endif
kp(1)=kp(1)+1
Up=x
return
else if (x.lt.0.) then
if (kp(4).eq.-32768) then
kp(4)=32767
Up=x
return
endif
kp(4)=kp(4)-1
if (kp(4).ne.-1) then
Up=x
return
endif
if (kp(3).eq.-32768) then
kp(3)=32767
Up=x
return
endif
kp(3)=kp(3)-1
if (kp(3).ne.-1) then
Up=x
return
endif
if (kp(2).eq.-32768) then
kp(2)=32767
Up=x
return
endif
kp(2)=kp(2)-1
if (kp(2).ne.-1) then
Up=x
return
endif
```

```
kp(1)=kp(1)-1                          Up=x
Up=x                                   return
return                                 endif
else                                   end
```

The lower bound of a number s is obtained simply using the function Up as

$$\text{Down} = -\text{Up}(-s) \ .$$

Appendix 8: Computation of the Diophantine Constant

In this appendix we prove that the golden mean

$$\omega = \omega_g \equiv \frac{\sqrt{5} - 1}{2}$$

satisfies the diophantine inequality

$$|\omega - \frac{p}{q}| \geq \frac{1}{\gamma q^2} \ , \qquad \forall p, q \in \mathbb{Z}, \ q \neq 0 \ , \tag{10.38}$$

with a constant

$$\gamma \equiv \frac{3 + \sqrt{5}}{2} \ . \tag{10.39}$$

Let us review some properties of continued fractions (see [Kh]).
Let ω be a positive irrational number and let $[a_0; a_1, a_2, ...]$, $a_k \in \mathbb{N}$, its continued fraction expansion, namely

$$\omega \equiv a_0 + \cfrac{1}{a_1 + \cfrac{1}{a_2 + ...}} \ ;$$

let

$$\frac{p_k}{q_k} \equiv [a_0; a_1, ..., a_k] \ , \qquad r_k \equiv [a_k; a_{k+1}, ...] \ .$$

Then, the following relations hold (see [Kh]):

$$p_k = p_{k-1} a_k + p_{k-2} \ , \qquad q_k = q_{k-1} a_k + q_{k-2} \ ,$$

for any $k \geq 1$, where $p_{-1} \equiv 1$, $q_{-1} \equiv 0$, $p_0 \equiv a_0$, $q_0 \equiv 1$;

$$p_{k-1} q_{k-2} - p_{k-2} q_{k-1} = (-1)^k \quad \forall k \geq 1 \ ; \qquad \frac{p_{2k}}{q_{2k}} \nearrow \omega \searrow \frac{p_{2k+1}}{q_{2k+1}} \ ; \tag{10.40}$$

$$\frac{1}{q_k(q_k + q_{k+1})} < |\omega - \frac{p_k}{q_k}| < \frac{1}{q_k q_{k+1}} \ ; \qquad \omega = \frac{r_k p_{k-1} + p_{k-2}}{r_k q_{k-1} + q_{k-2}} \ . \tag{10.41}$$

Lemma 1: Let $\Phi : [1, \infty) \to [1, \infty)$ be a continuous non decreasing function. Then from the inequality

$$|\omega q_k - p_k| \geq \frac{1}{\Phi(q_k)} \ , \qquad \forall k \geq 0 \ ,$$

it follows

$$|\omega q - p| \geq \frac{1}{\Phi(q)}, \qquad \forall q \neq 0.$$

Proof: If $\frac{p}{q} = \frac{p_k}{q_k}$ for some k then there is nothing to prove. Hence, assume that $\frac{p}{q} \neq \frac{p_k}{q_k}$ $\forall k \geq 0$. Then three cases are possible:

$$(i) \quad \frac{p}{q} < \frac{p_0}{q_0} \equiv a_0,$$

$$(ii) \quad \frac{p}{q} > \frac{p_1}{q_1},$$

$$(iii) \quad \frac{p}{q} \in I_k,$$

where $I_k \equiv (\frac{p_{k+1}}{q_{k+1}}, \frac{p_{k-1}}{q_{k-1}})$ for k odd and $I_k \equiv (\frac{p_{k-1}}{q_{k-1}}, \frac{p_{k+1}}{q_{k+1}})$ for k even.
In case (i):

$$|\omega q - p| \geq |\omega - \frac{p}{q}| > |\omega - a_0| = |\omega - \frac{p_0}{q_0}| \geq \frac{1}{\Phi(q_0)} = \frac{1}{\Phi(1)} \geq \frac{1}{\Phi(q)}.$$

In case (ii):

$$|\frac{p}{q} - \omega| > |\frac{p}{q} - \frac{p_1}{q_1}| \geq \frac{1}{q q_1} \quad \Rightarrow \quad |p - \omega q| > \frac{1}{q_1} = \frac{1}{a_1};$$

since $|\omega - a_0| \leq \frac{1}{a_1}$, one has

$$|\omega q - p| > |\omega - p_0| = |\omega q_0 - p_0| \geq \frac{1}{\Phi(q_0)} \geq \frac{1}{\Phi(q)}.$$

In case (iii), by (10.40):

$$\frac{1}{q q_{k-1}} \leq |\frac{p}{q} - \frac{p_{k-1}}{q_{k-1}}| |\frac{p_{k+1}}{q_{k+1}} - \frac{p_{k-1}}{q_{k-1}}| < |\frac{p_k}{q_k} - \frac{p_{k-1}}{q_{k-1}}| = \frac{1}{q_k q_{k-1}} \quad \Rightarrow q > q_k.$$

Again by (10.40),

$$|\omega - \frac{p}{q}| > |\frac{p_{k+1}}{q_{k+1}} - \frac{p}{q}| \geq \frac{1}{q q_{k+1}} \quad \Rightarrow |\omega q - p| \geq \frac{1}{q_{k+1}},$$

but, by (10.41) it is $|\omega q_k - p_k| \leq \frac{1}{q_{k+1}}$ and since $q > q_k$ Lemma 1 follows. $\square$

Lemma 2: For all $k \geq 0$

$$|\omega - \frac{p_k}{q_k}| = \frac{1}{\sigma_k q_k^2}$$

with $\sigma_k \equiv r_{k+1} + \frac{q_{k-1}}{q_k}$.

Proof: By (10.41) and (10.40)

$$|\omega - \frac{p_k}{q_k}| \equiv |\frac{r_{k+1} p_k + p_{k-1}}{r_{k+1} q_k + q_{k-1}} - \frac{p_k}{q_k}| \equiv \frac{1}{q_k (r_{k+1} q_k + q_{k-1})} =$$

$$= \frac{1}{q_k^2} \frac{1}{(r_{k+1} + \frac{q_{k-1}}{q_k})} \equiv \frac{1}{q_k^2 \sigma_k}.$$

By Lemma 1 one has to check (10.38) for $(p, q) = (p_k, q_k)$ and by Lemma 2 we can take $\gamma = \sup_{k \geq 0} \sigma_k$. Since

$$\frac{\sqrt{5} - 1}{2} = [0; 1, 1, 1, ...] \equiv [0; 1^\infty] \ ,$$

one finds

$$r_{k+1} = \frac{\sqrt{5} + 1}{2} \qquad (\forall k \geq 0) \ .$$

Finally, from

$$\frac{q_{-1}}{q_0} = 0 \ , \quad \frac{q_0}{q_1} = 1 \ , \quad \frac{q_k}{q_{k+1}} < 1 \quad (\forall k \geq 1)$$

(10.38) and (10.39) follow. Notice that one may have better estimates using the identity

$$\frac{q_k}{q_{k-1}} = [a_k; a_{k-1}, ..., a_1] \ , \qquad \forall k \geq 1 \ . \ \square$$

Appendix 9: Small-Divisor Series for Symplectic Maps

In this appendix we provide an upper bound, $s_l(\delta)$, on the small-divisor series,

$$[\sum_{n=1}^{\infty} (\frac{n^l}{\sin \frac{n\omega}{2}})^2 \ e^{-\delta n}]^{1/2} \ , \qquad l = 0, 1 \ , \tag{10.42}$$

arising in the theory of symplectic maps (cfr. §10).

We shall prove that, for any integers $N, l \geq 0$, the sum in (10.42) is bounded by:

$$s_l(\delta) \equiv [\sum_{n=1}^{N-1} (\frac{n^l}{\sin \frac{n\omega}{2}})^2 \ e^{-\delta n} + S_l^{(N)}]^{1/2} \ , \qquad N \in \mathbb{N} \ , \tag{10.43}$$

with

$$S_l^{(N)} \equiv (1 - e^{-\delta}) \frac{\pi^2 C^2}{12} \sum_{n=N}^{\infty} n^{2l+2} e^{-\delta n} \ ; \tag{10.44}$$

(for $N = 0$, the first sum in (10.43) is absent). For $l = 0, 1$ one can bound $S_l^{(N)}$ with:

$$S_0^{(N)} \leq \frac{\pi^2 C^2}{4} (1 - e^{-\delta}) \ e^{\frac{\delta}{2}} e^{-\alpha(N-1)} \frac{1}{\alpha^3} \ [2 + (2N + 1)\alpha + N^2 \alpha^2] \ ,$$

$$S_1^{(N)} \leq \frac{\pi^2 C^2}{4} (1 - e^{-\delta}) \ e^{\frac{\delta}{2}} e^{-\alpha(N-1)} \frac{1}{\alpha^5} \ [24 + (24N + 36)\alpha + (12N^2 + 24N + 14)\alpha^2$$
$$+ (4N^3 + 6N^2 + 4N + 1)\alpha^3 + N^4 \alpha^4] \ ,$$

$$\tag{10.45}$$

where $\alpha \equiv \delta(1 + \omega)$.

Proof: Let $b_n \equiv \sum_{N \leq k \leq n} \frac{1}{\sin^2(\frac{k\omega}{2})}$ for $n \geq N$ and $b_{N-1} = 0$; then, since $b_n - b_{n-1} = \frac{1}{\sin^2(\frac{n\omega}{2})}$, it follows that

$$\sum_{n=N}^{\infty} \left(\frac{n^l}{\sin(\frac{n\omega}{2})}\right)^2 e^{-\delta n} = \sum_{n=N}^{\infty} n^{2l} e^{-\delta n}(b_n - b_{n-1}) =$$

$$= \sum_{n=N}^{\infty} n^{2l} e^{-\delta n} b_n - \sum_{n=N}^{\infty} (n+1)^{2l} e^{-\delta n} e^{-\delta} b_n \leq$$

$$\leq (1 - e^{-\delta}) \sum_{n=N}^{\infty} n^{2l} e^{-\delta n} b_n \leq$$

$$\leq (1 - e^{-\delta}) \frac{\pi^2 C^2}{12} \sum_{n=N}^{\infty} n^{2l+2} e^{-\delta n} ,$$

where in the last inequality we used Rüssmann's estimate (cfr [R2]):

$$|b_n| = \sum_{N \leq k \leq n} \frac{1}{\sin^2(\frac{k\omega}{2})} \leq \sum_{1 \leq k \leq n} \frac{1}{4 \min_{l \in \mathbb{Z}} |\frac{\omega}{2\pi} k - l|} \leq$$

$$\leq \frac{\pi^2 C^2}{12} n^2 .$$

Now using that for any $k \geq 0$

$$\sum_{n=N}^{\infty} n^k e^{-\delta n} = (-1)^k \frac{d^k}{d\delta^k} \frac{e^{-\delta N}}{1 - e^{-\delta}}$$

and the estimate

$$\frac{e^{-\delta}}{1 - e^{-\delta}} < \frac{1}{\delta} \qquad \forall \delta > 0 ,$$

one obtains the claim. □

Appendix 10: KAM Statements for Symplectic Maps

Here we formulate the KAM algorithm and theorem for symplectic maps (see §10). The proofs are easily obtained by mimicking the arguments leading to §4 and §5 (and makes a good exercise); alternatively we refer the (tired) reader to [CC4] and [CC2].

KAM Algorithm for the Maps in (10.1)

The notations are as in §10; the style as in §4. Let $v \equiv v^{(0)}$ be a non-degenerate approximate solution of (10.6), let $\varepsilon(\theta) \equiv \varepsilon^{(0)}(\theta)$ be the associated error-function (see (10.13)); let v, ε be real-analytic in Δ_ξ^1 (or possibly in $\Delta_\xi^1 \times \mathcal{P}$) and let $\{\delta_j\}$ be a sequence of positive numbers such that $\sum_{j \geq 0}^{\infty} \delta_j < \xi_0 \equiv \xi$. Let $v^{(j)}$, $\varepsilon^{(j)}$,... be the functions constructed by iteratively applying Lemma 10.3, provided, of course, one has the needed control on $(1 + v^{(j-1)})^{-1}$.

We define, now, the norm-parameters and the KAM algorithm. Let $\mathcal{D}^\xi \equiv \{x = \theta + v(\theta) \mid \theta \in \Delta_\xi^1\}$ and $\mathcal{D}_\rho^\xi \equiv \{x = x_0 + x_1 \mid x_0 \in \mathcal{D}^\xi, \ x_1 \in \mathbb{C}, |x_1| \leq \rho\}$. Let $j \geq 0$

and let

$$\mathcal{N}_j \equiv \{M_j, \ \overline{M}_j, \ V_j, \ V_{1j}, \ E_j, \ \rho_j, \ F_{2j}\} \qquad (10.46)$$

be the set of positive numbers controlling the norms of $v^{(j)}$ on $\Delta^1_{\xi_j}$ and of f on $\mathcal{D}^\xi_{\rho_j}$ [i.e., $M_j \geq |1 + v^{(j)}_\theta|_{\xi_j} \equiv |1 + v^{(j)}_\theta|_j$, $\overline{M}_j \geq |(1 + v^{(j)}_\theta)^{-1}|_{\xi_j}$, $V_j \geq |v^{(j)}|_{\xi_j}$, $V_{1j} \geq |v^{(j)}_\theta|_{\xi_j}$, $E_j \geq |\mathscr{E}(v^{(j)})|_{\xi_j}$, $F_{2j} \geq |f_{xx}|_{\mathcal{D}^\xi_{\rho_j}} \equiv |f_{xx}|_j$]. Then, if $s_l(\delta_j)$ is an upper bound on the small-divisor series (see Appendix 9), the above norm-parameter can be defined as follows:

$$a_j \equiv (M_j \overline{M}_j s_0(\delta_j))^2 \ \{1 + (M_j \overline{M}_j)^2 \frac{s_0(2\xi_j)}{s_0(\delta_j)}\} \ ,$$

where $\xi_0 = \xi$ and (for $j \geq 1$) $\xi_j = \xi_{j-1} - \delta_{j-1}$,

$$W_j \equiv E_j \, a_j$$

and

$$W_{1j} \equiv E_j a_j \, (\frac{V_{1j}}{M_j} \delta_j^{-1} + \frac{s_1(\delta_j)}{s_0(\delta_j)}) \ .$$

Then one can take

$$M_{j+1} \equiv M_0 + \sum_{i=0}^{j} W_{1i} \ ,$$

$$\overline{M}_{j+1} \equiv \begin{cases} \overline{M}_j \cdot (1 - \overline{M}_j \sum_{i=0}^{j} W_{1i})^{-1} & \text{if } \sum_{i=0}^{j} W_{1i} < 1 \\[2mm] \infty & \text{if } \sum_{i=0}^{j} W_{1i} \geq 1 \ , \end{cases}$$

$$V_{j+1} \equiv V_0 + \sum_{i=0}^{j} W_i \ ,$$

$$V_{1(j+1)} \equiv V_{10} + \sum_{i=0}^{j} W_{1i}$$

and

$$E_{j+1} = (E_j)^2 \, a_j \, (\frac{a_j F_{2(j+1)}}{2} + \frac{\delta_j^{-1}}{M_j}) \ ,$$

where

$$F_2^{(j+1)} \equiv \sup_{\mathcal{D}^\xi_{\rho_{j+1}}} |f_{xx}| \ , \qquad \rho_0 \equiv 0, \qquad \rho_{j+1} \equiv \sum_{i=0}^{j} W_i \ .$$

Finally, the smallness condition (i.e., the analog of (5.1) in Theorem 5.1), which has been used in [CC4] to obtain the results discussed in §10 (for rotation numbers with

$\tau = 1$) is:

$$154 \cdot 10^{13} \; C^5 M^2 (M\overline{M})^{21/2} \xi^{-8} \; F_2 \; E \le 1 \tag{10.47}$$

where $F_2 \equiv \max\{1, \bar{F}_2\}$, $\bar{F}_2 \equiv \sup_{\mathcal{D}_r^{\xi}} |f_{xx}|$, $r = 1/67^4$.

Actually, the condition (10.47), which was deduced in [CC2] and used in [CC4], could be slightly improved using the techiques presented in this work.

References

[A1]   Arnold V.I., *Proof of a Theorem by A.N. Kolmogorov on the invariance of quasi-periodic motions under small perturbations of the Hamiltonian*, Russ. Math. Surveys **18**, 9 (1963)

[A2]   Arnold V.I. (ed.), *Encyclopedia of Math. Sciences*, Dynamical System III, Springer-Verlag **3** (1988)

[AL]   Aubry S., Le Daeron P.Y., *The discrete Frenkel-Kontorova model and its extensions I*, Physica **8D**, 381 (1983)

[BCCF]   Berretti A., Celletti A., Chierchia L., Falcolini C., *Analytic properties of conjugacy of invariant curves*, J. of Stat. Physics **66**, 1613 (1992)

[BC]   Berretti A., Chierchia L., *On the complex analytic structure of the golden invariant curve for the standard map*, Nonlinearity **3**, 39 (1990)

[B]   Bost J.-B., *Tores invariants des systèmes dynamiques Hamiltoniens*, Séminaire Bourbaki **639**, 113 (1984-85)

[C1]   Celletti A., *Analysis of resonances in the spin-orbit problem in Celestial Mechanics: The synchronous resonance (Part I)*, J. of Appl. Math. and Phys. (ZAMP) **41**, 174 (1990)

[C2]   Celletti A., *Analysis of resonances in the spin-orbit problem in Celestial Mechanics: Higher order resonances and some numerical experiments (Part II)*, J. of Appl. Math. and Phys. (ZAMP) **41**, 453 (1990)

[CC1]   Celletti A., Chierchia L., *Rigorous estimates for a computer-assisted KAM theory*, J. Math. Phys. **28**, 2078 (1987)

[CC2]   Celletti A., Chierchia L., *Construction of analytic KAM surfaces and effective stability bounds*, Commun. Math. Phys. **118**, 119 (1988)

[CC3]   Celletti A., Chierchia L., *A computer-assisted approach to small-divisors problems arising in hamiltonian mechanics*, IMA Volumes in Math. and its Applications, K.R. Meyer, D.S. Schmidt eds., **28**, 41 (1991)

[CC4]   Celletti A., Chierchia L., *Invariant curves for area-preserving twist maps far from integrable*, to appear in J. of Stat. Physics

[CF]   Celletti A., Falcolini C., *Capture probability of Mercury in spin-orbit resonances*, preprint CARR (1990)

[CFP]   Celletti A., Falcolini C., Porzio A., *Rigorous numerical stability estimates for the existence of KAM tori in a forced pendulum*, Ann. Inst. Henri Poincaré **47**, 85 (1987)

[CG]   Celletti A., Giorgilli A., *On the numerical optimization of KAM estimates by classical perturbation theory*, J. Appl. Mathem. and Phys. (ZAMP) **39**, 743 (1988)

[ChF]   Chierchia L., Falcolini C., *A non-linear programming problem arising in a KAM algorithm*, unpublished (1990)

[ChF2] Cierchia L., Falcolini C., *A direct proof of a theorem by Kolmogorov in Hamiltonian systems,* to appear in Annali Scuola Normale Superiore die Pisa Cl. Scienze.

[ChG]   Cierchia L., Gallavotti G., *Drift and diffusion in phase space,* Ann. Inst. Henri Poincaré **60**, 1 (1994)

[C]   Chirikov B.V., *A universal instability of many dimensional oscillator systems,* Physics Reports **52**, 263 (1979)

[D]   Danby J.M.A., *Fundamentals of Celestial Mechanics,* Macmillan, New York (1962)

[Do]   Douady R., *Une démonstration directe de léquivalence des théorèmes des tores invariants pour les difféomorphismes et les champs de vecteurs,* C.R. Acad. Sci. Paris **295**, 201 (1982)

[El]   Eliasson H., *Absolutely convergent series expansions for quasiperiodic motions,* preprint, Univ. of Stockholm (1987)

[Es]   Escande D.F., *Stochasticity in classical Hamiltonian systems: Universal aspects,* Physics Reports **121**, 165 (1985)

[ED]   Escande D.F., Doveil F., *Renormalization method for computing the threshold of the large-scale stochastic instability in two degrees of freedom Hamiltonian systems,* J. Stat. Physics **26**, 257 (1981)

[G1]   Gallavotti G., *The Elements of Mechanics,* Springer-Verlag, New York (1983)

[G2]   Gallavotti G., *Perturbation theory for classical Hamiltonian systems,* in Scaling and Self-Similarity in Physics, ed. J. Fröhlich, PPh. **7**, Birkhauser, Boston (1984)

[GP]   Goldreich P., Peale S., *Spin-orbit coupling in the solar system,* Astron. J. **71**, 425 (1966)

[Go]   Goroff D., unpublished (1983)

[Gr]   Greene J.M., *A method for determining a stochastic transition,* J. Math. Physics **20**, 1183 (1979)

[Ha]   Hamilton R.S., *The inverse function theorem of Nash and Moser,* Bull. Amer. Math. Soc. **7**, 65 (1982)

[He]   Henrard J., *Spin-orbit resonance and the adiabatic invariant,* in: "Resonances in the Motion of Planets, Satellites and Asteroids", S. Ferraz-Mello and W. Sessin eds., Sao Paulo, 19 (1985)

[Her1] Herman M., *Sur la conjugaison différentiable des difféomorphismes du cercle à des rotations,* Pub. I.H.E.S. **49**, 5 (1979)

[Her2] Herman M., *Sur le courbes invariantes par le difféomorphismes de l'anneau,* Vol. 1 Astérisque **103-104** (1983)

[Her3] Herman M., *Recent results and some open questions on Siegel's linearization theorems of germs of complex analytic diffeomorphisms of C^n near a fixed point,* preprint (1987)

[Kh]   Khintchine A., *Continued fractions,* Noordhoff Ltd. Groningen (1963)

[Ko]   Kolmogorov A.N., *On the conservation of conditionally periodic motions under small perturbation of the Hamiltonian,* Dokl. Akad. Nauk. SSR **98**, 469 (1954)

[L]   Lanford III O.E., *Computer assisted proofs in analysis,* Physics A **124**, 465 (1984)

[LR]   De La Llave R., Rana D., *Accurate strategies for K.A.M. bounds and their implementation,* IMA Volumes in Math. and its Applications, K.R. Meyer, D.S. Schmidt eds., **28**, 127 (1991)

[MK]   MacKay R.S., *Transition to chaos for area-preserving maps*, Lectures Notes in Physics **247**, 390 (1985)

[MKP]   MacKay R.S., Percival I.C., *Converse KAM: Theory and practice*, Comm. Math. Phys. **98**, 469 (1985)

[Ma]   Mather J.N., *Nonexistence of invariant circles*, Erg. theory and dynam. systems **4**, 301 (1984)

[Mo1]   Moser J., *On invariant curves of area-preserving mappings of an annulus*, Nach. Akad. Wiss. Göttingen, Math. Phys. Kl. II **1**, 1 (1962)

[Mo2]   Moser J., *A rapidly convergent iteration method and non-linear partial differential equations*, Ann. Scuola Norm. Sup. Pisa **20**, 265 (1966)

[Mo3]   Moser J., *Convergent series expansions for quasi-periodic motions*, Math. Annalen **169**, 136 (1967)

[Mo4]   Moser J., *Minimal solutions of variational problems on a torus*, Ann. Inst. Henri Poincaré **3**, 229 (1986)

[Mo5]   Moser J., *Break-down of stability*, in Lect. Notes in Physics **247**, 492 (1986)

[Mo6]   Moser J., *Monotone twist mappings and the calculus of variations*, Ergod. Th. & Dynam. Sys. **6**, 401 (1986)

[Pe]   Percival I.C., *Variational principles for invariant tori and cantori*, in *nonlinear dynamics and the beam-beam interaction*, AIP Conference Proceedings **57**, ed. M. Month, J.C. Herrera 302 (1980)

[Po]   Poincaré H., *Les Methodes Nouvelles de la Mechanique Celeste*, Gauthier-Villars, Paris (1892)

[R1]   Rüssmann H., *On optimal estimates for the solutions of linear partial differential equations of first order with constant coefficients on the torus*, Lectures Notes in Physics, **38**, 598 (1975)

[R2]   Rüssmann H., *On optimal estimates for the solutions of linear difference equations on the circle*, Celestial Mechanics **14**, 33 (1976)

[R3]   Rüssmann H., *Konvergente Reihenentwicklungen in der Störungstheorie der Himmelsmechanik*, Selecta Mathematica V, Heidelberg Taschenbücher **201**, 93 (1978)

[R4]   Rüssmann H., *Note on sums containing small divisors*, Comm. Pure and Appl. Math. **29**, 755 (1976)

[SZ]   Salamon D., Zehnder E., *KAM theory in configuration space*, Comment. Math. Helvetici **64**, 84 (1989)

[S]   Siegel C.L., *Iteration of analytic functions*, Ann. Math. **43**, 607 (1942)

[Vax]   (no author listed) *Vax Architecture handbook*, Digital Equipment Corporation (1981)

[W]   Wisdom J., Peale S.J., *The chaotic rotation of Hyperion*, Icarus **58**, 137 (1984)

[Z]   Zehnder E., *Generalized implicit function theorems with applications to some small divisor problems*, Comm. Pure Appl. Math., I **28**, 91 (1975), II **29**, 49 (1976)

Ergodicity in Hamiltonian Systems*

Carlangelo Liverani[1] *and Maciej P. Wojtkowski*[2]

[1]Mathematics Department, University of Rome II, Tor Vergata, Rome, Italy.
liverani@mat.utovrm.it
[2]Department of Mathematics, University of Arizona, Tucson, AZ 85721, USA.
maciejw@math.arizona.edu

Abstract. We discuss the Sinai method of proving ergodicity of a discontinuous Hamiltonian system with (nonuniform) hyperbolic behavior.

Symbols

α	amount of long leaves in a connecting square
$\mathscr{B}(p; r)$	Ball of radius r and center p
c	amount of overlap in neighboring squares
$\mathscr{C}$	sectors
d	distance
$k(c)$	maximal number of overlapping squares
L	linear map
$\mathscr{M}$	Symplectic manifold
$\mathscr{M}^{\pm}$	Symplectic boxes
μ	invariant measure
ω	symplectic form
$\mathscr{Q}$	quadratic form defining a sector
R	rectangles
$\mathscr{G}$	collection of rectangles
$\mathscr{S}^{\pm}$	singularity sets
T	map
U	big neighborhood in the smooth case
$\mathscr{U}(x)$	neighborhood of x
V	side of a sector
$\mathscr{W}$	linear symplectic space
W	stable and unstable manifolds

In the figures: The stable direction is vertical. The unstable direction is horizontal

---

* We would like to thank N. Chernov, L. Chierchia, V. Donnay, A. Katok, N. Simányi, D. Szász and L.-S. Young for helpful and enlightening discussions. The first author wishes to thank the Mathematics Department of the University of Arizona, Tucson and the Center for Applied Mathematics at Cornell University, in particular its director J. Guckenheimer, where he was visiting during part of this work, he also acknowledges the partial support received by CNR, grant n. 203.01.52, and by the GNFM. The second author gratefully acknowledges the hospitality of Forschungsinstitut für Mathematik at ETH Zürich, where the first draft of this paper was written. He also acknowledges the partial support from NSF Grant DMS–9017993.

0. Introduction

The notion of ergodicity was introduced by Boltzman as a property satisfied by a Hamiltonian flow on its energy manifold. The emergence of the KAM (Kolmogorov - Arnold - Moser) theory of quasiperiodic motions made it clear that very few Hamiltonian systems are actually ergodic. Moreover, those systems which seem to be ergodic do not lend themselves easily to rigorous methods.

Ergodicity is a rather weak property in the hierarchy of stochastic behavior of a dynamical system. The study of strong properties (mixing, K-property and Bernoulliness) in smooth dynamical systems began from the geodesic flows on surfaces of negative curvature. In particular, Hopf [H] invented a method of proving ergodicity, using horocycles, which turned out to be so versatile that it endured a lot of generalizations. It was developed by Anosov and Sinai [AS] and applied to Anosov systems with a smooth invariant measure. With the advances of the theory of Kolmogorov - Sinai entropy the Hopf method turned out to be also a basis for proving the K-property of Anosov systems.

The key role in this approach is played by the hyperbolic behavior in a dynamical system. By the hyperbolic behavior we mean the property of exponential divergence of nearby orbits. In the strongest form it is present in Anosov systems and Smale systems. It leads there to a rigid topological behavior. In weaker forms it seems to be a common phenomenon.

In his pioneering work on billiard systems Sinai [S] showed that already weak hyperbolic properties are sufficient to establish the strong mixing properties. Even the discontinuity of the system can be accommodated.

The Multiplicative Ergodic Theorem of Oseledets [O] makes Lyapunov exponents a natural tool to describe the hyperbolic behavior of a dynamical system with a smooth invariant measure.

Pesin [P] made the nonvanishing of Lyapunov exponents the starting point for the study of hyperbolic behavior. He showed that, if a diffeomorphism preserving a smooth measure has only nonvanishing Lyapunov exponents, then it has at most countably many ergodic components and (roughly speaking) on each component it has the Bernoulli property.

Pesin's work raised the question of sufficient conditions for ergodicity or, more modestly, for the openness (modulo sets of measure zero) of the ergodic components.

In his work, spanning two decades, on the system of colliding balls (gas of hard balls) Sinai developed a method of proving (local) ergodicity in discontinuous systems with nonuniform hyperbolic behavior. We will refer to it as the Sinai method. It was improved by Sinai and Chernov [CS] and by A.Krámli, N.Simányi and D.Szász [KSS]. In both papers the discussion is confined to the realm of semidispersing billiards.

The purpose of the present paper is to recover the Sinai method as a part of the theory of hyperbolic dynamical systems. In the process we have simplified some of the aspects of the method, and we have revealed its logical structure and limitations.

We rely on two developments. The first is the work of Katok and Strelcyn [KS] in which they generalized Pesin Theory to discontinuous systems. The other is the development of criteria for nonvanishing of Lyapunov exponents in Hamiltonian systems in papers [W1], [W2] and [W3]. In the language of these criteria Burns and Gerber [BG] found a sufficient condition for (local) ergodicity in the smooth case of lowest dimension (3 for flows preserving a smooth measure). It was later generalized by Katok [K1] to arbitrary dimension. As a byproduct of our general approach, which is aimed at discontinuous systems, we obtain a similar theorem (Main Theorem in the smooth case) and a new proof.

Let us give some advice to the reader on how to use our paper. The first three Sections demonstrate what the Sinai method is, and how it works. The discussion is conducted in the simplest possible environment of a linear discontinuous system on the two dimensional torus. It is reasonable to stop here, especially if the reader is only interested in two dimensional uniformly hyperbolic systems. But we do not recommend trying to read the heart of the paper without going through the first three Sections.

In Sections 4, 5 and 6 we develop the linear symplectic language in which we formulate our results. We suggest that the reader skips these sections and goes straight to Section 7 where we formulate the multitude of hypotheses and the two Main Theorems on local ergodicity, one for smooth systems and the other (much harder) for discontinuous systems. The reading of Section 7, and the following Sections, will require numerous trips back to Sections 4-6 for the necessary definitions and theorems.

If the reader does not care about the discontinuous case, she needs to read only Sections 8, 9 and 10 with significant leaps (since everything is simpler in the smooth case). Sections 11 and 12 contain almost the whole proof of the Main Theorem in the discontinuous case (it also relies on the results of Sections 8-10). The remaining part of the proof is contained in Section 13. It stands out by the level of technical complications.

Section 14 contains some classes of examples where all the hard work can be put to use, and one class where it cannot. The interest in this last example comes from the fact that it is multidimensional and all the Lyapunov exponents are different from zero. Unfortunately, it does not satisfy an important property (proper alignment of singularity sets). It points towards the need for a more flexible scheme.

1. A Model Problem

We will discuss here a very simple model problem in which the important features of the Sinai's method are not obscured by technical details. Our discussion will be very careful so that in the future when the technical details will cloud the horizon we will be able to refer the reader to these basic clarifications.

We consider a family of linear maps of the plane defined by

$$x'_1 = x_1 + ax_2$$
$$x'_2 = x_2,$$

where a is a real parameter. We use these linear maps to define (discontinuous) maps of the torus by restricting the formulas to the strip $\{0 \leq x_2 \leq 1\}$ and further taking them modulo 1. In this way we define a mapping T_1 of the torus $\mathbb{T}^2 = \mathbb{R}^2/\mathbb{Z}^2$ which is discontinuous on the circle $\{x_2 \in \mathbb{Z}\}$ (except when a is equal to an integer) and preserves the Lebesgue measure μ.

Similarly we define another family of maps depending on the same parameter a by restricting the formulas

$$x'_1 = x_1$$
$$x'_2 = ax_1 + x_2$$

to the strip $\{0 \leq x_1 \leq 1\}$ and then taking them modulo 1. Thus for each a we get a mapping T_2 of the torus which is discontinuous on the circle $\{x_1 \in \mathbb{Z}\}$ (except when a is equal to an integer) and preserves the Lebesgue measure μ.

Finally we introduce the composition of these maps $T = T_2 T_1$ which depends on one real parameter a. An alternative way of describing the map T is by introducing two fundamental domains for the torus $\mathcal{M}^+ = \{0 \leq x_1 + a x_2 \leq 1, \ 0 \leq x_2 \leq 1\}$ and $\mathcal{M}^- = \{0 \leq x_1 \leq 1, \ 0 \leq -a x_1 + x_2 \leq 1, \}$ (see Figure 1).

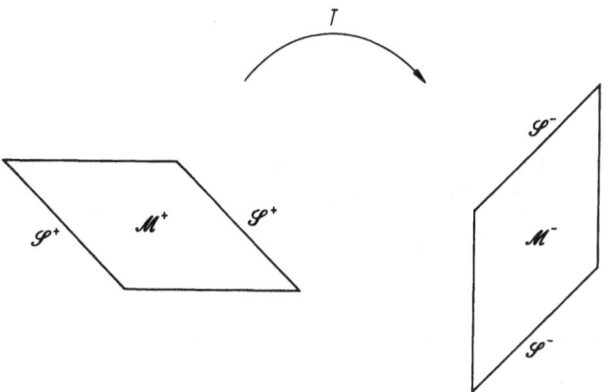

Figure 1. The map

The linear map defined by the matrix

$$\begin{pmatrix} 1 & a \\ a & 1+a^2 \end{pmatrix} = \begin{pmatrix} 1 & 0 \\ a & 1 \end{pmatrix} \begin{pmatrix} 1 & a \\ 0 & 1 \end{pmatrix}$$

takes $\mathcal{M}^+$ onto $\mathcal{M}^-$ thus defining a map of the torus which is discontinuous at most on the boundary of $\mathcal{M}^+$ and preserves the Lebesgue measure. This is our map T.

Let $\mathcal{S}^\pm = \partial \mathcal{M}^\pm$ be the boundary of $\mathcal{M}^\pm$. Except for integer values of a the mapping T is discontinuous on $\mathcal{S}^+$ and its inverse T^{-1} is discontinuous on $\mathcal{S}^-$. Let us stress that the map T is well defined in the closed domain $\mathcal{M}^+$ but two different points on the boundary $\mathcal{S}^+$ which correspond to the *same* point on the torus will be mapped onto two different points on the boundary $\mathcal{S}^-$ which correspond to two *different* points on the torus (except for the corner). We adopt the convention that the image under T of a point from $\mathcal{S}^+$ is the pair of image points in $\mathcal{S}^-$. With this convention we can apply T or any of its powers to any subset in the torus.

For integer values of $a \neq 0$ we have a hyperbolic algebraic automorphism of the torus, a prime example of an Anosov system. It is thus a Bernoulli system and has a nice Markov partition [AW]. We restrict ourselves to the study of ergodicity and we repeat the proof of ergodicity by the Hopf method, since the Sinai method is built upon it.

Let $f : \mathbb{T}^2 \to \mathbb{R}$ be a continuous function. We want to prove that for almost every $x \in \mathbb{T}^2$ the time averages

$$\frac{f(x) + f(Tx) + \cdots + f(T^{n-1}x)}{n}$$

converge as $n \to +\infty$ to the average value of f, i.e., $\int f d\mu$. Once this is established one can obtain the same property for all integrable functions by an approximation argument. From Birkhoff Ergodic Theorem (BET) we know that the time averages converge almost everywhere to a function $f^+ \in L^1(\mathbb{T}^2, \mu)$ which is invariant on the orbits of T, i.e., $f^+ \circ T = f^+$, and has the same average value as f, i.e., $\int f^+ d\mu = \int f d\mu$. Further applying BET to f and T^{-1} we obtain that the time averages in the past

$$\frac{f(x) + f(T^{-1}x) + \cdots + f(T^{-n+1}x)}{n}$$

converge almost everywhere as $n \to +\infty$ to $f^- \in L^1(\mathbb{T}^2, \mu)$ for which $f^- \circ T = f^-$ and $\int f^- d\mu = \int f d\mu$.

It is the usual magic of the ergodic theory which forces the functions f^+ and f^- to coincide almost everywhere. (Let us recall the argument: let

$$\mathcal{A}_+ = \{x \in \mathbb{T}^2 \mid f^+(x) > f^-(x)\};$$

by definition $\mathcal{A}_+$ is an invariant set, hence

$$\int_{\mathcal{A}_+} \left[f^+(x) - f^-(x)\right] d\mu(x) = \int_{\mathcal{A}_+} f(x)d\mu(x) - \int_{\mathcal{A}_+} f(x)d\mu(x) = 0$$

which implies $\mu(\mathcal{A}_+) = 0$ and $f^+ \leq f^-$ μ-almost everywhere. The same argument, this time applied to the set $\mathcal{A}_- = \{x \in \mathbb{T}^2 \mid f^-(x) > f^+(x)\}$, implies the converse inequality.)

For $a \neq 0$ the matrix

$$\begin{pmatrix} 1 & a \\ a & 1 + a^2 \end{pmatrix}$$

is a hyperbolic matrix with eigenvalues $\lambda = \lambda(a) > 1$ and $\frac{1}{\lambda} < 1$. For $x \in \mathbb{T}^2$ let us denote by $W^u(x)$ $(W^s(x))$ the line in $\mathbb{T}^2$ passing through x and having the direction of the unstable eigenvector (the stable eigenvector), i.e., the eigenvector with eigenvalue λ $(\frac{1}{\lambda})$. We call $W^u(x)$ $(W^s(x))$ the unstable (stable) leaf of x. The leaves of x have the following property. If $y \in W^u(x)$ $(y \in W^s(x))$ then the distance

$$d(T^n y, T^n x) = \lambda^{-|n|}d(y, x) \to 0 \quad \text{as} \quad n \to -\infty(+\infty).$$

Hence for $y, z \in W^{u(s)}(x)$

$$|f(T^n y) - f(T^n z)| \to 0 \quad \text{as} \quad n \to -\infty(+\infty).$$

It follows that for $y, z \in W^{u(s)}(x)$ either $f^\pm(y)$ and $f^\pm(z)$ are both defined and equal or they are both undefined. Lifting the functions f^+ and f^- to $\mathbb{R}^2$ and using the directions of the eigenvalues as coordinate directions we can say that f^+ is a function of one coordinate alone and f^- is a function of only the other coordinate. Since the two functions coincide almost everywhere they must be constant.

Let us examine what can be saved of this argument when a is not an integer. In such a case, we still have the stable and unstable directions but a line parallel to, say, the unstable direction is cut by $\mathcal{S}^-$ into pieces and if y and z belong to two different pieces the distance $d(T^n y, T^n z)$ does not decrease to zero as $n \to -\infty$. Since this last property is of crucial importance in the Hopf method, the unstable (and stable) leaves have to be much shorter than before. Here is how we construct them. For simplicity of notation we will formulate everything for the unstable leaves alone.

We proceed inductively. Thus, for $x \in int \mathcal{M}^-$, we define $W_1^u(x)$ as the open segment of the line through x with the direction of the unstable eigenvector which contains x and has both endpoints on $\mathcal{S}^-$. The preimage $T^{-1} W_1^u(x)$ is by a factor of λ shorter than $W_1^u(x)$ and, in general, is cut into two or three pieces by $\mathcal{S}^-$. We pick the piece which contains $T^{-1} x$ and take its image under T; this is our second approximate unstable leaf $W_2^u(x)$, i.e.,

$$W_2^u(x) = T\left(T^{-1} W_1^u(x) \cap W_1^u(T^{-1}(x))\right).$$

Unless $T^{-1} x \in \mathcal{S}^-$ the second approximate unstable leaf $W_2^u(x)$ is again an open segment containing x with endpoints on $\mathcal{S}^- \cup T\mathcal{S}^-$ and naturally $W_2^u(x) \subset W_1^u(x)$. Given $W_n^u(x)$, $n = 1, 2, \ldots$, we define the $n+1$ approximate unstable leaf of x $W_{n+1}^u(x)$ by

$$W_{n+1}^u(x) = T^n \left(T^{-n} W_n^u(x) \cap W_1^u(T^{-n}(x))\right).$$

If $x \notin \bigcup_{i=0}^{+\infty} T^i \mathcal{S}^-$ then this inductive procedure will yield a nested sequence of open segments containing x

$$W_1^u(x) \supset W_2^u(x) \supset \ldots$$

with endpoints on

$$\bigcup_{i=0}^{+\infty} T^i \mathcal{S}^-.$$

We can also describe this construction in the following way. First we consider a fairly long segment $W_1^u(x)$. Then we look at $T\mathcal{S}^-$, if it does not intersect $W_1^u(x)$ then we do not change it, if it splits $W_1^u(x)$ into several segments, then we keep the segment which contains x. We repeat it with $T^2 \mathcal{S}^-$ and further images of $\mathcal{S}^-$, so that the segment may be cut shorter infinitely many times. The property $x \notin \bigcup_{i=0}^{+\infty} T^i \mathcal{S}^-$ ensures that x stays always strictly inside the segment. It is quite remarkable that, for almost every x, this inductive process shortens the segment only finitely many times. More precisely we have

Proposition 1.1. *For almost all $x \in \mathcal{M}^- \setminus \bigcup_{i=0}^{+\infty} T^i \mathcal{S}^-$ the sequence of approximate unstable leaves of x stabilizes, i.e., there is a natural $N = N(x)$ such that*

$$\bigcap_{i=1}^{+\infty} W_i^u(x) = \bigcap_{i=1}^{N} W_i^u(x).$$

Proof. For $t > 0$, let

$$X_t = \{x \in \mathcal{M}^- \mid d(x, \mathcal{S}^-) \leq t\}$$

where $d(\cdot, \cdot)$ is the distance of a point form a set. Because $\mathcal{S}^-$ is a finite union of segments we have

$$\mu(X_t) \leq \text{const } t.$$

Choosing $t_n = \frac{1}{n^2}$ we get

$$\sum_{n=1}^{+\infty} \mu\left(X_{t_n}\right) < +\infty,$$

hence also

$$\sum_{n=1}^{+\infty} \mu\left(T^n X_{t_n}\right) < +\infty.$$

It follows by the Borel-Cantelli Lemma that almost every x belongs to only finitely many of the sets

$$TX_{t_1}, T^2 X_{t_2}, \ldots,$$

which means that except for finitely many values of n

$$d(T^{-n}x, \mathscr{S}^-) > \frac{1}{n^2}.$$

Choosing $c(x) > 0$ sufficiently small we can take care of the finite number of exceptional values of n so that

$$d(T^{-n}x, \mathscr{S}^-) > \frac{c(x)}{n^2}$$

for each $n = 1, 2, \ldots$. Each time $W_{n+1}^u(x)$ is shorter than $W_n^u(x)$ we must have

$$d(T^{-n}x, \mathscr{S}^-) < \frac{\text{length}\left(\mathcal{W}_n^u(x)\right)}{\lambda^n}.$$

But then

$$\frac{c(x)}{n^2} < \frac{\text{length}\left(\mathcal{W}_n^u(x)\right)}{\lambda^n} \leq \frac{\text{length}\left(W_1^u(x)\right)}{\lambda^n},$$

which can hold for at most finitely many values of n. $\square$

We define the unstable leaf only for points x in the set of full measure described in Proposition 1.1, by taking the intersection

$$W^u(x) = \bigcap_{i=1}^{+\infty} W_i^u(x).$$

In view of Proposition 1.1, for each $W^u(x)$, there are natural numbers $n_l(x)$ and $n_r(x)$ such that $T^{n_l(x)} W^u(x)$ has the left endpoint on $\mathscr{S}^-$ and $T^{n_r(x)} W^u(x)$ has the right endpoint on $\mathscr{S}^-$. Most importantly we have the exponential contraction of $W^u(x)$, i.e., for $y \in W^u(x)$ the distance

$$d(T^{-n}y, T^{-n}x) = \frac{d(y, x)}{\lambda^n} \to 0 \quad \text{as} \quad n \to +\infty.$$

Everything that we have done to construct the unstable leaves can be repeated for the stable leaves and they have analogous properties. Once we have the stable and unstable leaves we are ready to do the Hopf argument.

For any continuous function $f : \mathbb{T}^2 \to \mathbb{R}$ the forward ergodic average f^+ is constant on the stable leaves and the backward ergodic average f^- is constant on the unstable leaves. Let us call a point $x \in \mathbb{T}^2$ f-typical, if $f^+(x)$, $f^-(x)$, $W^u(x)$ and $W^s(x)$ are well defined and $f^+(x) = f^-(x)$. The set of f-typical points has full measure, so a stable (or an unstable) leaf contains a set of f-typical points of full arc-length, except for a family of leaves of total measure zero. If $W^s(x)$ is not one of those exceptional leaves, then the set

$$C_1 = \bigcup_{\substack{y \in W^s(x) \\ y \text{ is } f\text{-typical}}} W^u(y)$$

has positive measure and $f^- = f^+ = const$ on C_1. We can proceed by adding all the stable leaves through f-typical points in C_1 to obtain C_2, etc., but a priori there is no reason to expect that we will be able to cover all of the torus in this way. (Indeed one can imagine that there is a dividing line between two ergodic components of our system and that all the stable and unstable leaves stop short of crossing this line.) That is where the Hopf method breaks down. It can only tell us that the ergodic components have positive measure and, therefore, that there are at most countably many of them. (To be more precise, we cannot really claim that C_1 belongs to one ergodic component. To argue this we have to modify our argument by taking a sequence of continuous functions dense in L^1 and considering the set of points which are f-typical for all the functions f in the sequence. This set, as the intersection of countably many sets of full measure, has full measure. We can then use it in the definition of C_1 and claim that $f^- = f^+ = const$ on C_1 for all the functions in our dense sequence. This implies that such C_1 does belong to one ergodic component. It follows easily that every invariant subset of positive measure contains an ergodic component of positive measure. Hence all ergodic components have positive measure.)

2. The Sinai Method

We have seen, in the previous section, that the Hopf method is not sufficient to prove the ergodicity of a discontinuous map because the stable and unstable leaves may be short. The Sinai method amounts to establishing that most of the stable and unstable leaves are, in a certain sense, sufficiently long. The first (highly nontrivial) step in this method is to formulate precisely what is meant by "sufficiently long". As before, we do it only for the unstable leaves; the changes necessary in the case of stable leaves are automatic.

Let $\mathcal{U} \subset \mathbb{T}^2$ be a (small) square with the sides parallel to unstable and stable directions respectively (to make the geometry simpler let us think that the unstable direction is horizontal and the stable direction vertical). For any $0 < c < 1$ we construct a sequence $\mathcal{G}_n(c), n = 1, 2, \ldots,$ of coverings of $\mathcal{U}$ in the following way. Without loss of generality we can let

$$\mathcal{U} = \{(u, v) \mid -b < u < b, \ -b < v < b\}.$$

We consider the net $\mathcal{N}(n, c)$ defined by

$$\mathcal{N}(n, c) = \{\frac{c}{n}(m, k) \in \mathcal{U} \mid m, k \in \mathbb{Z}\}.$$

Now the covering $\mathcal{G}_n(c)$ is the collection of squares having centers at points from $\mathcal{N}(n, c)$ and sides, of length $\frac{1}{n}$, parallel to the sides of $\mathcal{U}$. If $c < \frac{1}{2}$ then $\mathcal{G}_n(c)$ is a covering of $\mathcal{U}$ (otherwise $\mathcal{G}_n(c)$ may cover only a smaller square). The parameter c will be chosen later to be very small, so that many squares in $\mathcal{G}_n(c)$ overlap. However, once c is fixed, a point in $\mathcal{U}$ may belong to at most $k(c)$ squares in $\mathcal{G}_n(c)$, where $k(c)$ is a fixed number independent of $n = 1, 2, \dots$; (one can easily establish that $k(c) \leq (\frac{1}{2c} + 1)^2$, but we will not use any explicit estimate).

We call two squares, in $\mathcal{G}_n(c)$, immediate neighbors if the distance between their centers is $\frac{c}{n}$. Two immediate neighbors overlap on $1 - c$ part of their areas.

One can naturally define a column of squares and a row of squares as special collections of squares in $\mathcal{G}_n(c)$ (see Figure 2). For example, a sequence $\{R_i\}_{i=1}^{l}$ of squares from $\mathcal{G}_n(c)$ is called a column of squares if, for every $i = 1, \dots, l - 1$, R_i and R_{i+1} are immediate neighbors, R_{i+1} is above R_i, and there is no square in $\mathcal{G}_n(c)$ below R_1 or above R_l.

For each square $R \in \mathcal{G}_n$ we introduce the stable, $\partial_s R$, and unstable, $\partial_u R$, boundaries of R; $\partial_s R$ is the union of the two boundary segments of R which have the stable (vertical) direction and $\partial_u R$ is the union of the two boundary segments of R which have the unstable (horizontal) direction. Given a point $x \in R$, the unstable leaf $W^u(x)$ may intersect both segments in $\partial_s R$ or it may be too short to reach one of them (or both). In the first case we say that $W^u(x)$ is long in R, or that it is connecting in R , in the second that it is short in R or that it is not connecting in R.

Definition 2.1. *Given α, $0 < \alpha < 1$, we call a square $R \in \mathcal{G}_n(c)$ α-connecting if the measure of the set of points $x \in R$ whose unstable leaf $W^u(x)$ is long in R is at least α part of the total area of R.*

Sinai formulates the property that most of unstable leaves are sufficiently long in the following way.

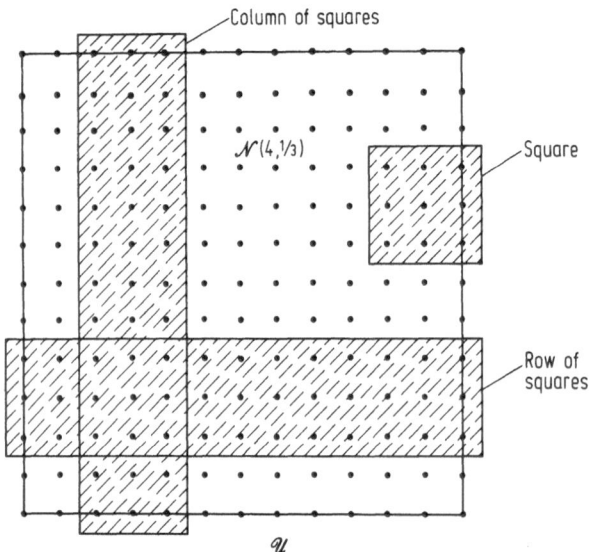

Figure 2. The covering

Sinai Theorem 2.2. *There is $\alpha_0 < 1$ such that for any α, $0 < \alpha \le \alpha_0$ and any c, $0 < c < 1$,*

$$\lim_{n \to +\infty} n \, \mu \left(\bigcup \{ R \in \mathcal{G}_n(c) \mid R \text{ is not } \alpha\text{-connecting} \} \right) = 0.$$

In other words, the theorem says that if α is sufficiently small, then the union of the squares in $\mathcal{G}_n(c)$ which are not α-connecting has measure $o(\frac{1}{n})$.

Before proving the Sinai Theorem let us show how it can be used to get information about ergodic components. Notice that Definition 2.1 and the Sinai Theorem can be repeated for stable leaves.

Proposition 2.3. *The square $\mathcal{U} \subset \mathbb{T}^2$ (for which the Sinai Theorem holds for both unstable leaves and stable leaves) belongs to one ergodic component of T.*

In view of the arbitrariness of the square $\mathcal{U}$ to which we can apply this Theorem we obtain immediately

Corollary 2.4. *The map T is ergodic.*

Proof of Proposition 2.3. Let us fix α sufficiently small so that the Sinai Theorem holds for α-connecting squares both in the unstable and stable versions. Next we fix c smaller than α. As a consequence two α-connecting squares in $\mathcal{G}_n(c)$, which are immediate neighbors, contain in their intersection a set of connecting leaves of positive measure. The reason is that immediate neighbors intersect over $1 - c$ part of their areas and hence the guaranteed α part of the square covered by connecting leaves cannot fit into the remaining c part of the square. In the following we will not change the values of α or c and, for simplicity, we will call an α-connecting square simply a connecting square. Thus a connecting square is α-connecting both with respect to stable and unstable leaves.

Consider any continuous function f on the torus. We call a point $y \in \mathbb{T}^2$ f-typical if the forward time average f^+ and the backward time average f^- are well defined at y and $f^+(y) = f^-(y)$. The set of f-typical points has full measure. We call a stable (unstable) leaf f-typical if its points, except for a subset of zero arc-length, are f-typical. The union of leaves which are not f-typical is a set of measure zero.

For any connecting square R let us define

$$W^{u(s)}(R) = \{ x \in R \mid W^{u(s)}(x) \text{ is } f\text{-typical and long in } R \}.$$

Although we cannot apply the Hopf argument to the whole torus we can use it in a connecting square R to claim that f^+ is constant on all of $W^s(R)$ and f^- is constant on all of $W^u(R)$ with the two constants coinciding. Note that we say here (and we mean it) "all of $W^{s(u)}$" and not almost all. Indeed, first of all f^+ is constant on each of the stable leaves in $W^s(R)$. Further let us fix an ustable leaf in $W^u(R)$. The stable leaves from $W^s(R)$ intersect this unstable leaf in f-typical points, except for a set of stable leaves of total measure zero. Hence excluding these exceptional stable leaves the value of f^+

on the stable leaves has to coincide with the constant value of f^- on the distinguished unstable leaf. We conclude that f^+ is constant almost everywhere on $W^s(R)$ and the constant is equal to the constant value of f^- on the unstable leaf. Since we could have used any other unstable leaf in $W^u(R)$ it follows that f^- is constant on all of $W^u(R)$. By symmetry f^+ is constant on all of $W^s(R)$. (The reader must have noticed the implicit use of the Fubini Theorem in the arguments above. It is only natural since the stable and unstable leaves are parallel segments. In the nonlinear case one has to use the "absolute continuity" of the foliations into stable and unstable manifolds. This property is all that we need, to make the present argument work.)

Further for two connecting squares R_1 and R_2 which are immediate neighbors f^+ is constant on $W^s(R_1) \cup W^s(R_2)$ and f^- is constant on $W^u(R_1) \cup W^u(R_2)$ with the two constants coinciding. Indeed, at least one of the intersections $W^u(R_1) \cap W^u(R_2)$ (if one square is above the other) or $W^s(R_1) \cap W^s(R_2)$ (if one square is next to the other) must have positive measure and hence is nonempty, forcing the constant value of f^+ or f^- to be the same for both squares.

After this observation we proceed to prove that the time average of f is almost everywhere constant in $\mathcal{U}$. To that end let $y, z \in \mathcal{U}$ be two f-typical points with f-typical leaves, $W^u(y)$ and $W^s(z)$ respectively. Our goal is to prove that $f^-(y) = f^+(z)$.

We say that $W^u(y)$ ($W^s(z)$) intersects completely a column (row) of squares in $\mathcal{G}_n(c)$ if it is connecting in one of the squares of the column (row). The Sinai Theorem allows us to claim that, for sufficiently large n, $W^u(y)$ intersects completely at least one column of connecting squares in $\mathcal{G}_n(c)$, i.e. a column in which all the squares are connecting, and $W^s(z)$ intersects completely at least one row of connecting squares. Indeed, suppose to the contrary that every column of squares in $\mathcal{G}_n(c)$ intersected completely by $W^u(y)$ contains at least one non-connecting square. Since the number of columns intersected completely by $W^u(y)$ grows linearly with n and the measure of one square in $\mathcal{G}_n(c)$ is $\frac{1}{n^2}$, we obtain that the measure of the union of non-connecting squares would be $O(\frac{1}{n})$ which contradicts the Sinai Theorem. (Here we have used the fact that the squares in $\mathcal{G}_n(c)$ cannot overlap more than $k(c)$ times.)

Let us fix a column and a row of connecting squares which are intersected completely by $W^u(y)$ and $W^s(z)$ respectively. Let R be the (unique) square which belongs both to the column and the row. Let further R_1 denote a square in which $W^u(y)$ is connecting and R_2 denote a square in which $W^s(z)$ is connecting. By the construction $y \in W^u(R_1)$ and f^- is constant on the union $W^u(R_1) \cup W^u(R)$. Similarly $z \in W^u(R_2)$ and f^+ is constant on $W^s(R_2) \cup W^s(R)$. It follows that $f^-(y) = f^+(z)$. In view of the arbitrariness in the choice of the f-typical leaves $W^u(y)$ and $W^s(z)$ we obtain that the time average of f must be constant in $\mathcal{U}$.

To finish the proof let us consider a T-invariant measurable subset A. Let g be the indicator function of A and

$$f_n \to g \quad \text{in } L^1(\mathbb{T}^2, \mu)$$

be a sequence of uniformly bounded continuous approximations to the indicator function. We will use the fact that the time average is continuous with respect to the L^1 norm to establish that the time average of g must be constant on $\mathcal{U}$. Indeed, if we denote by $\| \cdot \|_1$ the $L^1(\mathbb{T}^2, \mu)$ norm, then

$$\left\| f_n^+ - g^+ \right\|_1 = \left\| \lim_{N \to \infty} \frac{1}{N} \sum_{i=1}^{N} \left(f_n \circ T^i - g \circ T^i \right) \right\|_1$$

$$= \lim_{N \to \infty} \frac{1}{N} \left\| \sum_{i=1}^{N} \left(f_n \circ T^i - g \circ T^i \right) \right\|_1$$

by the Lebesgue Dominated Convergence Theorem.
Using the invariance of the measure we get

$$\left\| f_n^+ - g^+ \right\|_1 \le \lim_{N \to \infty} \frac{1}{N} \sum_{i=1}^{N} \left\| \left(f_n \circ T^i - g \circ T^i \right) \right\|_1 = \left\| f_n - g \right\|_1$$

Since the time averages f_n^+ of f_n are all constant (almost everywhere) on $\mathcal{U}$ the above inequality implies that the time average g^+ is constant (almost everywhere) on $\mathcal{U}$. But the invariance of A forces $g^+ = g$ so that either $\mathcal{U} \setminus A$ or $\mathcal{U} \cap A$ has measure zero. In view of the arbitrariness of the invariant set A it follows that $\mathcal{U}$ must belong to one ergodic component.

$\square$

3. Proof of the Sinai Theorem

The proof of the Sinai Theorem does not require a rigid geometric structure of the coverings $\mathcal{G}_n(c)$; it holds for any sequence of coverings by squares with side $\frac{1}{n}$ as long as there is a uniform bound on the number of squares covering one point. However, the lattice structure of the centers of the squares in $\mathcal{G}_n(c)$ allows to work with columns and rows of squares, as we did in the above application of the Sinai Theorem.

The first step in the proof is the choice of α_0. To that end we consider the smallest sector $\mathcal{C}$ in $\mathbb{R}^2$ symmetric about the horizontal (unstable) line which contains the lines with the two directions of the sides of $\mathcal{M}^-$, i.e., the directions of the segments in $\mathcal{S}^-$. Let

$$\mathcal{C} = \{ (\xi, \eta) \mid |\eta| \le \kappa(a) |\xi| \}.$$

It can be checked that $\kappa(a) < 1$ for any $a \ne 0$. We put $\alpha_0 = \frac{1}{2}(1 - \kappa(a))$. The reason for this choice is that, for any square with vertical and horizontal sides crossed by a line with the direction contained in $\mathcal{C}$, the shaded area in Figure 3 does not exceed $1 - 2\alpha$ part of the area of the square.

Let us observe that all of the segments in $\bigcup_{i=0}^{+\infty} T^i \mathcal{S}^-$ have directions contained in the sector $\mathcal{C}$. Indeed a linear hyperbolic map pushes lines towards the unstable direction except for the stable line, which stays put.

It follows from the construction of the unstable leaves (Proposition 1.1) that an unstable leaf has endpoints on forward images of $\mathcal{S}^-$ under T. Hence if an unstable leaf is short in a square then the square must be intersected by

$$\bigcup_{i=0}^{+\infty} T^i \mathcal{S}^-.$$

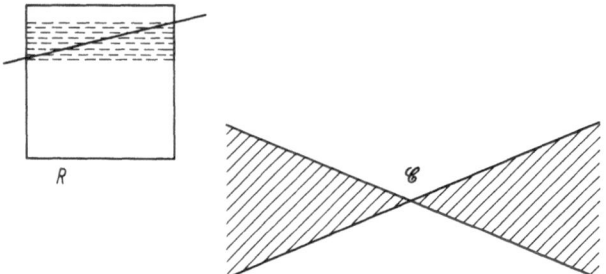

Figure 3. Leaves cut by a line with direction contained in the sector

Although this does not look like a severe restriction, since we can expect that the last set is dense, it has far reaching consequences. The reason being, heuristically, that the singularity lines $T^i \mathcal{S}^-$ become more and more horizontal as $i \to +\infty$ and they cannot cut effectively unstable leaves which are themselves horizontal.

We claim that, for any fixed $M \geq 1$, the singularity lines

$$\mathcal{S}_M^- = \bigcup_{i=0}^{M} T^i \mathcal{S}^-$$

by themselves can produce only few squares which are not α-connecting so that their total measure is $O(\frac{1}{n^2})$. To make this precise (and clear) we introduce an auxiliary notion of an M-bad square in a covering $\mathcal{G}_n(c)$. We say that a square $R \in \mathcal{G}_n(c)$ is M-bad if the measure of the set of points $y \in R$ such that the unstable leaf $W^u(y)$ has an endpoint in $R \cap \mathcal{S}_M^-$ (so that it is short in R) is greater than $1 - 2\alpha$ part of the measure of the square. (Loosely speaking a square is M-bad if it is not connecting because of the singularity lines in $\mathcal{S}_M^-$.)

If a square R intersects only one segment in $\mathcal{S}_M^-$ then the measure of points in R whose unstable leaves have endpoints on the intersection of this segment with R does not exceed $1 - 2\alpha_0 = \kappa(a)$ part of the measure of the square since the direction of the segment is in the sector $\mathcal{C}$. Hence an M-bad square has to intersect at least two segments in $\mathcal{S}_M^-$. But the singularity set $\mathcal{S}_M^-$ is a fixed finite collection of closed segments with only fixed finite number of intersection points (i.e., belonging to several segments). Away from the intersection points the segments are fairly wide apart and a small square cannot extend from one to another, see Figure 4. Hence, for sufficiently large n, an M-bad square in $\mathcal{G}_n(c)$ cannot be farther from one of the intersection points than $\frac{const}{n}$. It follows that the total measure of M-bad squares does not exceed $\frac{const}{n^2}$, where the constant depends only on a, c, α and M.

In this way we took care (in some sense) of the finite number of singularity lines in $\mathcal{S}_M^-$; we now face the problem of controlling the effects of the 'tail' $\bigcup_{i=M+1}^{+\infty} T^i \mathcal{S}^-$.

Let us suppose that a square $R \in \mathcal{G}_n(c)$ is not α-connecting and it is not M-bad. Hence at least α part of its area is covered by short leaves with endpoints in

$$R \cap \bigcup_{i=M+1}^{+\infty} T^i \mathcal{S}^-.$$

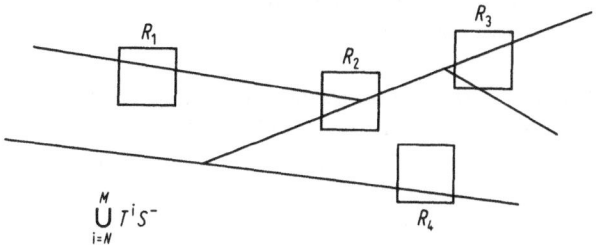

Figure 4. Singularity lines

Let $\mathcal{W}^u(y)$ be such a leaf short in R with an endpoint on $T^i \mathcal{G}^-$. Then

$$T^{-i}\left(W^u(y) \cap R\right) \subset X_{t_i}$$

where $t_i = n^{-1}\lambda^{-i}$ and, as before, $X_t = \{x \in \mathcal{M}^- \mid d(x, \mathcal{G}^-) \leq t\}$. Indeed, under the action of T^{-1}, an unstable leaf contracts by a factor of λ and the length of the part of $W^u(y)$ in R does not exceed $\frac{1}{n}$.

In view of this observation, we can claim that each square which is not α-connecting and which is not M-bad has at least α part of its area covered by

$$\bigcup_{i=M+1}^{+\infty} T^i X_{t_i}.$$

Since each point in $\mathcal{U}$ is covered by at most $k(c)$ squares from $\mathcal{G}_n(c)$, then the measure of the union of squares in $\mathcal{G}_n(c)$ which are not α-connecting and which are not M-bad does not exceed

$$k(c) \times \frac{1}{\alpha} \sum_{i=M+1}^{+\infty} \frac{const}{n\lambda^i} = \frac{1}{n}\left(\frac{k(c)}{\alpha} \sum_{i=M+1}^{+\infty} \frac{const}{\lambda^i}\right),$$

(here the constant is equal to the total length of $\mathcal{G}^-$). We have thus estimated the measure of the union of squares in $\mathcal{G}_n(c)$, which are not α-connecting and which are not M-bad, by the size of an individual square times the M-tail of a fixed convergent series. Some of the readers may have noticed that this completes the proof. For clarity, let us do it explicitly.

Let us take an arbitrary $\epsilon > 0$. We choose and fix $M = M(\epsilon)$ so large that the last series does not exceed $\frac{\epsilon}{2n}$, i.e.,

$$\frac{k(c)}{\alpha} \sum_{i=M+1}^{+\infty} \frac{const}{\lambda^i} < \frac{\epsilon}{2}.$$

Given M, we can still choose $n_0 = n_0(\epsilon, M)$ so large that, for any $n \geq n_0$, the measure of the union of M-bad squares in $\mathcal{G}_n(c)$ is less than $\frac{\epsilon}{2n}$. To estimate the measure of the union of squares in $\mathcal{G}_n(c)$, for $n \geq n_0$, which are not α-connecting we split them into those which are M-bad and those which are not. For both families of squares the measure of their union is less than $\frac{\epsilon}{2n}$. This proves our claim.   □

Remark 3.6. Let us point out that the property that the sector $\mathscr{C}$, defined by the directions of the segments in $\mathscr{S}^-$, is sufficiently narrow ($\kappa(a) < 1$) can be relaxed. For a general hyperbolic piecewise linear map it is sufficient that the segments in $\mathscr{S}^-$ are not parallel to the stable direction. In such a case we can find a natural N such that all the segments in $\bigcup_{i=N+1}^{+\infty} T^i \mathscr{S}^-$ have directions contained in a chosen narrow sector $\mathscr{C}$ (N is the number of iterates of T which do not put the singularity lines $\mathscr{S}^-$ into the chosen sector $\mathscr{C}$). Then the argument above applies to any square neighborhood $\mathscr{U}$ which does not intersect

$$\mathscr{S}_N^- = \bigcup_{i=0}^{N} T^i \mathscr{S}^- .$$

Similarly, in the version of the Sinai Theorem for the stable leaves we would have arrived at a natural N' such that the claim holds for any square $\mathscr{U}$ which does not intersect

$$\mathscr{S}_{N'}^+ = \bigcup_{i=0}^{N'} T^{-i} \mathscr{S}^+ .$$

Hence, it follows from Proposition 2.3 that any open square, with horizontal and vertical sides, which does not intersect $\mathscr{S}_N^- \cup \mathscr{S}_{N'}^+$ belongs to one ergodic component. This implies that the partition of $\mathbb{T}^2$ into ergodic components is coarser than the partition into (open) connected components of

$$\mathbb{T}^2 \setminus \left(\mathscr{S}_N^- \cup \mathscr{S}_{N'}^+ \right) .$$

Since $\mathscr{S}_N^- \cup \mathscr{S}_{N'}^+$ is a finite collection of segments, we obtain that there are at most finitely many ergodic components. To argue that there is only one component let us note that $\mathscr{S}_{N-1}^- \cup \mathscr{S}_{N'}^+$ and $T^N \mathscr{S}^-$ intersect in at most finitely many points which split the segments in $T^N \mathscr{S}^-$ into finitely many segments $\{I_k\}_{k=1}^{K_N}$ so that the interior of every I_k lies in the boundary of at most two connected components of $\mathbb{T}^2 \setminus \left(\mathscr{S}_N^- \cup \mathscr{S}_{N'}^+ \right)$, i.e., it has only one connected component on each side. Suppose that for such a segment I_k is in the boundary of two different ergodic components. Then $T I_k$ is also in the boundary of two different ergodic components. But $T I_k$ and $\mathscr{S}_N^- \cup \mathscr{S}_{N'}^+$ have only finitely many points of intersection, so that whole open sub-intervals of $T I_k$ must end up inside one connected component of $\mathbb{T}^2 \setminus \left(\mathscr{S}_N^- \cup \mathscr{S}_{N'}^+ \right)$ and thus it must have the same ergodic component on both sides. This contradiction implies that I_k does not take part in the splitting of $\mathbb{T}^2$ into ergodic components so we can drop it. In this way we can drop all of $T^N \mathscr{S}^-$ and claim that the partition into ergodic components is coarser than the partition into connected components of

$$\mathbb{T}^2 \setminus \left(\mathscr{S}_{N-1}^- \cup \mathscr{S}_{N'}^+ \right) .$$

It is now clear that we can proceed by dropping $T^{N-1} \mathscr{S}^-$ and $T^{-N'} \mathscr{S}^+$ as possible boundaries for the ergodic components and arriving eventually at $\mathscr{S}^+ \cup \mathscr{S}^-$ as the only possible boundaries we see that even these can be dropped. Hence there is only one ergodic component.

Let us spell out the property of T which is basic in this argument:

Although some points of $\mathscr{S}^-$ return to $\mathscr{S}^-$ under iterates of T, no interval in $\mathscr{S}^-$ can do it.

4. Sectors in a Linear Symplectic Space

For the convenience of the reader we will repeat here some of the material from [W3] and [LW].

Let $\mathscr{W}$ be a linear symplectic space of dimension $2d$ with the symplectic form ω. For instance we call $\mathscr{W} = \mathbb{R}^d \times \mathbb{R}^d$ the standard linear symplectic space if

$$\omega(w_1, w_2) = \langle \xi^1, \eta^2 \rangle - \langle \xi^2, \eta^1 \rangle,$$

where $w_i = (\xi^i, \eta^i)$, $i = 1, 2$, and $\langle \xi, \eta \rangle = \xi_1 \eta_1 + \cdots + \xi_d \eta_d$.

The symplectic group $Sp\,(d, \mathbb{R})$ is the group of linear maps of $\mathscr{W}$ ($2d \times 2d$ matrices if $\mathscr{W} = \mathbb{R}^d \times \mathbb{R}^d$) preserving the symplectic form i.e., $L \in Sp\,(d, \mathbb{R})$ if

$$\omega(Lw_1, Lw_2) = \omega(w_1, w_2)$$

for every $w_1, w_2 \in \mathscr{W}$.

By definition a Lagrangian subspace of a linear symplectic space $\mathscr{W}$ is a d-dimensional subspace on which the restriction of ω is zero (equivalently it is a maximal subspace on which ω vanishes).

Definition 4.1. *Given two transversal Lagrangian subspaces V_1 and V_2 we define the sector between V_1 and V_2 by*

$$\mathscr{C} = \mathscr{C}\,(V_1, V_2) = \{w \in \mathscr{W} \mid \omega(v_1, v_2) \geq 0 \text{ for } w = v_1 + v_2,\, v_i \in V_i,\, i = 1,\, 2\}$$

Equivalently, we define first the quadratic form $\mathfrak{Q}$ associated with an ordered pair of transversal Lagrangian subspaces,

$$\mathfrak{Q}(w) = \omega(v_1, v_2)$$

where $w = v_1 + v_2$, $v_i \in V_i$, $i = 1, 2$, is the unique decomposition of w. We have

$$\mathscr{C} = \{w \in \mathscr{W} \mid \mathfrak{Q}(w) \geq 0\}.$$

In the case of the standard symplectic space, $V_1 = \mathbb{R}^d \times \{0\}$ and $V_2 = \{0\} \times \mathbb{R}^d$ we get

$$\mathfrak{Q}\,((\xi, \eta)) = \langle \xi, \eta \rangle$$

and

$$\mathscr{C} = \{(\xi, \eta) \in \mathbb{R}^d \times \mathbb{R}^d \mid \langle \xi, \eta \rangle \geq 0\}.$$

We will refer to this $\mathscr{C}$ as the standard sector. Since any two pairs of transversal Lagrangian subspaces are symplectically equivalent we may consider only this case without any loss of generality. In the following we will alternate between the coordinate free geometric formulations and this special case. On the one hand, coordinate free formulations are important because we need to apply these concepts to the case of the derivative map which in general acts between two different tangent spaces, each one with its own sector. On the other hand, it turns out that many arguments are greatly simplified by resorting to these special coordinates.

It is natural to ask if a sector determines uniquely its sides. It is not a vacuous question since, for $d > 1$, there are many Lagrangian subspaces in the boundary of a sector. The answer is positive.

Proposition 4.2. *For two pairs of transversal Lagrangian subspaces V_1, V_2 and V'_1, V'_2 if*

$$\mathscr{C}(V_1, V_2) = \mathscr{C}(V'_1, V'_2)$$

then

$$V_1 = V'_1 \quad and \quad V_2 = V'_2.$$

Moreover V_1 and V_2 are the only isolated Lagrangian subspaces contained in the boundary of the sector $\mathscr{C}(V_1, V_2)$.

The proof of this Proposition can be found in [W3].

Based on the notion of the sector between two transversal Lagrangian subspaces (or the quadratic form $\mathfrak{Q}$) we define two monotonicity properties of a linear symplectic map. By $int'\mathscr{C}$ we denote the interior of the sector, i.e.,

$$int'\mathscr{C} = \{w \in \mathcal{W} | \mathfrak{Q}(w) > 0\}.$$

Definition 4.3. *Given the sector $\mathscr{C}$ between two transversal Lagrangian subspaces we call a linear symplectic map L monotone if*

$$L\mathscr{C} \subset \mathscr{C}$$

and strictly monotone if

$$L\mathscr{C} \subset int'\mathscr{C} \cup \{0\}.$$

A very useful characterization of monotonicity is given in the following

Theorem 4.4. *L is (strictly) monotone if and only if $\mathfrak{Q}(Lw) \geq \mathfrak{Q}(w)$ for every $w \in \mathcal{W}$ ($\mathfrak{Q}(Lw) > \mathfrak{Q}(w)$ for every $w \in \mathcal{W}$, $w \neq 0$).*

The fact that monotonicity implies the increase of the quadratic form defining the sector is a manifestation of a very special geometric structure of a sector and does not hold for cones defined by general quadratic forms. Before proving this theorem we will discuss several useful facts.

For a pair of transversal Lagrangian subspaces V_1 and V_2 and a linear map $L : \mathcal{W} \to \mathcal{W}$, we can define the following 'block' operators:

$$A : V_1 \to V_1, \ B : V_2 \to V_1$$
$$C : V_1 \to V_2, \ D : V_2 \to V_2.$$

They are uniquely defined by the requirement that for any $v_1 \in V_1$, $v_2 \in V_2$

$$L(v_1 + v_2) = Av_1 + Bv_2 + Cv_1 + Dv_2.$$

We will need the following Lemma.

Lemma 4.5. *If L is monotone with respect to the sector between V_1 and V_2 then LV_1 is transversal to V_2 and LV_2 is transversal to V_1.*

Proof. Suppose that there is $0 \neq \bar{v}_1 \in V_1$ such that $L\bar{v}_1 \in V_2$. We choose $\bar{v}_2 \in V_2$ so that

$$\mathfrak{Q}(\bar{v}_1 + \bar{v}_2) = \omega(\bar{v}_1, \bar{v}_2) > 0.$$

We have also

$$\omega(\bar{v}_1, \bar{v}_2) = \omega(L\bar{v}_1, L\bar{v}_2) = \omega(L\bar{v}_1, B\bar{v}_2 + D\bar{v}_2) = \omega(L\bar{v}_1, B\bar{v}_2).$$

Let $v_\epsilon = \bar{v}_1 + \epsilon\bar{v}_2$. We have that for $\epsilon > 0$ v_ϵ belongs to int$\mathscr{C}$. Hence also $\mathfrak{Q}(Lv_\epsilon) \geq 0$ for $\epsilon > 0$. On the other hand

$$\mathfrak{Q}(Lv_\epsilon) = \epsilon^2 \omega(B\bar{v}_2, D\bar{v}_2) - \epsilon\omega(L\bar{v}_1, B\bar{v}_2)$$

which is negative for sufficiently small positive ϵ.

This contradiction proves the Lemma. □

It follows from Lemma 4.5 that the operators $A : V_1 \to V_1$ and $D : V_2 \to V_2$ are invertible.

We switch now to the coordinate language. Let

$$L = \begin{pmatrix} A & B \\ C & D \end{pmatrix}$$

be a symplectic map of the standard symplectic space $\mathbb{R}^d \times \mathbb{R}^d$ monotone with respect to the standard sector. A, B, C, D are now just $d \times d$ matrices.

Let us describe those symplectic matrices which are monotone in the weakest sense, namely they preserve the quadratic form $\mathfrak{Q}$. We will call such matrices $\mathfrak{Q}$-isometries. Obviously a $\mathfrak{Q}$-isometry maps the sector onto itself. The converse is also true.

Proposition 4.6. *If L is a linear symplectic map and*

$$L\mathscr{C} = \mathscr{C}$$

then

$$L = \begin{pmatrix} A & 0 \\ 0 & A^{*-1} \end{pmatrix}.$$

In particular it preserves the quadratic form $\mathfrak{Q}$

$$\mathfrak{Q} \circ L = \mathfrak{Q}.$$

Proof. If $L\mathscr{C} = \mathscr{C}$ then L maps also the boundary of the sector $\mathscr{C}$ onto itself. It follows from Proposition 4.2 that both sides of the sector stay put under L. Hence $B = C = 0$. By symplecticity $D = A^{*-1}$. □

By Lemma 4.5 given a monotone L we can always factor out the following $\mathfrak{2}$-isometry on the left

$$L = \begin{pmatrix} A & B \\ C & D \end{pmatrix} = \begin{pmatrix} A & 0 \\ 0 & A^{*-1} \end{pmatrix} \begin{pmatrix} I & P \\ R & \cdot \end{pmatrix}$$

Symplecticity of L forces P and R to be symmetric and allows the further unique factorization

$$(4.7) \qquad\qquad L = \begin{pmatrix} A & 0 \\ 0 & A^{*-1} \end{pmatrix} \begin{pmatrix} I & 0 \\ P & I \end{pmatrix} \begin{pmatrix} I & R \\ 0 & I \end{pmatrix}.$$

Moreover monotonicity forces P and R to be positive semidefinite ($P \geq 0$, $R \geq 0$). Strict monotonicity means that P and R are positive definite ($P > 0$, $R > 0$). These claims follow from the following

Proof of Theorem 4.4. Using the above factorization we get for $w = (\xi, \eta)$

$$\mathfrak{2}(Lw) = \langle \xi, \eta \rangle + \langle R\eta, \eta \rangle + \langle P(\xi + R\eta), \xi + R\eta \rangle.$$

Putting $\eta = 0$ we obtain that $P \geq 0$. To show that also $R \geq 0$ let us consider an eigenvector η_0 of R with eigenvalue λ and let $\xi = a\eta_0$. We get that if $a \geq 0$ then $w = (\xi, \eta_0) \in \mathscr{C}$ so that $\mathfrak{2}(Lw) \geq 0$. It follows that

$$(a + \lambda)\langle \eta, \eta \rangle + (a + \lambda)^2 \langle P\eta, \eta \rangle \geq 0.$$

This implies immediately that $\lambda \geq 0$. This proves the monotone version of the Theorem. The strictly monotone version is obtained in a similar way. □

As a byproduct of the proof we get the following useful observation

Proposition 4.8. *A monotone map L is strictly monotone if and only if*

$$LV_i \subset int \, \mathscr{C} \cup \{0\}, \quad i = 1, 2.$$

□

The following Proposition simplifies computations with monotone maps.

Proposition 4.9. *If*

$$L = \begin{pmatrix} A & B \\ C & D \end{pmatrix}$$

is a strictly monotone map then by multiplying it by $\mathfrak{2}$-isometries on the left and on the right we can bring it to the form

$$\begin{pmatrix} I & I \\ T & I + T \end{pmatrix}$$

where T is diagonal and has the same eigenvalues as C^*B.

Proof. The factorization of the monotone map L yields

$$\begin{pmatrix} A & 0 \\ 0 & A^{*-1} \end{pmatrix} L = \begin{pmatrix} I & R \\ P & I + PR \end{pmatrix}$$

where $P > 0$, $R > 0$ and $PR = C^*B$.

We have further

$$\begin{pmatrix} R^{-\frac{1}{2}} & 0 \\ 0 & R^{\frac{1}{2}} \end{pmatrix} \begin{pmatrix} I & R \\ P & I+PR \end{pmatrix} \begin{pmatrix} R^{\frac{1}{2}} & 0 \\ 0 & R^{-\frac{1}{2}} \end{pmatrix} = \begin{pmatrix} I & I \\ K & I+K \end{pmatrix}$$

where $K = R^{\frac{1}{2}}PR^{\frac{1}{2}}$ has the same eigenvalues as $C^*B = PR$.

Finally if F is the orthogonal matrix which diagonalizes K, i.e., $F^{-1}KF$ is diagonal, then

$$\begin{pmatrix} F^{-1} & 0 \\ 0 & F^{-1} \end{pmatrix} \begin{pmatrix} I & I \\ K & I+K \end{pmatrix} \begin{pmatrix} F & 0 \\ 0 & F \end{pmatrix} = \begin{pmatrix} I & I \\ T & I+T \end{pmatrix}$$

has the desired form with $T = F^{-1}KF$ having the same eigenvalues as C^*B.   □

Let us note that in the last Proposition we can ask for the diagonal entries of T to be ordered because any permutation of the entries can be accomplished by an appropriate $\mathfrak{Q}$-isometry.

5. The Space of Lagrangian Subspaces Contained in a Sector

Let us fix a sector $\mathcal{C} = \mathcal{C}(V_1, V_2)$ between two transversal Lagrangian subspaces V_1 and V_2. We say that a Lagrangian subspace E is strictly contained in $\mathcal{C}$ if

$$E \subset int\,\mathcal{C} \cup \{0\}.$$

We denote by $Lag(\mathcal{C})$ the manifold of all such Lagrangian subspaces and by $\widehat{Lag}(\mathcal{C})$ its closure in the Lagrangian Grassmanian, i.e., $\widehat{Lag}(\mathcal{C})$ is the set of all Lagrangian subspaces contained in $\mathcal{C}$.

We will introduce a metric and a partial order into $Lag(\mathcal{C})$. Let

$$\pi_i : \mathcal{W} \to V_i, \ i = 1, 2,$$

be the natural projections, i.e.,

$$w = \pi_1 w + \pi_2 w \quad \text{for every } w \in \mathcal{W}.$$

If a Lagrangian subspace E is strictly contained in $\mathcal{C}$ then $\pi_i E = V_i$, $i = 1, 2$, so that $\pi_i|_E$ (the restriction of π_i to the subspace E) is a one to one map of E onto V_i.

With every subspace $E \in Lag(\mathcal{C})$ we can associate a positive definite quadratic form on V_1 obtained by the formula

$$\mathfrak{Q} \circ (\pi_1|_E)^{-1}.$$

It will turn out that this is actually a one-to-one correspondence between positive definite quadratic forms on V_1 and Lagrangian subspaces contained strictly in $\mathcal{C}$.

Definition 5.1. *For two Lagrangian subspaces $E_1, E_2 \in Lag(\mathcal{C})$ we define the relation $E_1 \leq E_2$ ($E_1 < E_2$) by the inequality of the corresponding quadratic forms*

$$\mathfrak{Q} \circ (\pi_1|_{E_1})^{-1} \leq (<)\mathfrak{Q} \circ (\pi_1|_{E_2})^{-1}.$$

We define the distance of two Lagrangian subspaces $E_1, E_2 \in Lag(\mathscr{C})$ by

$$d(E_1, E_2) = \frac{1}{2} \sup_{0 \neq v \in V_1} |\ln \mathfrak{Q} \circ (\pi_1|_{E_1})^{-1} (v) - \ln \mathfrak{Q} \circ (\pi_1|_{E_2})^{-1} (v)|.$$

It is easy to see that $d(\cdot, \cdot)$ is indeed a metric.

There are other ways to introduce the partial order and the metric. The coordinate free definitions simplify some of the arguments in the following. For equivalent definitions of the metric see [LW], [Ve]. Theses definitions are justified by the following theorem.

Theorem 5.2. *For two transversal Lagrangian subspaces $E_1, E_2 \in Lag(\mathscr{C})$,*

$$E_1 < E_2 \quad \text{if and only if} \quad \mathscr{C}(E_1, E_2) \subset \mathscr{C}(V_1, V_2).$$

Further, if $E_1 < E_2$ then for a Lagrangian subspace $E \in Lag(\mathscr{C})$,

$$E \subset \mathscr{C}(E_1, E_2) \quad \text{if and only if} \quad E_1 \leq E \leq E_2.$$

Corollary 5.3. *If $E_1, E_2 \in Lag(\mathscr{C})$ and $E_1 < E_2$ then the diameter of the set $\widehat{Lag}(\mathscr{C}(E_1, E_2))$ in $Lag(\mathscr{C})$ is equal to the distance of E_1 and E_2.* $\qquad\square$

We will prove Theorem 5.2 at the end of this Section.

Let us introduce a convenient parametrization of $Lag(\mathscr{C})$ by symmetric positive definite matrices. We consider the standard sector $\mathscr{C}$ in $\mathbb{R}^d \times \mathbb{R}^d$ with $V_1 = \mathbb{R}^d \times \{0\}$ and $V_2 = \{0\} \times \mathbb{R}^d$. Let $U : \mathbb{R}^d \to \mathbb{R}^d$ be a linear map and

$$gU = \{(\xi, \eta) \in \mathbb{R}^d \times \mathbb{R}^d \mid \eta = U\xi\}$$

be its graph. The linear subspace gU is a Lagrangian subspace if and only if U is symmetric and further for a symmetric U its graph $gU \subset \mathscr{C}$ if and only if $U \geq 0$. Every Lagrangian subspace in $Lag(\mathscr{C})$ is transversal to V_2 so that it is a graph of a linear map as above. We will find the following Lemma useful.

Lemma 5.4. *If a Langrangian subspace $E \subset \mathscr{C}(V_1, V_2)$ is transversal to both V_1 and V_2 then it is strictly contained in the sector.*

Proof. We use the coordinate description of the standard sector. Thus the Lagrangian subspace E being transversal to V_2 is the graph of a symmetric positive semidefinite matrix. Since E is also transversal to V_1 the matrix is nondegenerate and hence positive definite. It follows immediately that E is strictly contained in the sector. $\qquad\square$

We have obtained a one-to-one correspondence between Lagrangian subspaces in $Lag(\mathscr{C})$ and symmetric positive definite matrices. The quadratic form on V_1 introduced in Definition 5.1 becomes the form defined by the positive definite matrix. The partial order becomes the familiar partial order between symmetric matrices.

The image of a Lagrangian subspace under a symplectic linear map is again a Lagrangian subspace. Moreover monotone maps take Lagrangian subspaces strictly contained in $\mathscr{C}$ into Lagrangian subspaces strictly contained in $\mathscr{C}$. Hence a monotone map L defines a map of $Lag(\mathscr{C})$ into itself. We will denote it again by $L : Lag(\mathscr{C}) \to Lag(\mathscr{C})$.

To simplify notation we will also write U instead of gU. We have that

$$L = \begin{pmatrix} A & B \\ C & D \end{pmatrix}$$

acts on Lagrangian subspaces by the following Möbius transformation

$$LU = (C + DU)(A + BU)^{-1}.$$

In particular the action of a 2-isometry

$$L = \begin{pmatrix} A & 0 \\ 0 & A^{*-1} \end{pmatrix}$$

is given by

$$LU = A^{*-1}UA^{-1}.$$

By putting $A = U^{\frac{1}{2}}$ we see that any $U > 0$ can be mapped onto identity matrix I. Thus 2-isometries act transitively on $Lag(\mathscr{C})$. Moreover it is not hard to see that

Proposition 5.5. *The action of a 2-isometry on $Lag(\mathscr{C})$ preserves the partial order and the metric.* $\qquad\qquad\qquad\Box$

Let $E_0 = \{(\xi, \eta) \mid \xi = \eta\}$. By straightforward computations we find that

(5.6)
$$\mathscr{C}(V_1, E_0) = \{(\xi, \eta) \mid \langle \xi, \eta \rangle - \langle \eta, \eta \rangle \geq 0\},$$
$$\mathscr{C}(E_0, V_2) = \{(\xi, \eta) \mid \langle \xi, \eta \rangle - \langle \xi, \xi \rangle \geq 0\}.$$

We get that

(5.7)
$$\mathscr{C}(V_1, E_0) \subset \mathscr{C}(V_1, V_2),$$
$$\mathscr{C}(E_0, V_2) \subset \mathscr{C}(V_1, V_2),$$
$$\mathscr{C}(V_1, E_0) \cap \mathscr{C}(E_0, V_2) = E_0.$$

Because the group of 2-isometries acts transitively on $Lag(\mathscr{C})$ (5.7) holds not just for the special Lagrangian subspace E_0 from (5.6) but for any Lagrangian subspace from $Lag(\mathscr{C})$. (It just happens that the easiest way to establish (5.7) is to do the calculation in the standard sector.)

Proposition 5.8. *For two Lagrangian subspaces $E_1, E_2 \in Lag(\mathscr{C})$ the following are equivalent*

(1) $E_1 \leq E_2$,

(2) $E_2 \subset \mathscr{C}(E_1, V_2)$,

(3) $E_1 \subset \mathscr{C}(V_1, E_2)$.

Proof. We will be using the coordinate description of the standard sector. Since the group of 2-isometries acts transitively on $Lag(\mathscr{C})$ we can assume that E_1 is equal to E_0 from (5.6). Let U_2 be the positive definite matrix defining E_2. We get from (5.6) that $E_2 \subset \mathscr{C}(E_0, V_2)$ if and only if $U_2 \geq I$. Hence (1) is equivalent to (2). Similarly let E_2 be equal to E_0 and U_1 be the positive definite matrix defining E_1. Using (5.6) again we get that $E_1 \subset \mathscr{C}(V_1, E_0)$ if and only if $U_1 - U_1^2 \geq 0$ which is equivalent to $U_1 \leq I$. This proves the equivalence of (1) and (3). $\qquad\square$

Proof of Theorem 5.2. If $E_1 < E_2$ then, by Proposition 5.8 and Lemma 5.4, E_2 is strictly contained in $\mathscr{C}(E_1, V_2)$. Using (5.7) we get

$$\mathscr{C}(E_1, E_2) \subset \mathscr{C}(E_1, V_2) \subset \mathscr{C}(V_1, V_2).$$

Suppose now that $\mathscr{C}(E_1, E_2) \subset \mathscr{C}(V_1, V_2)$. By Proposition 5.8, it suffices to show that $E_2 \subset \mathscr{C}(E_1, V_2)$. If it is not so then there is $e_2 \in E_2$ which does not belong to $\mathscr{C}(E_1, V_2)$. Let us consider $v_1 = \pi_1 e_2$ where $\pi_1 : \mathcal{W} \to V_1$ is the projection onto V_1 in the direction of V_2. Let further e_1 be the unique element in E_1 such that $\pi_1 e_1 = v_1$ (i.e., $e_1 = (\pi_1|_{E_1})^{-1} v_1$). Clearly the difference between the two vectors $v_2 = e_2 - e_1$ belongs to V_2. Because $e_2 = e_1 + v_2$ and $e_2 \notin \mathscr{C}(E_1, V_2)$ we have $\omega(e_1, v_2) < 0$ so that $\omega(-e_1, e_2) > 0$. It follows that $v_2 = e_2 - e_1 \in int\, \mathscr{C}(E_1, E_2) \subset int\, \mathscr{C}(V_1, V_2)$. We have thus reached a contradiction, that v_2 belongs to V_2 and also to $int\, \mathscr{C}(V_1, V_2)$. This contradiction proves that indeed $E_2 \subset \mathscr{C}(E_1, V_2)$ which, by Proposition 5.8, implies that $E_1 < E_2$ (remember that E_1 and E_2 are assumed to be transversal). The first part of the Theorem is proven.

To prove the second part, let $E_1 < E_2$ and $E \subset \mathscr{C}(E_1, E_2)$. By Proposition 5.8, we get $E_2 \subset \mathscr{C}(E_1, V_2)$. It follows, in view of (5.7), that $\mathscr{C}(E_1, E_2) \subset \mathscr{C}(E_1, V_2)$, and hence $E \subset \mathscr{C}(E_1, V_2)$ which is equivalent (again by Proposition 5.8) to $E_1 \leq E$. Similarly, we get $E \leq E_2$.

In the opposite direction, if $E_1 \leq E < E_2$ then, by Proposition 5.8, E_1 and E are strictly contained in $\mathscr{C}(V_1, E_2)$ and $E_1 \subset \mathscr{C}(V_1, E)$. Applying now the equivalence of (2) and (3) in Proposition 5.8 to the case of $E_1, E \in Lag(\mathscr{C}(V_1, E_2))$, we get immediately $E \subset \mathscr{C}(E_1, E_2)$. The case of $E_1 \leq E \leq E_2$ can be now treated by continuity. $\qquad\square$

Let us consider a special family of Lagrangian subspaces in the standard sector: the graphs of multiples of the identity matrix, i.e., for a real number u let

$$Z_u = \{(\xi, \eta) \mid \eta = e^u \xi\}.$$

We have that

$$d(Z_{u_1}, Z_{u_2}) = \frac{1}{2}|u_1 - u_2|.$$

Lemma 5.9. *Let us fix two Lagrangian subspaces, Z_{u_1} and Z_{u_2}, $u_1 < u_2$. If for a Lagrangian subspace $E \in Lag(\mathscr{C})$*

$$d(Z_{u_1}, E) \leq \frac{1}{2}(u_2 - u_1)$$

then

$$E \leq Z_{u_2}.$$

Proof. Let the Lagrangian subspace E be the graph of a positive definite matrix U. For every nonzero $\xi \in \mathbb{R}^d$, we have

$$\ln\langle \xi, U\xi \rangle - \ln\langle \xi, e^{u_1}\xi \rangle \leq u_2 - u_1.$$

It follows that, for every nonzero $\xi \in \mathbb{R}^d$,

$$\ln \frac{\langle \xi, U\xi \rangle}{\langle \xi, \xi \rangle} \leq u_2.$$

We conclude that $U \leq e^{u_2} I$.   □

We will use the following consequence of the last Lemma.

Proposition 5.10. *Let* $E_1 < E_2$ *be two Lagrangian subspaces contained strictly in* $\mathscr{C}(V_1, V_2)$. *There is a symplectic map which maps the sector* $\mathscr{C}(V_1, V_2)$ **onto** *the standard sector* $\mathscr{C}$ *and the sector* $\mathscr{C}(E_1, E_2)$ **into** *the sector* $\mathscr{C}(Z_{-u}, Z_u)$ *if and only if* $d(E_1, E_2) \leq u$.

Proof. By a symplectic map we can map the subspace V_1 onto $\mathbb{R}^d \times \{0\}$, the subspace V_2 onto $\{0\} \times \mathbb{R}^d$ and E_1 onto Z_{-u} (because $\mathfrak{D}$-isometries act transitively on $Lag(\mathscr{C})$). It follows from Lemma 5.9 that the sector $\mathscr{C}(E_1, E_2)$ will be then automatically mapped into $\mathscr{C}(Z_{-u}, Z_u)$.

The converse follows from Corollary 5.3.   □

For aesthetical reasons we will be using Proposition 5.10 in a different coordinate system.

Let us introduce the family of sectors

$$\mathscr{C}_\rho = \{(\xi, \eta) \mid \|\eta\| \leq \rho\|\xi\|\}$$

for any real $\rho > 0$.

Proposition 5.11. *Let* $E_1 < E_2$ *be two Lagrangian subspaces contained strictly in* $\mathscr{C}(V_1, V_2)$. *There is a symplectic map which maps the sector* $\mathscr{C}(V_1, V_2)$ **onto** *the sector* $\mathscr{C}_{\rho^{-1}}$ *and the sector* $\mathscr{C}(E_1, E_2)$ **into** *the sector* $\mathscr{C}_\rho$ *if and only if*

$$d(E_1, E_2) \leq \ln \frac{1 + \rho^2}{1 - \rho^2},$$

with $0 < \rho < 1$.

Proof. Let us define a symplectic map L by

$$\xi' = \frac{1}{\sqrt{2}} (\rho^{-\frac{1}{2}} \xi - \rho^{\frac{1}{2}} \eta),$$

$$\eta' = \frac{1}{\sqrt{2}} (\rho^{-\frac{1}{2}} \xi + \rho^{\frac{1}{2}} \eta).$$

A direct computation shows that, if $\rho < 1$, $L\mathscr{C}_{\rho^{-1}} = \mathscr{C}$ and $L\mathscr{C}_\rho = \mathscr{C}(Z_{-u}, Z_u)$, with $u = \log \frac{1+\rho^2}{1-\rho^2}$. The result follows then from Proposition 5.10.   □

6. Unbounded Sequences of Linear Monotone Maps

In this section we fix a sector $\mathscr{C} = \mathscr{C}(V_1, V_2)$ between two Lagrangian subspaces. One can think that $\mathscr{C}$ is the standard sector. We start by computing the coefficient of expansion of $\mathfrak{D}$ under the action of a monotone symplectic map.

For a linear symplectic map L, monotone with respect to the sector $\mathscr{C}$, we define the coefficient of expansion at $w \in int\mathscr{C}$ by

$$\beta(w, L) = \sqrt{\frac{\mathfrak{Q}(Lw)}{\mathfrak{Q}(w)}}.$$

Further, we define the least coefficient of expansion by

$$\sigma_\mathscr{C}(L) = \inf_{w \in int\mathscr{C}} \beta(w, L).$$

Let us note that, for two monotone maps L_1 and L_2,

$$\sigma_\mathscr{C}(L_2 L_1) \geq \sigma_\mathscr{C}(L_2)\,\sigma_\mathscr{C}(L_1),$$

i.e., the coefficient of expansion $\sigma_\mathscr{C}$ is supermultiplicative.

We will omit the index $\mathscr{C}$ in $\sigma_\mathscr{C}(L)$ when it is clear which sector we have in mind.

We want to find the value of the expansion coefficient in coordinates. We will use the fact that this infimum does not change if L is multiplied on the left or on the right by $\mathfrak{Q}$-isometries. So let

$$L = \begin{pmatrix} A & B \\ C & D \end{pmatrix}$$

be a monotone matrix. By the factorization (4.7), $C^*B = PR$ is equal to the product of two positive semidefinite matrices, and so it has only real non-negative eigenvalues. Let us denote them by $0 \leq t_1 \leq \cdots \leq t_d$. The monotone map L is strictly monotone if and only if $t_1 > 0$.

Proposition 6.1. *For a monotone map L*

$$\sigma(L) = \sqrt{1 + t_1} + \sqrt{t_1} = \exp \sinh^{-1} \sqrt{t_1}.$$

Moreover, if L is strictly monotone

$$\sigma(L) = \beta(w, L)$$

for some $w \in int\ \mathscr{C}$.

Proof. Let us put

$$m(L) = \sqrt{1 + t_1} + \sqrt{t_1} = \min_{1 \leq i \leq d}\left(\sqrt{1 + t_i} + \sqrt{t_i}\right).$$

First we prove the inequality $\beta(w, L) \geq m(L)$ for $w \in int\mathscr{C}$. Since both $\beta(w, L)$ and $m(L)$ are continuous functions of L it is sufficient to prove the inequality for strictly monotone maps only. In view of Proposition 4.9 we can take

$$L = \begin{pmatrix} I & I \\ T & I+T \end{pmatrix}$$

with diagonal T and $t_1, \ldots, t_d$ on the diagonal. We compute directly, for $w = (\xi, \eta)$ such that $\mathfrak{Q}(w) = 1$,

$$
\begin{aligned}
(\beta(w, L))^2 &= \sum_{i=1}^{d} \left(t_i \xi_i^2 + (1 + 2t_i) \xi_i \eta_i + (1 + t_i) \eta_i^2 \right) \\
&= \sum_{i:\xi_i \eta_i \geq 0} \left(\left(\sqrt{t_i} \xi_i - \sqrt{1+t_i} \eta_i \right)^2 + \left(\sqrt{1+t_i} + \sqrt{t_i} \right)^2 \xi_i \eta_i \right) \\
&\quad + \sum_{i:\xi_i \eta_i < 0} \left(\left(\sqrt{t_i} \xi_i + \sqrt{1+t_i} \eta_i \right)^2 + \left(\sqrt{1+t_i} - \sqrt{t_i} \right)^2 \xi_i \eta_i \right) \geq \\
&\geq \sum_{i:\xi_i \eta_i \geq 0} \left(\sqrt{1+t_i} + \sqrt{t_i} \right)^2 \xi_i \eta_i + \sum_{i:\xi_i \eta_i < 0} \left(\sqrt{1+t_i} + \sqrt{t_i} \right)^{-2} \xi_i \eta_i \geq \\
&\geq (1 + \delta) m(L)^2 - \delta m(L)^{-2} \geq m(L)^2
\end{aligned}
$$

where

$$
\delta = \left(\sum_{i:\xi_i \eta_i \geq 0} \xi_i \eta_i \right) - 1 = - \sum_{i:\xi_i \eta_i < 0} \xi_i \eta_i \geq 0
$$

and all the inequalities become equalities for

$$
\xi_1 = \left(\frac{1+t_1}{t_1} \right)^{\frac{1}{4}}, \quad \eta_1 = \left(\frac{t_1}{1+t_1} \right)^{\frac{1}{4}}, \quad \xi_i = 0, \eta_i = 0, \ i = 2, \ldots, d.
$$

Thus the Proposition is proven for strictly monotone matrices, and for all monotone matrices we get the inequality $\sigma(L) \geq m(L)$. To get the equality $\sigma(L) = m(L)$ for all monotone matrices, we proceed as follows. For any $\epsilon > 0$, we choose a strictly monotone matrix L_ϵ, so close to the identity that $m(L_\epsilon L) < m(L) + \epsilon$. Since $L_\epsilon L$ is strictly monotone and our Proposition has been proven for strictly monotone matrices, there is $w_\epsilon \in \text{int}\mathscr{C}$ such that

$$
\beta(w_\epsilon, L_\epsilon L) = m(L_\epsilon L) = \sigma(L_\epsilon L).
$$

But $\beta(w, L_\epsilon L) > \beta(w, L)$, for any $w \in \text{int}\mathscr{C}$. Hence

$$
m(L) \leq \sigma(L) \leq \beta(w_\epsilon, L) < \beta(w_\epsilon, L_\epsilon L) = m(L_\epsilon L) < m(L) + \epsilon,
$$

which ends the proof.   □

For a given sector $\mathscr{C} = \mathscr{C}(V_1, V_2)$ let $\mathscr{C}' = \mathscr{C}(V_2, V_1)$ be the complementary sector. We have

Proposition 6.2. *If L is (strictly) monotone with respect to $\mathscr{C}$ then L^{-1} is (strictly) monotone with respect to $\mathscr{C}'$ and $\sigma_\mathscr{C}(L) = \sigma_{\mathscr{C}'}(L^{-1})$.*

Proof. We have that the union

$$
\mathscr{C}(V_1, V_2) \cup \text{int}\mathscr{C}(V_2, V_1)
$$

is equal to the whole linear symplectic space $\mathcal{W}$. Hence if

$$L\mathcal{C}(V_1, V_2) \subset \mathcal{C}(V_1, V_2)$$

then

$$\mathcal{C}(V_1, V_2) \subset L^{-1}\mathcal{C}(V_1, V_2)$$

and finally

$$L^{-1}\text{int}\mathcal{C}(V_2, V_1) \subset \text{int}\mathcal{C}(V_2, V_1).$$

The last property is easily seen to be equivalent to the monotonicity of L^{-1}.

To obtain the equality of the coefficients of least expansion we will use the standard sector and the block description of L. Let (see (4.7))

$$L = \begin{pmatrix} A & 0 \\ 0 & A^{*-1} \end{pmatrix} \begin{pmatrix} I & 0 \\ P & I \end{pmatrix} \begin{pmatrix} I & R \\ 0 & I \end{pmatrix}.$$

The linear symplectic map $\begin{pmatrix} 0 & I \\ -I & 0 \end{pmatrix}$ takes the standard sector $\mathcal{C}$ onto $\mathcal{C}'$ and further

$$L_1 = \begin{pmatrix} 0 & -I \\ I & 0 \end{pmatrix} L^{-1} \begin{pmatrix} 0 & I \\ -I & 0 \end{pmatrix}$$

has the same least coefficient of expansion with respect to $\mathcal{C}$ as L^{-1} with respect to $\mathcal{C}'$. Since

$$L^{-1} = \begin{pmatrix} I & -R \\ 0 & I \end{pmatrix} \begin{pmatrix} I & 0 \\ -P & I \end{pmatrix} \begin{pmatrix} A^{-1} & 0 \\ 0 & A^* \end{pmatrix}$$

we get

$$L_1 = \begin{pmatrix} I & P \\ R & I + RP \end{pmatrix} \begin{pmatrix} A^* & 0 \\ 0 & A^{-1} \end{pmatrix}.$$

Our claim follows now from the formula in Proposition 6.1 and the fact that PR has the same eigenvalues as RP.   □

The next Proposition is a useful addition to the Corollary 5.3.

Proposition 6.3. *For a strictly monotone map L*

$$d(LV_1, LV_2) = \ln \frac{\sigma(L)^2 + 1}{\sigma(L)^2 - 1}.$$

Proof. Since $\mathfrak{D}$ − isometries preserve the distance between Lagrangian subspaces it follows from Proposition 4.9 that we can restrict our calculations to

$$L = \begin{pmatrix} I & I \\ T & I + T \end{pmatrix}$$

with diagonal T. By the Definition 5.1 we have

$$d(LV_1, LV_2) = \frac{1}{2} \sup_{0 \neq \xi \in \mathbb{R}^d} | \ln\langle \xi, T\xi \rangle - \ln\langle \xi, (T + I)\xi \rangle |$$

$$= \frac{1}{2} \sup_{0 \neq \xi \in \mathbb{R}^d} \ln \frac{\langle \xi, (I + T^{-1})\xi \rangle}{\langle \xi, \xi \rangle} = \max_i \frac{\ln \left(1 + t_i^{-1}\right)}{2} = \frac{\ln \left(1 + t_1^{-1}\right)}{2}$$

where $t_1 \leq t_2 \leq \cdots \leq t_d$ are the eigenvalues of T. The desired formula is now obtained by a straightforward calculation. $\square$

We introduce now an important property of a sequence of monotone maps. Let us consider a sequence of linear symplectic monotone maps $\{L_i\}_{i=1}^{+\infty}$. To simplify notation let us put $L^n = L_n \ldots L_1$.

Definition 6.4. *A sequence $\{L_1, L_2, \ldots\}$ of monotone maps is called unbounded if, for all $w \in int\mathscr{C}$,*

$$\mathfrak{Q}(L^n w) \rightarrow +\infty \quad as \quad n \rightarrow +\infty.$$

It is called strictly unbounded if, for all $w \in \mathscr{C}, w \neq 0$,,

$$\mathfrak{Q}(L^n w) \rightarrow +\infty \quad as \quad n \rightarrow +\infty.$$

Theorem 6.5. *A sequence $\{L_1, L_2, \ldots\}$ of maps, monotone with respect to $\mathscr{C}$, is unbounded if and only if*

$$\bigcap_{n=1}^{+\infty} L_1^{-1} L_2^{-1} \ldots L_n^{-1} \mathscr{C}' = one \ Lagrangian \ subspace,$$

where $\mathscr{C}'$ is the complementary sector.

Corollary 6.6. *If a sequence of monotone maps $\{L_1, L_2, \ldots\}$ is unbounded then the sequence $\{L_2, L_3, \ldots\}$ is also unbounded.* $\square$

We were not able to find a proof of Corollary 6.6 independent of Theorem 6.5.

Proof of Theorem 6.5. We note that $\{L_1, L_2, \ldots\}$ is unbounded if and only if for any strictly monotone L the sequence $\{L, L_1, L_2, \ldots\}$ is unbounded.

The next step is to prove that $\{L_1, L_2, \ldots\}$ is unbounded if and only if for every strictly monotone L

$$(6.7) \qquad\qquad \sigma_\mathscr{C}\left(L^n L\right) \rightarrow +\infty \quad as \quad n \rightarrow +\infty.$$

Indeed the last property implies immediately that $\{L, L_1, L_2, \ldots\}$ is unbounded and so, if it holds for all strictly monotone L, then also $\{L_1, L_2, \ldots\}$ is unbounded. To prove the converse we will need the following fact from point set topology:

Lemma. *Let $f_1 \leq f_2 \leq \ldots$, be a nondecreasing sequence of real-valued continuous functions defined on a compact Hausdorff space X. If for every $x \in X$*

$$\lim_{n \rightarrow +\infty} f_n(x) = +\infty,$$

then

$$\lim_{n \to +\infty} \inf_{x \in X} f_n(x) = +\infty.$$

If $\{L_1, L_2, \ldots\}$ is unbounded and L is strictly monotone then we have

$$\sigma_{\mathscr{C}}\left(L^n L\right) = \inf_{w \in int\mathscr{C}} \frac{\sqrt{\mathfrak{Q}(L^n L w)}}{\sqrt{\mathfrak{Q}(w)}} \geq \inf_{0 \neq w \in \mathscr{C}} \frac{\sqrt{\mathfrak{Q}(L^n L w)}}{\sqrt{\mathfrak{Q}(Lw)}} \sigma_{\mathscr{C}}(L).$$

Applying the Lemma to

$$f_n(w) = \frac{\sqrt{\mathfrak{Q}(L^n L w)}}{\sqrt{\mathfrak{Q}(Lw)}}, \quad n = 1, 2, \ldots,$$

which can be considered as a sequence of functions on the compact space of rays in $\mathscr{C}$ we obtain (6.7).

Now we will be proving that (6.7) is equivalent to

$$\bigcap_{n=1}^{+\infty} L^{-1} L_1^{-1} L_2^{-1} \ldots L_n^{-1} \mathscr{C}' = \text{one Lagrangian subspace},$$

where $\mathscr{C}' = \mathscr{C}(V_2, V_1)$ is the complementary sector. The sectors

$$\mathscr{C}'_n = L^{-1} L_1^{-1} L_2^{-1} \ldots L_n^{-1} \mathscr{C}' = L^{-1} \left(L^n\right)^{-1} \mathscr{C}' = \mathscr{C}(L^{-1} \left(L^n\right)^{-1} V_2, L^{-1} \left(L^n\right)^{-1} V_1)$$

$n = 1, 2, \ldots$, form a nested sequence. We consider the space $Lag(\mathscr{C}')$ of all Lagrangian subspaces contained strictly in $\mathscr{C}'$ with the metric defined in Section 5. The sequence of subsets $\widehat{Lag}(\mathscr{C}'_n) \subset Lag(\mathscr{C}'), n = 1, 2, \ldots, \ldots$, is a nested sequence of compact subsets. Hence its intersection contains exactly one point ($=$ Lagrangian subspace) if and only if their diameters converge to zero. By Corollary 5.3 the diameter of $\widehat{Lag}(\mathscr{C}'_n)$ is equal to the distance of the Lagrangian subspaces $L^{-1} \left(L^n\right)^{-1} V_2$ and $L^{-1} \left(L^n\right)^{-1} V_1$. By Proposition 6.3 this distance is equal to

$$\ln \frac{s_n^2 + 1}{s_n^2 - 1},$$

where $s_n = \sigma_{\mathscr{C}'}\left(L^{-1}(L^n)^{-1}\right)$. But, by Proposition 6.2,

$$\sigma_{\mathscr{C}'}\left(L^{-1}(L^n)^{-1}\right) = \sigma_{\mathscr{C}}(L^n L).$$

This shows that indeed the set

$$\bigcap_{n=1}^{+\infty} \widehat{Lag}(C'_n)$$

contains exactly one point if and only if (6.7) holds.   □

We will use the following characterization of strict unboundedness.

Theorem 6.8. *Let $\{L_i\}_{i=1}^{+\infty}$ be a sequence of linear symplectic monotone maps. The following are equivalent.*

(1) *The sequence $\{L_i\}_{i=1}^{+\infty}$ is strictly unbounded,*

(2) $\displaystyle\inf_{0\neq w\in\mathscr{C}}\frac{\sqrt{\mathfrak{Q}(L^n w)}}{\|w\|} \to +\infty$ *as* $n\to +\infty$,

(3) $\sigma(L^n) \to +\infty$ *as* $n\to +\infty$,

(4) *the sequence $\{L_i\}_{i=1}^{+\infty}$ is unbounded and L^{n_0} is strictly monotone for some* $n_0 \geq 1$.

Proof. The Lemma used in the proof of Theorem 6.5 applied to the sequence of functions

$$f_n(w) = \frac{\sqrt{\mathfrak{Q}(L^n w)}}{\|w\|}, \quad n = 1, 2, \ldots,$$

shows that (1) $\Rightarrow$ (2). Further (2) $\Rightarrow$ (3) because

$$\sigma(L^n) = \inf_{w\in int\mathscr{C}}\frac{\sqrt{\mathfrak{Q}(L^n w)}}{\sqrt{\mathfrak{Q}(w)}} \geq \inf_{0\neq w\in\mathscr{C}}\frac{\sqrt{\mathfrak{Q}(L^n w)}}{\|w\|}\inf_{w\in int\mathscr{C}}\frac{\|w\|}{\sqrt{\mathfrak{Q}(w)}}.$$

The implication (3) $\Rightarrow$ (4) is obvious ($\sigma(L^n) > 1$ if and only if L^n is strictly monotone, cf. Proposition 6.1). Finally, let the sequence $\{L_i\}_{i=1}^{+\infty}$ be unbounded, and let L^{n_0} be strictly monotone. By Corollary 6.6, the sequence $\{L_{n_0+1}, L_{n_0+2}, \ldots\}$ is also unbounded. It follows that $\{L_i\}_{i=1}^{+\infty}$ is strictly unbounded. $\qquad\square$

The following example plays a role in the study of special Hamiltonian systems.

Example. Let

$$L_n = \begin{pmatrix} A_n & 0 \\ 0 & A_n^{*-1} \end{pmatrix}\begin{pmatrix} I & 0 \\ P_n & I \end{pmatrix}\begin{pmatrix} I & R_n \\ 0 & I \end{pmatrix},$$

$n = 1, 2, \ldots$, be a sequence of monotone symplectic matrices with nonexpanding A_n, i.e., $\|A_n \xi\| \leq \|\xi\|$ for all ξ. We assume further that the symmetric matrices R_n satisfy

$$\tau'_n I \geq R_n \geq \tau_n I \quad\text{and}\quad \frac{\tau'_n}{\tau_n} \leq C,$$

for some positive constants C and τ_n, τ'_n, $n = 1, 2, \ldots$. We do not make any assumptions about P_n (beyond $P_n \geq 0$ which is forced by the monotonicity of L_n). Note that if a symmetric matrix R satisfies $\tau I \leq R \leq \tau' I$ then $\tau\|\eta\| \leq \|R\eta\| \leq \tau'\|\eta\|$. Indeed

$$\langle R\eta, R\eta \rangle = \frac{\left\langle RR^{\frac{1}{2}}\eta, R^{\frac{1}{2}}\eta \right\rangle}{\left\langle R^{\frac{1}{2}}\eta, R^{\frac{1}{2}}\eta \right\rangle}\langle R\eta, \eta \rangle,$$

which yields the estimate.

Proposition 6.9. *If $\sum_{n=1}^{+\infty}\tau_n = +\infty$ then the sequence $\{L_1, L_2, \ldots\}$ is unbounded.*

Proof. Let $w_1 = (\xi_1, \eta_1) \in int\mathscr{C}$ and $w_{n+1} = (\xi_{n+1}, \eta_{n+1}) = L_n w_n$, $n = 1, 2, \ldots$. Our goal is to show that

$$q_n = \mathfrak{Q}(w_n) \to +\infty \quad\text{as}\quad n\to +\infty.$$

We have $\xi_{n+1} = A_n (\xi_n + R_n \eta_n)$, so that

(6.10) $\qquad \|\xi_{n+1}\| \leq \|\xi_n\| + \|R_n \eta_n\| \leq \|\xi_n\| + \tau_n' \|\eta_n\| \leq \|\xi_1\| + \sum_{i=1}^{n} \tau_i' \|\eta_i\|.$

At the same time, $q_n = \langle \xi_n, \eta_n \rangle \leq \|\xi_n\| \|\eta_n\|$, so that

(6.11) $\qquad\qquad\qquad\qquad \|\eta_n\| \geq \dfrac{q_n}{\|\xi_n\|},$

and hence (see also the proof of Theorem 4.4)

$$q_{n+1} \geq q_n + \langle R_n \eta_n, \eta_n \rangle \geq q_n + \tau_n \|\eta_n\|^2 \geq q_n + \tau_n \|\eta_n\| \frac{q_n}{\|\xi_n\|}.$$

Using (6.10), we obtain from the last inequality

(6.12) $\qquad \dfrac{q_{n+1}}{q_n} \geq 1 + \dfrac{\tau_n \|\eta_n\|}{\|\xi_1\| + \sum_{i=1}^{n-1} \tau_i' \|\eta_i\|} \geq 1 + \dfrac{1}{C} \dfrac{\tau_n' \|\eta_n\|}{\|\xi_1\| + \sum_{i=1}^{n-1} \tau_i' \|\eta_i\|}.$

If $\sum_{i=1}^{+\infty} \tau_i' \|\eta_i\| < +\infty$ then by (6.10) the sequence $\|\xi_n\|$ is bounded from above, and hence by (6.11) the sequence $\|\eta_n\|$ is bounded away from zero which is a contradiction (in view of $\sum_{i=1}^{+\infty} \tau_i' = +\infty$).

Hence

$$\sum_{i=1}^{+\infty} \tau_i' \|\eta_i\| = +\infty.$$

Now the claim follows from (6.12) and the following

Lemma 6.13. *For a sequence of positive numbers* $a_0, a_1, \ldots,$ *if*

$$\sum_{n=1}^{+\infty} a_n = +\infty \quad then \quad \sum_{n=1}^{+\infty} \frac{a_n}{\sum_{i=0}^{n-1} a_i} = +\infty.$$

Proof of the Lemma. We have for $1 \leq k \leq l$

$$\sum_{n=k}^{l} \frac{a_n}{\sum_{i=0}^{n-1} a_i} \geq \frac{\sum_{n=k}^{l} a_n}{\sum_{n=0}^{l} a_n} \to 1 \quad as \quad l \to +\infty.$$

$\square$ $\square$

7. Properties of the System and the Formulation of the Results

In this section we describe in detail the class of systems to which the present paper applies. We divide the conditions that the systems must satisfy into several groups. The multitude of conditions is justified by the fact that we want to include discontinuous systems (there is only one way to be continuous but many ways to be discontinuous !). In the case of a symplectomorphism of a compact symplectic manifold most of these conditions are vacuous. Because of that we will single out this case and we will refer to it as the smooth case. The bulk of our effort is devoted to the discontinuous case.

A. The Phase Space

In the smooth case the phase space $\mathcal{M}$ is a smooth compact symplectic manifold.

In the discontinuous case it is a disjoint union of nice subsets of the linear symplectic space. More precisely, let us consider the standard linear symplectic space $\mathcal{W} = \mathbb{R}^d \times \mathbb{R}^d$ equipped with a Riemannian metric uniformly equivalent to the standard Euclidean scalar product. For simplicity we assume that the Riemannian volume element is equal to the symplectic volume element (measure) μ.

By a submanifold of $\mathcal{W}$ we mean an embedded submanifold of $\mathcal{W}$. Further, we define a piece of a submanifold $\mathcal{N}$ to be a compact subset of $\mathcal{N}$ which is the closure of its interior (in the relative topology of the submanifold $\mathcal{N}$). A piece X of a submanifold has a well defined boundary which we will denote by ∂X (it is the set of boundary points with respect to the relative topology of the submanifold). Notice that at every point of a piece of a submanifold, including a boundary point, we have a well defined tangent subspace.

A submanifold carries the measure defined by the Riemannian volume element. The measure of the boundary of a piece of a submanifold is not necessarily zero.

The phase space is made up of pieces of $\mathcal{W}$ which have regular boundaries in the sense of the following definition.

Definition 7.1. *A compact subset $X \subset \mathcal{W}$ is called regular if it is a finite union of pieces $X_i, i = 1, \ldots, k$, of $2d - 1$-dimensional submanifolds*

$$X = X_1 \cup \cdots \cup X_k.$$

The pieces overlap at most on their boundaries, i.e.,

$$X_i \cap X_j \subset \partial X_i \cup \partial X_j, \ i, j = 1, \ldots k;$$

and the boundary ∂X_i of each piece X_i, $i = 1, \ldots k$, is a finite union of compact subsets of $2d - 2$-dimensional submanifolds.

To picture such sets one can think of the boundary of a $2d$-dimensional cube. The faces are pieces of $2d - 1$-dimensional submanifolds and they clearly overlap only at their boundaries. The boundary of each face is a union of pieces of $2d - 2$ dimensional submanifolds (actually it is a union of $2d - 2$ dimensional cubes). Let us stress that in the definition of a regular set we do not impose any requirements on the $2d - 2$ dimensional subsets in the boundary. Due to the generality of the definition one cannot even claim that the union of two regular sets is regular.

As a consequence of Definition 7.1 the natural measures on the pieces $X_i, i = 1, \ldots, k$, of any regular subset X can be concocted to give a well defined measure μ_X on X (the $2d - 1$ dimensional Riemannian volume). It is so because the boundaries of the pieces being themselves finite unions of subsets of submanifolds of lower dimension have zero measure. Hence if we put

$$\partial X = \bigcup_{i=1}^{k} \partial X_i,$$

then

(7.2) $\mu_X(\partial X) = 0$.

Moreover, by the regularity of the measure μ_X, it follows from (7.2) that, if we denote by $(\partial X)^\delta$ the δ-neighborhood of ∂X in X, then

(7.3) $$\lim_{\delta \to 0} \mu_X\left((\partial X)^\delta\right) = 0.$$

Further we have the following Proposition.

Proposition 7.4. *For a subset Y of $X \subset \mathcal{W}$ let the δ-neighborhood of Y in $\mathcal{W}$ be denoted by Y^δ, i.e.,*

$$Y^\delta = \{x \in \mathcal{W} \mid d(x, Y) \le \delta\}.$$

If X is a regular $(2d-1$-dimensional) subset of $\mathcal{W}$ and $Y \subset X$ is **closed** *then*

$$\lim_{\delta \to 0} \frac{\mu(Y^\delta)}{2\delta} = \mu_X(Y).$$

We will use in the following only the weaker property

(7.5) $$\limsup_{\delta \to 0} \frac{\mu(Y^\delta)}{\delta} \le const \, \mu_X(Y).$$

We leave the proof of Proposition (7.4) or of the weaker property (7.5) to the reader.

Definition 7.6. *A compact subset $\mathcal{M} \subset \mathcal{W}$ is called a symplectic box if the boundary $\partial \mathcal{M}$ of $\mathcal{M}$ is a regular subset of $\mathcal{W}$ and the interior $int \mathcal{M}$ of $\mathcal{M}$ is connected and dense in $\mathcal{M}$.*

We can now formulate the requirements on the phase space of a discontinuous system. *The phase space of our system is a finite disjoint union of symplectic boxes.*

To simplify notation we assume that the phase space is just one symplectic box $\mathcal{M}$. It will be quite obvious how to generalize the subsequent formulations to the case of several symplectic boxes.

B. The Map T (the Dynamical System)

In the smooth case the map T is a symplectomorphism $T : \mathcal{M} \to \mathcal{M}$.

In the discontinuous case we assume that the symplectic box $\mathcal{M}$ is partitioned in two ways into unions of equal number of symplectic boxes

$$\mathcal{M} = \mathcal{M}_1^+ \cup \cdots \cup \mathcal{M}_m^+ = \mathcal{M}_1^- \cup \cdots \cup \mathcal{M}_m^-.$$

Two boxes of one partition can overlap at most on their boundaries, i.e.,

$$\mathcal{M}_i^\pm \cap \mathcal{M}_j^\pm \subset \partial \mathcal{M}_i^\pm \cap \partial \mathcal{M}_j^\pm, \quad i, j = 1, \ldots, m.$$

The map T is defined separately on each of the symplectic boxes $\mathcal{M}_i^+$, $i = 1, \ldots, m$. It is a symplectomorphism of the interior of each $\mathcal{M}_i^+$ onto the interior $\mathcal{M}_i^-$, $i = 1, \ldots, m$ and a homomorphism of $\mathcal{M}_i^+$ onto $\mathcal{M}_i^-$, $i = 1, \ldots, m$. We assume that the derivative DT is well behaved near the boundaries of the symplectic boxes. Namely, we assume that it satisfies the Katok-Strelcyn conditions so that we can apply their results [KS] on the existence of (un)stable manifolds and the property of absolute continuity.

We will say that T is a (discontinuous) symplectic map of $\mathcal{M}$. Formally T is not well defined on the set of points which belong to the boundaries of several plus-boxes: it has several values. We adopt the convention that the image of a subset of $\mathcal{M}$ under T contains all such values.

Let us introduce the singularity sets $\mathcal{S}^+$ and $\mathcal{S}^-$.

$$\mathcal{S}^\pm = \{p \in \mathcal{M} \mid p \text{ belongs to at least two of the boxes } \mathcal{M}_i^\pm, i = 1, \ldots, m\}.$$

The plus-singularity set $\mathcal{S}^+$ is a closed subset and T is continuous on its complement. Similarly T^{-1} is continuous on the complement of $\mathcal{S}^-$.

We have that $\mathcal{S}^+ \cup \partial\mathcal{M}$ is the union of all the boundaries of the plus-boxes and $\mathcal{S}^- \cup \partial\mathcal{M}$ is the union of all the boundaries of the minus-boxes, i.e.,

$$\mathcal{S}^\pm \cup \partial\mathcal{M} = \bigcup_{i=1}^{m} \partial\mathcal{M}_i^\pm.$$

Note that most of the points in the boundary $\partial\mathcal{M}$ of $\mathcal{M}$ do not belong to $\mathcal{S}^-$ or $\mathcal{S}^+$.

We assume that the singularity sets $\mathcal{S}^\pm$ and the union of boundaries $\bigcup_{i=1}^{m} \partial\mathcal{M}_i^\pm$ are regular sets.

An important role in our discussion will be played by the singularity sets of the higher iterates of T. We define for $n \geq 1$

$$\mathcal{S}_n^+ = \mathcal{S}^+ \cup T^{-1}\mathcal{S}^+ \cup \cdots \cup T^{-n+1}\mathcal{S}^+.$$

and

$$\mathcal{S}_n^- = \mathcal{S}^- \cup T\mathcal{S}^- \cup \cdots \cup T^{n-1}\mathcal{S}^-.$$

We have that T^n is continuous on the complement of $\mathcal{S}_n^+$ and T^{-n} is continuous on the complement of $\mathcal{S}_n^-$.

Regularity of singularity sets. *We assume that for every $n \geq 1$ the singularity sets $\mathcal{S}_n^+$ and $\mathcal{S}_n^-$ are regular.*

We will formulate, in Lemma 7.7, an abstract condition on the first power of T alone that guarantees the regularity of the singularity sets but it requires that the map is a diffeomorphism on every symplectic box up to and including its boundary i.e., it can be extended to a diffeomorphism of an open neighborhood of $\mathcal{M}_i^+$ onto an open neighborhood of $\mathcal{M}_i^-$, $i = 1, \ldots, m$.

Hence it is very appealing to restrict the discussion to such maps. Unfortunately, such a restriction would leave out important examples: billiard systems where the derivative may blow up at the boundary. The conditions in the work of Katok and Strelcyn [KS] were tailored for such systems.

Nevertheless the reader is invited to be generous with the restrictions on the regularity of T, this will make it easier to follow the main line of the argument.

C. Monotonicity of T

In the smooth case we assume that two continuous bundles of transversal Lagrangian subspaces are chosen in an open subset $U \subset \mathcal{M}$ (U is not necessarily dense). We denote them by $\{V_1(p)\}_{p \in \mathcal{U}}$ and $\{V_2(p)\}_{p \in \mathcal{U}}$ respectively.

In the discontinuous case we assume that two continuous bundles of transversal Lagrangian subspaces are chosen in the interior of the symplectic box $\mathcal{M}$. Their limits (if they exist at all) at the boundary $\partial \mathcal{M}$ are allowed to be nontransversal (to have nonzero intersection).

We consider the bundle of sectors (see Definition 4.1) defined by these Lagrangian subspaces

$$\mathcal{C}(p) = \mathcal{C}(V_1(p), V_2(p)).$$

Let

$$\mathcal{C}'(p) = \mathcal{C}(V_2(p), V_1(p))$$

be the complementary sector.

We require that the derivative of the map and its iterates, where defined, is monotone, if only monotonicity is well defined (cf. Definition 4.3).

More precisely, in the smooth case we require that, if $p \in U$ and $T^k p \in U$ for $k \geq 1$, then

$$D_p T^k \mathcal{C}(p) \subset \mathcal{C}(T^k p).$$

In the discontinuous case we assume that

$$D_p T \mathcal{C}(p) \subset \mathcal{C}(Tp)$$

for points p in the interior of every symplectic box $\mathcal{M}_i^+$, $i = 1, \ldots, m$.

We call a point $p \in int\mathcal{M}$ ($p \in U$ in the smooth case) *strictly monotone in the future* if there is $n \geq 1$ such that $D_p T^n$ is defined and it is strictly monotone (in the smooth case we require, naturally, that $T^n p \in U$), i.e.,

$$D_p T^n \mathcal{C}(p) \subset int \mathcal{C}(T^n p) \cup \{0\}.$$

Similarly a point p is called *strictly monotone in the past* if there is $n \geq 1$ such that $D_p T^{-n}$ is strictly monotone with respect to the complementary sectors, i.e.,

$$D_p T^{-n} \mathcal{C}'(p) \subset int \mathcal{C}'(T^{-n} p) \cup \{0\}.$$

It is clear that if p is strictly monotone in the future then its preimages are also strictly monotone in the future. By Proposition 6.2 we also have that if p is strictly monotone in the future then there is $n \geq 1$ such that $T^n p$ is strictly monotone in the past.

Strict monotonicity almost everywhere. *We assume that almost all points in $\mathcal{M}$ (U in the smooth case) are strictly monotone.*

This property implies that all Lyapunov exponents are non-zero almost everywhere in $\mathcal{M}$ (in U in the smooth case). The proof of this fact is quite simple and can be found in [W1]. It will also follow easily from our Proposition 8.4. Thus by the work of Pesin [P] in the smooth case and of Katok and Strelcyn [KS] in the discontinuous case, through almost every point there are local stable and unstable manifolds of dimension d and the foliations into these manifolds are absolutely continuous.

The sectors $\mathcal{C}(p)$ contain the unstable Lagrangian subspaces (tangent to the unstable manifolds) and the complementary sectors $\mathcal{C}'(p)$ contain the stable Lagrangian subspaces (tangent to the stable manifolds). The sectors can be viewed as a priori approximations to the unstable and stable subspaces. We will refer to the sectors as unstable sector and stable sector respectively.

This ends the list of required properties for the smooth case. The last three properties of our system are introduced only for the discontinuous case.

D. Alignment of Singularity Sets

For a codimension one subspace in a linear symplectic space its characteristic line is, by definition, the skeworthogonal complement (which is a one dimensional subspace).

Proper alignment of $\mathcal{S}^-$ and $\mathcal{S}^+$. *We assume that the tangent subspace of $\mathcal{S}^-$ at any $p \in \mathcal{S}^-$ has the characteristic line contained strictly in the sector $\mathcal{C}(p)$ and that the tangent subspace of $\mathcal{S}^+$ at any $p \in \mathcal{S}^+$ has the characteristic line contained strictly in the complementary sector $\mathcal{C}'(p)$. We say that the singularity sets $\mathcal{S}^-$ and $\mathcal{S}^+$ are properly aligned.*

Let us note that if a point in $\mathcal{S}^\pm$ belongs to several pieces of submanifolds then we require that the tangent subspaces to *all* of these pieces have characteristic lines in the interior of the sector.

It will be clear from the way in which the proper alignment of singularity sets is used in Section 12 that it is sufficient to assume that there is N such that $T^N \mathcal{S}^-$ and $T^{-N} \mathcal{S}^+$ are properly aligned. We will show, in section 14, that for the system of falling balls even this weaker property fails. Hence the study of ergodicity of this system would require some further relaxation of this property.

Let us note that it is helpful in establishing the regularity of singularity sets $\mathcal{S}_n^\pm$ if the boundaries of $\mathcal{M}$ have tangent subspaces with characteristic lines contained in the boundary of the sector. It is so in some examples. More precisely we have the following lemma.

Lemma 7.7. *If the map T is a diffeomorphism up to and including the boundaries of the symplectic boxes $\mathcal{M}_1^+, \ldots, \mathcal{M}_m^+$, satisfies properties C, D and the boundary $\partial \mathcal{M}$ of $\mathcal{M}$ has all the tangent subspaces with characteristic lines contained in the boundary of the sectors then the sets $\mathcal{S}_n^\pm, n \geq 1$, are regular (i.e. the property **B** is automatically verified).*

Proof. Let us recall that, by assumption, $\mathcal{S}^-$ and $\bigcup_{i=1}^m \partial \mathcal{M}_i^+$ are regular subsets. Further the intersection of any properly aligned regular subset X (the characteristic lines of its tangent subspaces are contained strictly in the unstable sector $\mathcal{C}$) with any of the symplectic boxes $\mathcal{M}_1^+, \ldots, \mathcal{M}_m^+$ is a regular subset. Indeed let $X_1, \ldots, X_p$ be the pieces of $2d - 1$ dimensional manifolds which make up X ($X = \bigcup_{i=1}^p X_i$) and $Y_1, \ldots, Y_q$ be the pieces of $2d - 1$ dimensional manifolds which make up the boundary of say $\mathcal{M}_1^+$ ($\partial \mathcal{M}_1^+ = \bigcup_{j=1}^q Y_j$). By the proper alignment of the pieces we can assume that any X_i and Y_j are pieces of transversal submanifolds. Hence, the intersection of the submanifolds is a submanifold of dimension $2d - 2$, and therefore $X_i \cap Y_j$ is a compact subset of the $2d - 2$-dimensional manifold. It follows that the intersection of X_i with $\mathcal{M}_1^+$ is a finite union of pieces of the $2d - 1$ dimensional manifold and also a regular subset. The same can be repeated for the other symplectic boxes $\mathcal{M}_2^+, \ldots, \mathcal{M}_m^+$.

Moreover we have that any $(X_i \cap \mathcal{M}_1^+) \cup \partial \mathcal{M}_1^+$, $i = 1, \ldots, p$, is a regular subset and further $(X \cap \mathcal{M}_1^+) \cup \partial \mathcal{M}_1^+$ is a regular subset. It follows that $T\left((X \cap \mathcal{M}_1^+) \cup \partial \mathcal{M}_1^+\right) = (TX \cap \mathcal{M}_1^-) \cup \partial \mathcal{M}_1^-$ is a regular subset and, after repeating the argument for the other symplectic boxes, we get that for any regular and properly aligned subset X the sets $TX \cup \bigcup_{i=1}^m \partial \mathcal{M}_i^-$ and $TX \cup \mathcal{S}^-$ are regular. Moreover $TX \cup \mathcal{S}^-$ is properly aligned.

Now the proof can clearly be completed by induction since

$$\mathcal{S}_{n+1}^- = T\mathcal{S}_n^- \cup \mathcal{S}^-.$$

The argument for $\mathcal{S}^+$ is completely analogous.   □

The last two properties are rather technical. They are used only in Section 13 in the proof of the 'tail bound'. It remains an open question if one can do without them.

E. Noncontraction Property

There is a constant a, $0 < a \leq 1$, such that, for any $n \geq 1$ and every $p \in \mathcal{M} \setminus \mathcal{S}_n^+$,

$$\|D_p T^n v\| \geq a\|v\|,$$

for every vector v in the sector $\mathscr{C}(p)$.

This condition is hard (or impossible) to check unless we know of a special norm (or any other homogeneous function on the tangent vectors in the sector uniformly equivalent to the Riemannian metric) which does not decrease under DT. In the theory of Anosov systems a metric with such a property can be constructed, and it is called a Lyapunov metric.

In the case of semi-dispersing billiards the noncontraction property does not hold in general. Chernov and Sinai [CS] were using a semi-norm (the configuration norm) which does not decrease on vectors in the unstable sector, but it is not equivalent to the Riemannian metric. The reason why it suffices is quite intricate. It turns out that the (d-dimensional) subspace of null vectors of the semi-norm (on which the seminorm vanishes) is contained in the boundary of the sectors and in the subspaces tangent to $\partial \mathcal{M}$. Moreover, all of $\mathcal{S}^- (\mathcal{S}^+)$ is in the image (preimage) of $\partial \mathcal{M}$. Because of these properties, Chernov and Sinai were able to establish that the measure of a δ-neighborhood of the singularity set in this semi-norm is $\mathcal{O}(\delta)$. It is possible to formulate such properties axiomatically, but we found it too cumbersome.

Finally, let us note that the noncontraction property is automatically satisfied, if T is differentiable up to and including the boundaries of symplectic boxes, the bundle of sectors is defined in all of $\mathcal{M}$ (including the boundaries), and DT is strictly monotone everywhere. Indeed, in such a case we can measure vectors in $\mathscr{C}$ by the form $\mathfrak{Q}$ defined by the bundle of sectors $DT^{-1}\mathscr{C}$ which are uniformly larger than $\mathscr{C}$ by compactness. The form $\mathfrak{A}$ does not decrease under DT, and, by a compactness argument, it is equivalent to the Riemannian metric in the sectors $\mathscr{C}$.

F. Sinai-Chernov Ansatz

This is a property of the derivatives of the iterates of T on the singularity set itself, of T^{-1} on $\mathscr{S}^+$ and of T on $\mathscr{S}^-$. Namely, we require that, for almost every point in $\mathscr{S}^-$ with respect to the measure $\mu_{\mathscr{S}}$ ($\mu_{\mathscr{S}}$ is the $2d-1$ dimensional Riemannian volume on $\mathscr{S}^- \cup \mathscr{S}^+$), all iterates of T are differentiable and for almost every point in $\mathscr{S}^+$ all iterates of T^{-1} are differentiable. Note that the last requirement holds automatically under the assumptions of Lemma 7.7. Moreover,

we assume that for almost every point $p \in \mathscr{S}^-$ with respect to the measure $\mu_{\mathscr{S}}$, the sequence of derivatives $\{D_{T^n p}T\}_{n \geq 0}$ is strictly unbounded (cf. Definition 6.4). Analogous property must hold for $\mathscr{S}^+$ and T^{-1}.

By Theorem 6.8 the forward part of Sinai - Chernov Ansatz is equivalent to the following property. For almost every point $p \in \mathscr{S}^-$ with respect to the measure $\mu_{\mathscr{S}}$

$$\lim_{n \to +\infty} \sigma(D_p T^n) = +\infty,$$

where the coefficient σ is defined at the beginning of Section 6.

In several examples unboundedness holds for all orbits by virtue of Proposition 6.9, but strict monotonicity is hard to establish.

We have completed the formulation of the conditions. Under these conditions we will prove the following two theorems.

Main Theorem (Smooth case). *For any $n \geq 1$ and any $p \in U$ such that $T^n p \in U$ and $\sigma(D_p T^n) > 1$ (i.e., p is strictly monotone) there is a neighborhood of p which is contained in one ergodic component of T.*

It follows from this theorem that if U is connected and every point in it is strictly monotone then $\bigcup_{i=-\infty}^{+\infty} T^i U$ belongs to one ergodic component. Such a theorem was first proven by Burns and Gerber [BG] for flows in dimension 3. It was later generalized by Katok [K1] to arbitrary dimension, and recently also to a non-symplectic framework [K2]. Our proof is a byproduct of the preparatory steps in the proof of the following

Main Theorem (Discontinuous case). *For any $n \geq 1$ and for any $p \in \mathcal{M} \setminus \mathscr{S}_n^+$ such that $\sigma(D_p T^n) > 3$ there is a neighborhood of p which is contained in one ergodic component of T.*

Let us note that the conditions of the last theorem are satisfied for almost all points $p \in \mathcal{M}$. Indeed let

$$\mathcal{M}_{n,\epsilon} = \{p \in \mathcal{M} \mid \sigma(D_p T^n) > 1 + \epsilon\}.$$

Since almost all points are strictly monotone, then

$$\bigcup_{n=1}^{+\infty} \bigcup_{\epsilon>0} \mathcal{M}_{n,\epsilon}$$

has full measure. By the Poincare Recurrence Theorem and the super-multiplicativity of the coefficient σ, we conclude that

$$\bigcup_{n=1}^{+\infty} \mathcal{M}_{n,2}$$

has also full measure.

Hence the theorem implies, in particular, that all ergodic components are essentially open. The theorem allows also to go further since we assume that only finitely many iterates of T are differentiable at p, so that we can apply it to orbits that end up on the singularity sets both in the future and in the past (e.g. $p \in \mathcal{S}^-$ and $T^n p \in \mathcal{S}^+$). We need though a specific amount of hyperbolicity on this finite orbit ($\sigma(D_p T^n) > 3$); note that in the smooth case any amount of hyperbolicity ($\sigma(D_p T^n) > 1$) is sufficient.

This theorem gives a fairly explicit description of points which can lie in the boundary of an ergodic component. By checking that there are only few such points (e.g. that they form a set of codimension 2) one can conclude that a given system is ergodic.

In all the examples that we know, any point with an infinite orbit (in the future or in the past) has the unbounded sequence of derivatives (in the sense of Definition 6.4). In such a case, it follows from Theorem 6.8 that for any strictly monotone point with the infinite orbit in the future the condition $\sigma(D_p T^n) > 3$ is satisfied automatically, if only n is sufficiently large.

There is no need to formulate the Main Theorem separately for a point p which has only the backward orbit ($p \in \mathcal{S}^+$). We can simply apply the theorem to T_p^{-n} (one can appreciate now the convenience of Proposition 6.2).

Let us finish this Section with an example where the role of the proper alignment of singularity sets is exposed. The well known Baker's Transformation maps the unit square as shown in Figure 5a and it is ergodic. Let us consider a variation of this construction where the square is stretched and squeezed as before but now the middle one half is left at the bottom and the quarters on the left and right are translated to the top as shown in Figure 5b. This time the map T is not ergodic. The ergodic components are separated by the vertical segment in the middle although for any point p on this segment we have that

$$\sigma(D_p T^2) = 4.$$

Of all the conditions formulated above only the proper alignment of singularity sets is violated. Namely, part of $\mathcal{S}^-$ has stable (vertical) direction (all of $\mathcal{S}^+$ has stable direction, which is fine), see Figure 6 where $\mathcal{S}^\pm$ are indicated by bold lines. For the standard Baker's transformation the condition of the proper alignment is clearly satisfied.

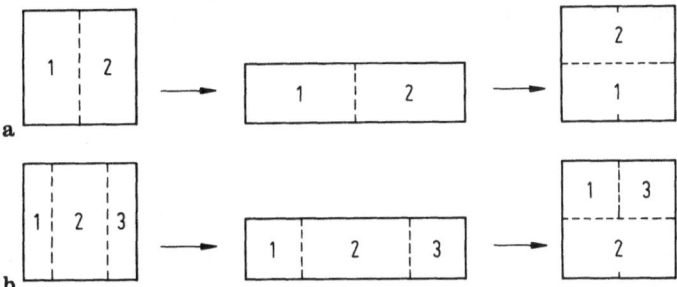

Figure 5. The Baker Map and the Modified Baker Map

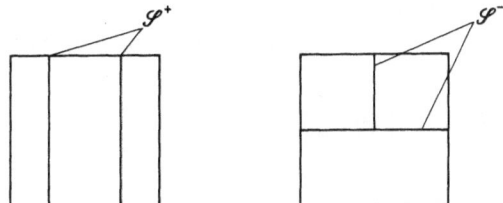

Figure 6. The discontinuity lines of the Modified Baker Map

8. Construction of the Neighborhood and the Coordinate System

We will construct a convenient coordinate system in a neighborhood of a strictly monotone point $p \in \mathcal{M}$. There are two cases: strict monotonicity in the past and strict monotonicity in the future but they are completely symmetric. Therefore, we will discuss only
one of them. Namely, we assume that there is $N \geq 1$ such that

$$
\begin{array}{lll}
& i) & T^{-N} \text{ is differentiable at } p \,:\, p \notin \mathcal{S}_N^- \cup \partial\mathcal{M}, \; (\textit{discontinuous case}) \\
(8.1) & & T^{-N}p \in U, \; (\textit{smooth case}) \\
& ii) & D_p T^{-N} \text{ is strictly monotone.}
\end{array}
$$

We will find a neighborhood $\mathcal{U}(p)$ in which there is an abundance of "long" stable
and unstable manifolds. Let us emphasize that we have assumed only that p (and its N
preimages) does not belong to $\mathcal{S}^-$ but it may very well belong to $\mathcal{S}^+$. Such a level of
generality is crucial in obtaining local ergodicity also for points in the singularity sets
$\mathcal{S}^\pm$.

Our first requirement on the neighborhood is that T^{-N} is a diffeomorphism of $\mathcal{U}(p)$
onto a neighborhood of $\bar{p} = T^{-N}p$ (and in the smooth case both neighborhoods are
contained in U).

By the Darboux theorem a symplectic manifold looks locally like a piece of the
standard linear symplectic space. Hence reducing $\mathcal{U}(p)$ further, if necessary, we can

identify it with a neighborhood $\mathcal{U}$ of the standard linear symplectic space $\mathbb{R}^d \times \mathbb{R}^d$

$$\mathcal{U} = \mathcal{U}_a = \mathcal{V}_a \times \mathcal{V}_a,$$

where

$$\mathcal{V}_a = \{x = (x^1, \ldots, x^d) \in \mathbb{R}^d \mid |x^i| < a, i = 1, \ldots, d\}.$$

(In the discontinuous case we have assumed from the very beginning that a symplectic box is a subset in $\mathbb{R}^d \times \mathbb{R}^d$). We assume that the point p becomes the zero point and the symplectic structure is the standard one. In particular all the tangent spaces in $\mathcal{U}(p)$ can be identified with $\mathbb{R}^d \times \mathbb{R}^d$. The choice of a cube for the shape of the neighborhood is important only for some of the arguments in Section 11, otherwise we want to stress that our neighborhood $\mathcal{U}$ is the cartesian product of neighborhoods $\mathcal{V}_a$ in the d-dimensional linear space and we will not use any special directions there.

Let us further introduce for any positive ρ the following sectors in the tangent space of $\mathcal{U}$.

$$\mathcal{C}_\rho = \{(\xi, \eta) \in \mathbb{R}^d \times \mathbb{R}^d \mid \|\eta\| \leq \rho\|\xi\|\}$$

and the complementary sector

$$\mathcal{C}'_\rho = \{(\xi, \eta) \in \mathbb{R}^d \times \mathbb{R}^d \mid \|\xi\| \leq \rho^{-1}\|\eta\|\}.$$

By the assumption (8.1) the sector $D_{\bar{p}}T^N\mathcal{C}(\bar{p})$ is strictly inside the sector $\mathcal{C}(p)$. We change coordinates in $\mathcal{U}$ in such a way that for some $\bar{\rho} < 1$

$$\mathcal{C}'(p) = \mathcal{C}'_{\bar{\rho}^{-1}}$$

and

$$D_{\bar{p}}T^N\mathcal{C}(\bar{p}) \subset \mathcal{C}_{\bar{\rho}}.$$

By Propositions 5.11 and 6.3 this can be done with $\bar{\rho} = (\sigma(D_{\bar{p}}T^N))^{-1}$.

We pick ρ, $\bar{\rho} < \rho < 1$. By the continuity of the sector bundle $\mathcal{C}(z)$, $z \in \mathcal{U}$, and of the derivative $D_y T^N$, $y \in T^{-N}\mathcal{U}$, if we reduce the size of $\mathcal{U}$ appropriately, we can achieve that for any $z \in \mathcal{U}$ (see Figure 7)

(8.2) $$\mathcal{C}'(z) \subset \mathcal{C}'_{\rho^{-1}}$$

and for any $y \in T^{-N}\mathcal{U}$

(8.3) $$D_y T^N\mathcal{C}(y) \subset \mathcal{C}_\rho.$$

The properties (8.2) and (8.3) can be viewed as a uniform separation of sectors $D_y T^N\mathcal{C}(y)$ and $\mathcal{C}(T^N y)$ over $\mathcal{U}$ ($y \in T^{-N}\mathcal{U}$). This property seems to be asymmetric in time, i.e., T plays here a different role than T^{-1}. Nevertheless we can obtain from (8.2) and (8.3) the following fundamental Proposition which is perfectly symmetric in time.

We will say that a point $z \in \mathcal{U}$ has k spaced returns in a given time interval if there are k moments of time in this interval

$$i_1 < i_2 < \cdots < i_k$$

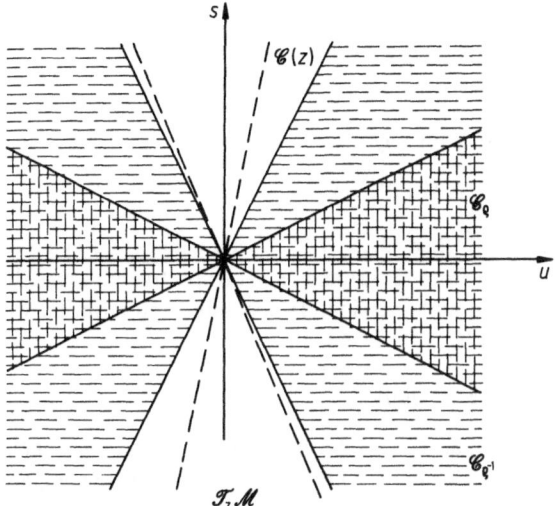

Figure 7. The sectors

at which z visits $\mathcal{U}$, i.e.,

$$T^{i_j}z \in \mathcal{U} \text{ for } j = 1, \ldots, k,$$

and the visits are spaced by at least time N, i.e.,

$$i_{j+1} - i_j \geq N \text{ for } j = 1, \ldots, k - 1.$$

Proposition 8.4. *If T^n is differentiable at $z \in \mathcal{U}$ for $n \geq N$ and $z' = T^n z \in \mathcal{U}$ then*

(8.5u) $$D_z T^n \mathscr{C}_{\rho-1} \subset \mathscr{C}_\rho$$

and

(8.5s) $$D_{z'} T^{-n} \mathscr{C}'_\rho \subset \mathscr{C}'_{\rho-1}.$$

Moreover for $(\xi', \eta') = D_z T^n (\xi, \eta)$ if $(\xi, \eta) \in \mathscr{C}_\rho$ then

(8.6u) $$\|\xi'\| \geq b\rho^{-k} \|\xi\|,$$

and if $(\xi', \eta') \in \mathscr{C}'_{\rho-1}$ then

(8.6s) $$\|\eta\| \geq b\rho^{-k} \|\eta'\|,$$

where k is the maximal number of spaced returns of z in the time interval from N to n and

$$b = \sqrt{1 - \rho^4}.$$

Proof. It follows from (8.2) that for any $x \in \mathcal{U}$

$$\mathscr{C}_\rho \subset \mathscr{C}_{\rho-1} \subset \mathscr{C}(x).$$

Hence

$$D_z T^{n-N} \mathcal{C}_{\rho^{-1}} \subset \mathcal{C}(T^{n-N}z).$$

Now (8.5u) follows from (8.3).

Let us further note that (8.3) implies that for any $x \in \mathcal{U}$

$$D_x T^{-N} \mathcal{C}'_\rho \subset \mathcal{C}'(T^{-N}x).$$

We obtain (8.5s) by applying first $D_{z'} T^{-N}$, then $D_{T^{-N}z'} T^{-n+N}$ and using (8.2) again.

The properties (8.6u) and (8.6s) follow from (8.5u) and (8.5s) respectively in exactly the same way. We will prove only the unstable version. To measure vectors in $\mathcal{C}_\rho$ we use the form $\mathcal{Q}$ associated with the sector $\mathcal{C}_{\rho^{-1}}$. It is equal to

$$\rho^{-1}\|\xi\|^2 - \rho\|\eta\|^2$$

and on every spaced return to $\mathcal{U}$ the value of this form on vectors from $\mathcal{C}_{\rho^{-1}}$ gets increased by at least the factor ρ^{-2}, cf. Propositions 5.11 and 6.3 . It remains to compare the value of this form at $(\xi, \eta) \in \mathcal{C}_\rho$ with $\|\xi\|^2$. We have

$$\rho^{-1}\|\xi\|^2 \geq \rho^{-1}\|\xi\|^2 - \rho\|\eta\|^2 \geq (\rho^{-1} - \rho^3)\|\xi\|^2$$

which immediately yields (8.6u).

□

Having achieved the symmetry with respect to the direction of time we will restrict the discussion in the next section to the case of unstable manifolds using the unstable version of Proposition 8.4. It can be then repeated for the stable manifolds with the use of the stable version.

Remark 8.7. *If p is not a periodic point then by reducing the neighborhood $\mathcal{U}$ we can guarantee that any successive visits to $\mathcal{U}$ are spaced by at least the time N. In such a case the number of spaced returns becomes simply the number of returns to $\mathcal{U}$. It is so also if $N = 1$.*

9. Unstable Manifolds in the Neighborhood $\mathcal{U}$

Let us repeat the properties of T and $\mathcal{U}$ established in the previous section which we will rely upon. Note that the original point p does not appear explicitly.

There is a positive number $\rho < 1$ such that for any $z \in \mathcal{U}$

(9.1) $\mathcal{C}_\rho \subset \mathcal{C}_{\rho^{-1}} \subset \mathcal{C}(z)$

and for any $y \in T^{-N}\mathcal{U}$

(9.2) $D_y T^N \mathcal{C}(y) \subset \mathcal{C}_\rho.$

It follows that if $z \in \mathcal{U}$ and $T^n z \in \mathcal{U}$ for $n \geq N$ then

(9.3) $D_z T^n \mathcal{C}_{\rho^{-1}} \subset \mathcal{C}_\rho.$

Moreover if

$$(\xi, \eta) \in \mathscr{C}_\rho \text{ and } (\xi', \eta') = D_z T^n(\xi, \eta)$$

then

(9.4) $$\|\xi'\| \geq b\rho^{-k}\|\xi\|$$

where k is the maximal number of spaced returns to $\mathcal{U}$ between the times N and n and $b = \sqrt{1 - \rho^4}$.

By the Pesin theory [P] in the smooth case and the Katok-Strelcyn theory [KS] in the general case, for almost all $z \in \mathcal{U}$, we have a local unstable manifold $W_{loc}^u(z)$ through z. Further, the tangent spaces of $W_{loc}^u(z) \cap \mathcal{U}$ are Lagrangian subspaces contained in $\mathscr{C}_\rho$. But the general theory does not give us a good hold on their size.

Let $\pi_i : \mathcal{V} \times \mathcal{V} \to \mathcal{V}$, $i = 1, 2$, be the projection on the first and second component respectively. We denote by $\mathscr{B}(c; r)$ the open ball with the center at c and the radius r.

Definition 9.5 . *We say that an unstable manifold in $\mathcal{U}$ of a point $z = (z_1, z_2) \in \mathcal{U}$ has size ε if it contains the graph of a smooth mapping from $\mathscr{B}(z_1; \varepsilon)$ to $\mathcal{V}$. We denote such a graph by $W_\varepsilon^u(z)$ and we will call it the unstable manifold of size ε.*

By the definition of an unstable manifold $W_\varepsilon^u(z)$ of size ε its projection onto the first component is the open ball with the center at $\pi_1 z$ and radius ε.

Lemma 9.6. *The projection onto the second component of an unstable manifold through $z = (z_1, z_2) \in \mathcal{U}$ of size ε lies in the open ball with the center at z_2 and the radius $\rho\varepsilon$, i.e.,*

$$\pi_2\left(W_\varepsilon^u(z)\right) \subset \mathscr{B}(z_2; \rho\varepsilon).$$

Proof. Let $W_\varepsilon^u(z)$ be the graph of

$$\psi : \mathscr{B}(z_1; \varepsilon) \to \mathcal{V}.$$

The subspace $\{(\xi, D\psi\xi)|\xi \in \mathbb{R}^d\}$ is tangent to $W_\varepsilon^u(z)$ and hence is contained in $\mathscr{C}_\rho$. It follows that

$$\|D\psi\| \leq \rho.$$

By the mean value theorem if $z' = (z_1', z_2') \in W_\varepsilon^u(z)$ then

$$\|z_2' - z_2\| = \|\psi(z_1') - \psi(z_1)\| \leq \sup \|D\psi\|\|z_1' - z_1\| < \rho\varepsilon.$$

$\square$

In contrast to the model problem at the beginning where we had fairly long initial unstable leaves and then we cut them because of the discontinuity of our system, we start here with small unstable manifolds and "grow" them until they are large or until they hit the singularity, whichever comes first. This is done in the proof of the following Theorem.

Theorem 9.7. *For any $\delta > 0$ almost every point z in $\mathcal{U}_\delta^1$,*

$$\mathcal{U}_\delta^1 = \mathcal{U}_{a_1(\delta)}$$

where $a_1(\delta) = a - b^{-1}\delta$ ($\mathfrak{U}_a$ is defined in §8 and $b = \sqrt{1 - \rho^4}$), either has an unstable manifold of size δ or it has an unstable manifold of size $\delta' < \delta$ such that the closure of $W^u_{\delta'}(z)$ intersects $\bigcup_{j>N} T^j \mathscr{S}^-$.

Proof. Let $\mathscr{A}(\varepsilon) \subset \mathfrak{U}^1_\delta$ be the set of points which have unstable manifolds of size ε. By the Katok-Strelcyn theory almost all points in $\mathfrak{U}^1_\delta$ belong to $\bigcup_{\varepsilon>0} \mathscr{A}(\varepsilon)$. Let us fix $\mathscr{A}(\varepsilon)$ of positive measure and let k be the smallest natural number such that

$$b\rho^{-k}\varepsilon \geq \delta.$$

Almost all points in $\mathscr{A}(\varepsilon)$ have k spaced returns to $\mathscr{A}(\varepsilon)$ in the past. Let z be such a point and let

$$-N \geq -i_1 > \cdots > -i_k = -n$$

be the k times of spaced returns of this point, i.e.,

$$T^{-i_j}z \in \mathscr{A}(\varepsilon), \quad j = 1, \ldots, k.$$

The geometric idea for growing unstable manifolds is to take the unstable manifold of size ε through the point $T^{-n}z$ and map it forward under T^n. The expansion property (9.4) guarantees then that the image contains the unstable manifold of size δ. There are two complications in this argument. First it may happen that T^n is not continuous on the unstable manifold $W^u_\varepsilon(T^{-n}z)$, that is

$$W^u_\varepsilon(T^{-n}z) \cap \mathscr{S}^+_n \neq \emptyset.$$

The other problem occurs when parts of the images of the unstable manifold are outside of $\mathfrak{U}$ where the expansion property (9.4) may fail.

To present clearly the core of the argument, we ignore for the time being these two difficulties and assume that T^n is differentiable on $W^u_\varepsilon(T^{-n}z)$ and that

$$T^{n-i_j}W^u_\varepsilon(T^{-n}z) \subset \mathfrak{U}, \quad j = 0, \ldots, k,$$

here we set $i_0 = 0$. We can prove then that z has an unstable manifold of size δ. Indeed let $W^u_\varepsilon(T^{-n}z)$ be the graph of

$$\psi : \mathscr{B}(\pi_1(T^{-n}z); \varepsilon) \to \mathscr{V}$$

and let us consider the map

$$\varphi : \mathscr{B}(\pi_1(T^{-n}z); \varepsilon) \to \mathscr{V}$$

defined by $\varphi(x) = \pi_1(T^n(x, \psi x))$. By (9.4) this map is an expanding map with the coefficient of expansion not less than $b\rho^{-k}$, i.e.,

$$\|D\varphi\xi\| \geq b\rho^{-k}\|\xi\|.$$

Hence the image of $\mathscr{B}(\pi_1(T^{-n}z); \varepsilon)$ by φ contains the ball $\mathscr{B}(\pi_1 z; \delta)$. Additional complication is caused by the fact that φ is not necessarily one-to-one. But since φ is a local diffeomorphism, we can define φ^{-1} on $\mathscr{B}(\pi_1 z; \delta)$ as the branch of the inverse for which

$\varphi^{-1}\pi_1 z = \pi_1(T^{-n}z)$. Therefore, $T^n W_\varepsilon^u(T^{-n}z)$ contains the graph of the map

$$\pi_2 \circ T^n \circ (id \times \psi) \circ \varphi^{-1}$$

which defines $W_\delta^u(z)$.

Let us now address the general case. We will construct the maximal subset of $W_\varepsilon^u(T^{-n}z)$ on which T^n is differentiable and its images at the return times to $\mathcal{U}$ are contained in $\mathcal{U}$. Our first step is to consider the connected component of

$$W_\varepsilon^u(T^{-n}z) \setminus \mathcal{S}_n^+$$

which contains $T^{-n}z$ and denote it by $\widetilde{W}_\varepsilon^u(T^{-n}z)$. Further, the connected component of

$$\bigcap_{j=0}^{k} T^{i_j-n}\left(T^{n-i_j}\widetilde{W}_\varepsilon^u(T^{-n}z) \cap \mathcal{U}\right)$$

which contains $T^{-n}z$ will be denoted by $\widetilde{\widetilde{W}}_\varepsilon^u(T^{-n}z)$. It is the part of the unstable manifold which has the desired properties.

Now we consider the image

$$T^n \widetilde{\widetilde{W}}_\varepsilon^u(T^{-n}z)$$

and we let δ' be the largest positive number such that $W_{\delta'}^u(z)$ is well defined and contained in $T^n \widetilde{\widetilde{W}}_\varepsilon^u(T^{-n}z)$.

If $\delta' \geq \delta$ then we are done. Let us hence assume that $\delta' < \delta$.

It follows from the maximality of δ', that the boundary of $W_{\delta'}^u(z)$ must contain a point from the boundary of $T^n \widetilde{\widetilde{W}}_\varepsilon^u(T^{-n}z)$. Let z' be such a point. If z' belongs to $\bigcup_{i \geq N}^{n-1} T^i \mathcal{S}^-$, then we are again done. If not then T^{-n} is differentiable at z', and hence $T^{-n}z'$ belongs to the boundary of $\widetilde{\widetilde{W}}_\varepsilon^u(T^{-n}z)$ and does not belong to $\mathcal{S}_n^+$. It follows now from the construction of $\widetilde{\widetilde{W}}_\varepsilon^u(T^{-n}z)$ that $T^{-n}z'$ must belong to the boundary of $W_\varepsilon^u(T^{-n}z)$ or for some $j, 0 \leq j \leq k$, $T^{-i_j}z'$ belongs to the boundary of $\mathcal{U}$.

We will obtain now a contradiction by using the expansion property (9.4). Let $W_{\delta'}^u(z)$ be the graph of

$$\chi : \mathcal{B}(\pi_1 z; \delta') \to \mathcal{V}$$

and let

$$\gamma_0 : [0, 1) \to \mathcal{B}(\pi_1 z; \delta')$$

be the segment connecting $\pi_1 z$ and $\pi_1 z'$. We consider the preimages of the curve

$$\{(\gamma_0(t), \chi\gamma_0(t)) \mid 0 \leq t < 1\} \subset W_{\delta'}^u(z),$$

and obtain $\gamma_j : [0, 1) \to \mathcal{V}, \; j = 0, \ldots, k$ by the formula

$$\gamma_j(t) = \pi_1\left(T^{-i_j}(\gamma_0(t), \chi\gamma_0(t))\right).$$

It follows from (9.4) that the length of γ_0 is not smaller than the length of γ_j times $b\rho^{-j}$. If $T^{-n}z'$ belongs to the boundary of $W_\varepsilon^u(T^{-n}z)$ then the length of γ_k is at least ε and we get the contradiction

$$\delta' \geq b\rho^{-k}\varepsilon \geq \delta.$$

Finally if $T^{-i_j}z'$ belongs to the boundary of $\mathcal{U}$ for some $j, 0 \leq j \leq k$, then γ_j which connects $\pi_1(T^{-i_j}z) \in \mathcal{U}_\delta^1$ and $\pi_1(T^{-i_j}z')$ must have the length at least $b^{-1}\delta$. We get again the contradiction

$$\delta' \geq b\rho^{-j}b^{-1}\delta \geq \delta.$$

$\square$

Definition 9.8. *We say that the unstable manifold $W_\delta^u(z)$ of size δ is cut by $T^i\mathcal{S}^-$, $i \geq 0$, if its boundary contains a point from $T^i\mathcal{S}^-$.*

By Theorem 9.7 to guarantee that at least some points (and in the case of a smooth map almost all points) have unstable manifolds of size δ we need to step away from the boundary of $\mathcal{U}$ by at least $b^{-1}\delta$. In the following we fix a sufficiently small δ_0 and restrict our discussions to $\mathcal{U}^1 = \mathcal{U}_{\delta_0}^1$. We can then claim that in $\mathcal{U}^1$ almost every point has a uniformly large unstable manifold (of size δ_0) or a smaller unstable manifold cut by some image of the singularity set $\mathcal{S}^-$.

By $\bar{\mathcal{B}}(c; r)$ we denote the closed ball with the center at c and the radius r. We define a rectangle $R(z; \delta)$ with the center at $z = (z_1, z_2)$ and the size δ as the Cartesian product of closed balls

$$R(z; \delta) = \bar{\mathcal{B}}(z_1; \frac{\delta}{2}) \times \bar{\mathcal{B}}(z_2; \frac{\delta}{2}).$$

Definition 9.10. *We say that the unstable manifold $W_{\delta'}^u(z')$ of $z' = (z_1', z_2')$ of size δ' is connecting in the rectangle $R(z; \delta)$ with the center at $z = (z_1, z_2)$ and size δ if*

$$\bar{\mathcal{B}}(z_1; \frac{\delta}{2}) \subset \mathcal{B}(z_1'; \delta')$$

and

$$\pi_2\left(W_{\delta'}^u(z') \cap R(z; \delta)\right) \subset \mathcal{B}(z_2; \frac{\delta}{2}).$$

We can say equivalently, that an unstable manifold $W_{\delta'}^u(z')$ is connecting in the rectangle $R(z; \delta)$, if the intersection of $W_{\delta'}^u(z')$ with the rectangle is the graph of a smooth mapping from the closed ball $\bar{\mathcal{B}}(\pi_1 z; \frac{\delta}{2})$ to the open ball $\mathcal{B}(\pi_2 z; \frac{\delta}{2})$. Clearly it is necessary that $\delta' > \frac{\delta}{2}$.

Definition 9.11. *For a given rectangle $R(z; \delta)$ with the center at $z = (z_1, z_2)$ and size δ we define its unstable core as the subset of those points $z' = (z_1', z_2') \in R(z; \delta)$ for which*

$$\rho\|z_1' - z_1\| + \|z_2' - z_2\| < (1 - \rho)\frac{\delta}{2}.$$

The role of an unstable core is revealed in the following Lemma.

Lemma 9.12. *If an unstable manifold $W_{\delta'}^u(z')$ of size $\delta' > \|\pi_1 z' - \pi_2 z\| + \frac{\delta}{2}$ intersects the unstable core of a rectangle $R(z; \delta)$, then it is connecting in the rectangle.*

Proof. Let $z = (z_1, z_2)$ and $z' = (z_1', z_2')$, let $W_{\delta'}^u(z')$ be the graph of $\psi : \mathcal{B}(z_1'; \delta') \to \mathcal{V}$, and let $(x_1', \psi x_1')$ be a point in the unstable core of the rectangle. By the condition on δ'

$$\bar{\mathcal{B}}(z_1; \frac{\delta}{2}) \subset \mathcal{B}(z_1'; \delta').$$

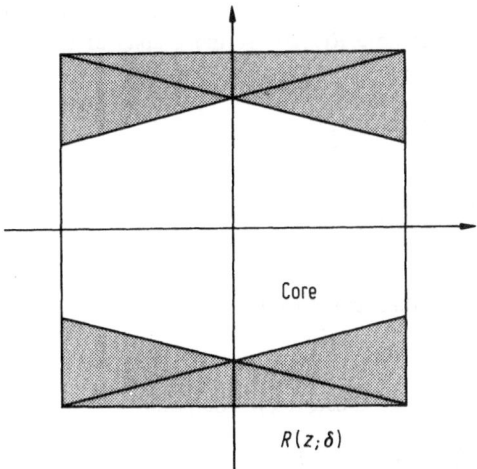

Figure 8. The core of a rectangle

We have to check only that if $x_1 \in \bar{\mathcal{B}}(z_1; \frac{\delta}{2})$ then

$$\|\psi x_1 - z_2\| < \frac{\delta}{2}.$$

We have

$$\|\psi x_1 - z_2\| \leq \|\psi x_1 - \psi x_1'\| + \|\psi x_1' - z_2\|$$
$$\leq \sup \|D\psi\| \|x_1 - x_1'\| + \|\psi x_1' - z_2\|$$
$$\leq \rho \|x_1 - z_1\| + \rho \|x_1' - z_1\| + \|\psi x_1' - z_2\|$$
$$< \rho \frac{\delta}{2} + (1 - \rho) \frac{\delta}{2} = \frac{\delta}{2}.$$

□

The point of the above lemma is that even a large unstable manifold may fail to be connecting in a rectangle, if it intersects the rectangle too close to the boundary.

10. Local Ergodicity in the Smooth Case

Contrary to the title of this section, we will consider here several propositions valid in the general case. Incidentally, they will suffice to obtain local ergodicity in the smooth case (Main Theorem).

It is important to remember that all of Section 9 can be repeated for stable manifolds. In this section we will be using both stable and unstable manifolds.

Lemma 10.1. *If an unstable manifold and a stable manifold are connecting in a rectangle, then there is a unique point of intersection of these manifolds in the rectangle, and it belongs to the interior of the rectangle.*

Proof. Let the rectangle have the center at $z = (z_1, z_2)$ and size δ. The intersections of the unstable and stable manifolds with the rectangle $R(z; \delta)$ are the graphs of smooth mappings

$$\psi^u : \mathcal{B}(z_1; \frac{\delta}{2}) \to \mathcal{B}(z_2; \frac{\delta}{2})$$

and

$$\psi^s : \mathcal{B}(z_2; \frac{\delta}{2}) \to \mathcal{B}(z_1; \frac{\delta}{2}),$$

respectively.

Since both ψ^u and ψ^s are contractions, so is their composition

$$\psi^s \psi^u : \mathcal{B}(z_1; \frac{\delta}{2}) \to \mathcal{B}(z_1; \frac{\delta}{2}).$$

Hence, the composition has a unique fixed point $x_1 \in \mathcal{B}(z_1; \frac{\delta}{2})$. The point

$$(x_1, \psi^u x_1) = (\psi^s \psi^u x_1, \psi^u x_1)$$

is the desired intersection point. $\qquad\qquad\qquad\qquad\qquad\qquad\qquad\qquad\qquad\qquad$ $\square$

For a rectangle R we denote by $W^{(u)s}(R)$ the union of the intersections with R of all (un)stable manifolds connecting in R, i.e.,

$$W^{(u)s}(R) = \bigcup \{R \cap W^{(u)s}_{\delta'}(z') \mid W^{(u)s}_{\delta'}(z') \text{ is connecting in } R\}.$$

The union of the unstable core and the stable core of a rectangle will be in the following called simply the core of the rectangle.

Proposition 10.2. *For any rectangle $R \subset \mathcal{U}^1$, if the sets $W^s(R)$ and $W^u(R)$ have positive measure, then $W^s(R) \cup W^u(R)$ belongs to one ergodic component of T.*

Proof. The proof is done by the Hopf method as described in Sections 1 and 2.

Let us fix a continuous function defined on our phase space. For all points in one (un)stable manifold the (backward) forward time averages are the same. As shown in Section 1 the forward and backward time averages have to coincide almost everywhere. Our goal is to show that they are constant almost everywhere in $W^s(R) \cup W^u(R)$.

There is a technical difficulty stemming from the fact that the foliations into stable and unstable manifolds are not smooth in general. One has to use the absolute continuity of the foliations which was proven in [KS] under the conditions which fit our scheme. (It is by far the hardest fact to prove in their theory.)

It follows from absolute continuity of the foliation into unstable manifolds that except for the union of unstable manifolds from $W^u(R)$ of total measure zero almost every point (with respect to the Remannian volume in the manifold) in an unstable manifold from $W^u(R)$ has equal forward and backward time averages. Let us take such a typical unstable manifold. Again by the property of absolute continuity the union of stable manifolds in $W^s(R)$ which intersect this chosen unstable manifold at points where the forward and backward time averages exist and are equal differs from $W^s(R)$ by a set of zero measure. Hence the time average of our function is constant almost everywhere in $W^s(R)$. Similarly, the time average of our function is constant almost everywhere in $W^u(R)$.

Finally, using the property of absolute continuity for the third time, we can claim that $W^u(R)$ and $W^s(R)$ intersect on a subset of positive measure. Hence the time average of our function is constant almost everywhere in $W^s(R) \cup W^u(R)$.

To prove that $W^s(R) \cup W^u(R)$ belongs to one ergodic component, we proceed in the same way as at the end of Section 2.   □

We are ready to prove the local ergodicity in the smooth case

Proof of Main Theorem (smooth case). All the constructions started in Section 9 apply to our point p. We will prove that a neighborhood $\mathcal{U}^2$, only slightly smaller than $\mathcal{U}^1$, belongs to one ergodic component. Indeed, according to Lemma 9.12 all the points in the (un)stable core of a rectangle $R \subset \mathcal{U}^1$ which have an (un)stable manifold of sufficiently large size belong to $W^{(u)s}(R)$. By Theorem 9.7, in the smooth case almost every point in $\mathcal{U}^1$ has both the unstable manifold and the stable manifold of size δ_0. Hence by Lemma 9.12, for any rectangle $R \subset \mathcal{U}^1$ of size $\delta < \delta_0$, the set $W^s(R)$ contains at least the stable core of R and $W^u(R)$ contains at least the unstable core of R. Clearly then the sets $W^s(R)$ and $W^u(R)$ have positive measure, and we can apply Proposition 10.2.

To end the proof, we consider a family of rectangles of size $\delta \leq \delta_0$ contained in $\mathcal{U}^1$ whose cores cover a slightly shrunk neighborhood $\mathcal{U}^2 \subset \mathcal{U}^1$. By Proposition 10.2, we can claim that each core belongs to one ergodic component. Since the cores form an open cover of the connected set $\mathcal{U}^2$, we can conclude that $\mathcal{U}^2$ belongs to one ergodic component.   □

Actually we can claim that, under the assumptions of the Main Theorem, the whole neighborhood $\mathcal{U}$ constructed in Section 8 belongs to one ergodic component. Indeed, by taking $\delta \to 0$ the above argument applies to $\mathcal{U}^2 \to \mathcal{U}^1$ so that actually $\mathcal{U}^1$ belongs to one ergodic component. If we now recall that the δ_0 in the definition of $\mathcal{U}^1$ can be chosen arbitrarily small, we can see that also the whole neighborhood $\mathcal{U}$ belongs to one ergodic component. This does not strengthen the theorem, but it demonstrates the usefulness of coverings with rectangles of size $\delta \to 0$. It will be crucial in the treatment of the discontinuous case.

Let us outline the plan for proving local ergodicity in the general case. We cover the neighborhood $\mathcal{U}^2$ with rectangles of size δ. At least for some rectangles R the sets $W^s(R)$ and $W^u(R)$ will have positive measure. We will be actually interested in the property that these sets cover certain fixed (but otherwise arbitrarily small) portion of the core of the rectangle, and we will call such rectangles connecting. One may then expect to have more and more connecting rectangles as $\delta \to 0$. The precise formulation of such a property is the subject of Sinai Theorem. The method of the proof requires that the size ρ of the sector is less than $\frac{1}{3}$. In applying Sinai Theorem it is convenient to work with more structured coverings, namely the centers of the rectangles will belong to a lattice with vertices so close that the cores of nearest neighbors will overlap almost completely. Consequently, if both nearest neighbors R_1 and R_2 are connecting then the union of $W^s(R_1) \cup W^u(R_1)$ and $W^s(R_2) \cup W^u(R_2)$ belongs to one ergodic component (as in the proof of Proposition 2.3). It will follow from Sinai Theorem that the network of connecting rectangles becomes more and more dense as $\delta \to 0$, so that we will be able to claim that one ergodic component reaches from any place in the neighborhood $\mathcal{U}^1$ to any other place. We will conclude by using the Lebesgue Density Theorem to show that $\mathcal{U}^2$ belongs to one ergodic component.

11. Local Ergodicity in the Discontinous Case

Given $\delta > 0$, we consider a shrunk neighborhood $\mathcal{U}_\delta^2$ defined by the requirement that a rectangle with the center in $\mathcal{U}_\delta^2$ and size δ lies completely in $\mathcal{U}^1$. (One can easily see that $\mathcal{U}_\delta^2 = \mathcal{U}_{a_2(\delta)}$ where $a_2(\delta) = a_1(\delta_0) - \frac{\delta}{2}$.) Let us note that $\mathcal{U}_\delta^2 \to \mathcal{U}^1$ as $\delta \to 0$.
    Let $\mathcal{N}(\delta, c)$ be the net defined by

$$\mathcal{N}(\delta, c) = \{c\delta(m, k) \in \mathcal{U}_\delta^2 \mid m, k \in \mathbb{Z}^d\}.$$

We consider the family $\mathcal{G}_\delta$ of all rectangles with the centers in $\mathcal{N}(\delta, c)$ and size δ

$$\mathcal{G}_\delta = \{R(z; \delta) \mid z \in \mathcal{N}(\delta, c)\}.$$

If c is sufficiently small the family $\mathcal{G}_\delta$ is a covering of $\mathcal{U}_\delta^2$. The parameter c will be chosen later to be very small, so that many rectangles in $\mathcal{R}_\delta$ overlap. But once c is fixed, a point may belong to at most a fixed number of rectangles, which we denote by $k(c)$ (it does not depend on δ).

Definition 11.1. *Given α, $0 < \alpha < 1$, we call a rectangle $R \in \mathcal{R}_\delta$ α-connecting in the (un)stable direction (or briefly connecting) if at least the α part of the measure of the (un)stable core of R is covered by $W^{(u)s}(R)$.*

Sinai Theorem 11.2. *If $\rho < \frac{1}{3}$ then there is $\alpha, 0 < \alpha < 1$, such that for any c*

$$\lim_{\delta \to 0} \delta^{-1} \mu \left(\bigcup \{R \in \mathcal{G}_\delta \mid R \text{ is not } \alpha\text{-connecting }\} \right) = 0,$$

i.e., the union of rectangles which are not α-connecting in either the stable or the unstable direction has measure $o(\delta)$

    It is very important for the application of this theorem that, given $\rho < \frac{1}{3}$, we get a certain α (which may be very small if ρ is close to $\frac{1}{3}$), and we are free to choose c (which determines the overlap of the rectangles in $\mathcal{G}_\delta$) as small as we may need.
    We will prove Sinai Theorem in Sections 12 and 13. In the remainder of this Section we will show how to obtain the Main Theorem in the discontinuous case from Sinai Theorem.
    We start with some auxiliary abstract facts. The first one is in Measure Theory. For any finite subset A, we will denote by $|A|$ the number of elements in A.

Lemma 11.3. *Let $\{Y_a \mid a \in A\}$ be a finite family of measurable subsets of equal measure m in the measure space (X, ν), such that no point in X belongs to more than k elements of the family. For any subfamily $\{Y_a \mid a \in A_1\}$, $A_1 \subset A$, we have*

$$\frac{m}{k}|A_1| \leq \nu \left(\bigcup_{a \in A_1} Y_a \right) \leq m|A_1|.$$

Further, if for a measurable subset $Y \subset X$ and some $\alpha, 0 < \alpha < 1$,

$$\nu(Y_a \cap Y) \geq \alpha\nu(Y_a) \quad \text{for} \quad a \in A_1,$$

then

$$\nu\left(\bigcup_{a \in A_1} Y_a \cap Y\right) \geq \frac{\alpha}{k}\nu\left(\bigcup_{a \in A_1} Y_a\right).$$

□

The second fact is in Combinatorics. Let us consider the lattice $\mathbf{Z}^d$ and its finite pieces

$$L_n = L_n(d) = \{0, 1, \ldots, n - 1\}^d \subset \mathbf{Z}^d.$$

Let $K \subset L_n$ be an arbitrary subset, which we call a configuration. We think of elements of K as occupied sites and elements of $L_n \setminus K$ as empty sites.

For a given configuration $K \subset L_n$, we consider the graph obtained by connecting by straight segments all pairs of occupied sites which are nearest neighbors. Let $gK \subset K$ be the family of sites in the largest connected component of the graph.

Proposition 11.4. *Let $K_n \subset L_n(d), n = 1, 2, \ldots,$ be a sequence of configurations. If*

$$\frac{|L_n \setminus K_n|}{|L_n|} \to 0 \quad \text{as} \quad n \to +\infty,$$

then

$$\frac{|gK_n|}{|L_n|} \to 1 \quad \text{as} \quad n \to +\infty.$$

Proof. The Proposition will follow immediately from the following combinatorial Lemma.

Lemma 11.5. *Let $K \subset L_n(d)$ be an arbitrary configuration. If*

$$\frac{|L_n \setminus K|}{n^{d-1}} < a < 1,$$

then

$$\frac{|gK|}{n^d} \geq 1 - (d - 1)a.$$

Proof. The proof is by induction on d. For $d = 1$ the statement is obvious. Suppose it is true for some d. We will establish it for $d + 1$.

We partition $L_n(d + 1)$ into subsets $L_n(d) \times \{i\}, i = 0, \ldots, n - 1$ and we call them floors. We pick the floor with the fewest number of empty sites. Clearly the number of empty sites there does not exceed an^{d-1}, so that we can apply to it the inductive assumption. We obtain in this floor a connected graph with at least $(1 - (d - 1)a)n^d$ elements.

Now we partition $L_n(d + 1)$ into subsets $\{z\} \times \{0, \ldots, n - 1\}, z \in L_n(d)$, and we call them columns. A column is called an elevator, if all of its elements are occupied. The number of elevators is at least $(1 - a)n^d$. Hence the number of elevators which intersect the connected graph in the floor considered above is at least $(1 - da)n^d$. Adding these elevators to the graph, we obtain a connected graph with at least $(1 - da)n^{d+1}$ elements, which ends the proof of the inductive step.   □ □

Proof of Main Theorem (Discontinuous case). All the constructions of Sections 8 through 10 apply with some $\rho < \frac{1}{3}$. We will be proving that the neighborhood $\mathcal{U}^1$ belongs to one ergodic component.

The Sinai Theorem gives us $\alpha < 1$ which depends only on ρ, and may have to be very small if ρ is very close to $\frac{1}{3}$. Let us consider the lattice $\mathcal{N}(\delta, c)$ and the covering $\mathcal{G}_\delta$. We choose c so small that, if the centers of two rectangles in $\mathcal{G}_\delta$ are nearest neighbors in $\mathcal{N}(\delta, c)$, then their unstable cores (and then automatically also stable cores) overlap on more than $1 - \alpha$ part of their measure. Note that such a property depends on c, but is independent of the value of δ. This choice of c has the following consequence. If two rectangles R_1 and R_2 with centers at nearest neighbors in $\mathcal{N}(\delta, c)$ are α-connecting in the unstable direction, then $W^u(R_1)$ and $W^u(R_2)$ intersect on a subset of positive measure. If in addition we also know that $W^s(R_1)$ and $W^s(R_2)$ have positive measure, then using Proposition 10.2 we obtain that

$$W^u(R_1) \cup W^u(R_2) \cup W^s(R_1) \cup W^s(R_2)$$

belongs to one ergodic component.

We consider the configuration $\mathcal{K}(\delta)$ in the lattice $\mathcal{N}(\delta, c)$ which consists of the centers of all rectangles in $\mathcal{G}_\delta$ which are α-connecting both in the stable and unstable directions. As in the discussion proceeding Proposition 11.4, we consider the graph obtained by connecting with straight segments all pairs of nearest neighbors in $\mathcal{K}(\delta)$. Let, as before, $g\mathcal{K}(\delta)$ be the collection of vertices in the largest connected component of this graph. By our construction, the set

$$Y(\delta) = \bigcup \{W^u(R(z; \delta)) \cup W^s(R(z; \delta)) \mid z \in g\mathcal{K}(\delta)\}$$

belongs to one ergodic component. This set is crucial in our proof that $\mathcal{U}^1$ belongs to one ergodic component. It may be very small in measure (if α is small), but it covers at least certain fixed α' part of the measure of each of the rectangles with centers in $g\mathcal{K}(\delta)$, i.e.,

(11.6) $\mu(R(z; \delta) \cap Y(\delta)) \geq \alpha' \mu(R(z; \delta))$

for any $z \in g\mathcal{K}(\delta)$ (α' is smaller than α, since α is only the part of the measure of the (un)stable core covered by the connecting (un)stable manifolds). It remains to show that the points in $g\mathcal{K}(\delta)$ reach into all parts of $\mathcal{U}^1$. It will follow from Sinai Theorem.

By Sinai Theorem, the total measure covered by rectangles which are not α-connecting is $o(\delta)$. Using Lemma 11.3, we can translate this estimate as

$$k(c)^{-1} |\mathcal{N}(\delta, c) \setminus \mathcal{K}(\delta)| \delta^{2d} = o(\delta).$$

Since in addition

$$\frac{|\mathcal{N}(\delta, c)|}{(c\delta)^{2d}} = O(1),$$

we see that the assumptions of Proposition 11.4 are satisfied and we can claim that

(11.7) $\dfrac{|g\mathcal{K}(\delta)|}{|\mathcal{N}(\delta, c)|} \to 1 \quad \text{as} \quad \delta \to 0.$

We are ready to finish the proof by a contradiction. Suppose there are two T invariant disjoint subsets E_1 and E_2 which have intersections with $\mathcal{U}^1$ of positive measure. Let us pick two Lebesgue density points p_1 and p_2 for $E_1 \cap \mathcal{U}^1$ and $E_2 \cap \mathcal{U}^1$, respectively.

Next, we fix cubes C_1 and C_2 with centers at p_1 and p_2 so small that

$$\mu(C_i \cap E_i) \geq \left(1 - \frac{\alpha'}{2k(c)}\right) \mu(C_i), \ i = 1, 2.$$

It follows from (11.7) that

$$\frac{|(\mathcal{N}(\delta, c) \setminus g\mathcal{H}(\delta)) \cap C_i|}{|\mathcal{N}(\delta, c)|} \to 0 \text{ as } \delta \to 0, \ i = 1, 2.$$

Since

$$\frac{|\mathcal{N}(\delta, c)|}{|\mathcal{N}(\delta, c) \cap C_i|} = O(1), \ i = 1, 2,$$

we conclude that

$$\frac{|(\mathcal{N}(\delta, c) \cap C_i) \setminus g\mathcal{H}(\delta)|}{|\mathcal{N}(\delta, c) \cap C_i|} \to 0 \text{ as } \delta \to 0, \ i = 1, 2.$$

Now we get immediately that

(11.8) $\quad \mu\left(\left(\bigcup\{R(z; \delta) | z \in g\mathcal{H}(\delta) \cap C_i\}\right) \Delta C_i\right) \to 0 \text{ as } \delta \to 0, \ i = 1, 2,$

where Δ denotes the symmetric difference, i.e., for any two sets A and B

$$A \Delta B = (A \setminus B) \cup (B \setminus A).$$

By (11.6) and Lemma 11.3

$$\mu\left(\bigcup\{R(z; \delta) | z \in g\mathcal{H}(\delta) \cap C_i\} \cap Y(\delta)\right) \geq \frac{\alpha'}{k(c)} \mu\left(\bigcup\{R(z; \delta) | z \in g\mathcal{H}(\delta) \cap C_i\}\right),$$

$i = 1, 2$.

Comparing this with (11.8) and remembering how dense E_i is in C_i, $i = 1, 2$, we conclude that, for sufficiently small δ, the set $Y(\delta)$ must intersect both E_1 and E_2 over subsets of positive measure, which contradicts the fact that it belongs to one ergodic component. $\qquad \square$

12. Proof of the Sinai Theorem

We will be proving only the unstable version of the theorem, i.e., we will estimate the measure of the union of rectangles which are not α-connecting in the unstable direction. Everything can be then repeated for the stable manifolds.

For a point y in the core of a rectangle $R(z; \delta)$ there are two possibilities:

(1) the point y has an unstable manifold of size $\delta' > \|\pi_1 y - \pi_1 z\| + \frac{\delta}{2}$ (which is connecting in $R(z; \delta)$ by Lemma 9.12),

(2) the point y has an unstable manifold of size $\delta' \leq \|\pi_1 y - \pi_1 z\| + \frac{\delta}{2}$ cut by $\bigcup_{i \geq 0} T^i \mathcal{S}^-$.

If a rectangle $R(z; \delta)$ is not connecting, then the second possibility must occur for at least $1 - \alpha$ part of its core.

The neighborhood $\mathcal{U}$ was chosen so small that $\mathcal{S}_N^- = \bigcup_{i=0}^{N-1} T^i \mathcal{S}^-$ is disjoint from $\mathcal{U}$. It follows that, for points in $\mathcal{U}^1$, the unstable manifolds of size $\delta' < \delta_0$ cannot be cut by this singularity set. For any $M \geq N$ let us introduce the following special case of the second property:

(2_M) *the point y has an unstable manifold of size $\delta' \leq \|\pi_1 y - \pi_1 z\| + \frac{\delta}{2}$ cut by $\bigcup_{i=N}^{M} T^i \mathcal{S}^-$.*

Further, we introduce the auxiliary notion of a M-nonconnecting rectangle. Roughly speaking, it is a rectangle which is not connecting because of the singularity set $\bigcup_{i=N}^{M} T^i \mathcal{S}^-$.

Definition 12.1. *Given $\alpha < \frac{1}{2}$, we say that a rectangle R of size δ is M-nonconnecting, if at least $1 - 2\alpha$ part of the measure of the unstable core of R consists of points which satisfy the property (2_M).*

The plan of the proof is the following. We divide the argument into two parts. We will prove that, for every $\varepsilon > 0$ there is $M = M(\varepsilon)$ and δ_ε such that, for all $\delta < \delta_\varepsilon$, the total measure of all rectangles in $\mathcal{G}_\delta$ which are not α-connecting and are not M-nonconnecting is less than $\delta\varepsilon$. This is the subject of the 'tail bound' (section 13), and it is by far the hardest part of the proof. It will require global considerations (i.e., outside of $\mathcal{U}$). The particular value of α is immaterial there.

We will start with the easier part proving that, for a given $\rho < \frac{1}{3}$, there is α such that, for any $\varepsilon > 0$ and any $M \geq N$ there is δ_ε such that, for all $\delta < \delta_\varepsilon$, the total measure of all M-nonconnecting rectangles of size δ is less than $\delta\varepsilon$. Let us formulate it in a separate Proposition. Its proof will be completely confined to the neighborhood $\mathcal{U}$.

Proposition 12.2. *For any $\rho < \frac{1}{3}$, there is $\alpha, 0 < \alpha < 1$, such that, for any $M \geq N$,*

$$\lim_{\delta \to 0} \delta^{-1} \mu \left(\bigcup \{ R \mid R \text{ has size } \delta \text{ and is } M\text{-nonconnecting } \} \right) = 0.$$

Proof. We rely on our assumption that $\mathcal{S}^-$ and its images are sufficiently 'nice'. More precisely we have required that the singularity set $\mathcal{S}_{M+1}^- = \bigcup_{i=0}^{M} T^i \mathcal{S}^-$ is regular. The definition of regularity was tailored to the needs of this proof. In particular, the singularity set $\mathcal{S}_{M+1}^-$ is a finite union of pieces of submanifolds I_k of codimension one, with boundaries $\partial I_k, k = 1, \ldots, p$. The boundaries $\partial I_k, k = 1, \ldots, p$ are themselves finite unions of compact subsets of submanifolds of codimension 2 . What is more

$$I_k \cap I_l \subset \partial I_k \cup \partial I_l \text{ for any } k, l.$$

In each of the pieces of submanifolds $I_k, k = 1, \ldots, p$, we consider the open neighborhood of the boundary of radius r, and we denote by J_r the union of these neighborhoods, i.e.,

$$J_r = \bigcup_{k=1}^{p} \{ p \in I_k \mid d(p, \partial I_k) < r \}.$$

For each δ let $r(\delta)$ be the smallest r such that, for any $k \neq l$, the distance of $I_k \setminus J_r$ and $I_l \setminus J_r$ is not less than 2δ. (In particular, for any $k \neq l$, the sets $I_k \setminus J_r$ and $I_l \setminus J_r$ are disjoint compact subsets, and their distance is at least 2δ.) Clearly

$$\lim_{\delta \to 0} r(\delta) = 0.$$

Hence, by the property (7.3)

(12.3) $$\lim_{\delta \to 0} \mu_{\mathscr{S}}(J_{r(\delta)}) = 0$$

where $\mu_{\mathscr{S}}$ is the natural volume element on $\mathscr{S}_{M+1}^-$.

Let us note that, if a rectangle $R = R(z; \delta)$ contains a point with the unstable manifold of size $\delta' < \delta$ cut by S_{M+1}^-, then it intersects the 2δ-neighborhood of S_{M+1}^-, but it does not necessarily intersect the singularity set itself. For technical reasons we prefer to blow up every rectangle, so that the blown up rectangle must intersect S_{M+1}^- itself, and not only its neighborhood. For a fixed $b_0 < \frac{1}{3}$, to be chosen later, and for any rectangle $R = R(z; \delta)$, we introduce the blown up rectangle

$$\tilde{R} = \mathscr{B}(\pi_1 z, \ (1 + 2b_0)\frac{\delta}{2}) \times \mathscr{B}(\pi_2 z, \ \frac{\delta}{2}).$$

The diameter of $\tilde{R}$ is less than 2δ, since we assume that $b_0 < \frac{1}{3}$.

Let y belong to the core of $R = R(z; \delta)$, satisfy the property (2_M), and

$$\|\pi_1 y - \pi_1 z\| \leq b_0 \frac{\delta}{2}.$$

This implies that the unstable manifold $W_\delta^u(y)$ is contained in $\tilde{R}$, so that $\tilde{R}$ intersects $\bigcup_{i=N}^{M} T^i \mathscr{S}^-$. For a fixed b_0, if α is sufficiently small, then any M-nonconnecting rectangle must contain such a point y. We conclude that, for α sufficiently small, if a rectangle R of size δ is M-nonconnecting, then $\tilde{R}$ intersects at least one of the pieces of submanifolds $I_k, k = 1, \ldots, p$. If for a rectangle R of size δ the blown up rectangle $\tilde{R}$ intersects two pieces of submanifolds I_k and $I_l, k \neq l$ then, by definition of $r(\delta)$, it must intersect $J_{r(\delta)}$, and so it must be contained in the neighborhood of $J_{r(\delta)}$ of radius 2δ. By (12.3) and Proposition 7.4 (or (7.5)) the measure of the neighborhood of $J_{r(\delta)}$ of radius 2δ is $o(\delta)$ (i.e., when divided by δ, it tends to zero as δ tends to zero). It remains to consider those blown up rectangles which intersect only one of the submanifolds $I_k, k = 1, \ldots, p$.

The proof will be finished when we prove that there is α so small that, for any sufficiently small δ, if a blown up rectangle $\tilde{R}$ of size δ intersects only one of the submanifolds $I_k, k = 1, \ldots, p$, then the rectangle R is **not** M-nonconnecting.

Our first observation is that there is a constant K depending only on the manifolds $I_k, k = 1, \ldots, p$, such that for any $x, x' \in I_k$ there is v in the tangent space to I_k at x ($v \in \mathscr{T}_x I_k$) for which

(12.4) $$\|x' - x - v\| \leq K\|x' - x\|^2$$

Here we consider the tangent space $\mathscr{T}_x I_k$ of I_k at x as a subspace in $\mathbb{R}^d \times \mathbb{R}^d$. This property is a formulation of the fact that smooth submanifolds are locally close to their tangent subspaces and follows easily from the Taylor expansion.

Further, in view of the proper alignment of the singularity manifolds, the tangent subspaces $\mathscr{T}_x I_k, x \in I_k \cap \mathscr{U}^1$ must have their characteristic lines in $\mathscr{C}_\rho$.

Let us now take a rectangle $R = R(z; \delta)$ such that the blown up rectangle $\tilde{R}$ intersects I_k. We will show that $\pi_2(I_k \cap \tilde{R})$ is contained in a fairly narrow layer. To show this, let $x = (x_1, x_2), x' = (x_1', x_2') \in I_k \cap \tilde{R}$ and let $v = (\xi, \eta) \in T_x I_k$ be the vector for which (12.4) holds. We pick a nonzero vector $v_0 = (\xi_0, \eta_0) \in T_x I_k$ with the direction of the characteristic line. For convenience, we scale it so that $\|\xi_0\| = 1$. We have, by the definition of a characteristic line,

$$\omega(v, v_0) = \langle \xi, \eta_0 \rangle - \langle \eta, \xi_0 \rangle = 0.$$

It follows that

$$|\langle \eta, \xi_0 \rangle| = |\langle \xi, \eta_0 \rangle| \le \rho \|\xi_0\| \|\xi\| = \rho \|\xi\|.$$

Replacing v by $x - x'$ in the last inequality and using (12.4), we get

$$\left| \langle \xi_0, x_2' - x_2 \rangle \right| \le \rho(\|x_1' - x_1\| + K\|x' - x\|^2) + K\|x' - x\|^2.$$

Since both x and x' are in $\widetilde{R}$, we have that

$$\|x_1' - x_1\| < (1 + 2b_0)\delta$$

and

$$\|x' - x\| < 2\delta.$$

Therefore, for any $x, x' \in I_k \cap \widetilde{R}$, we obtain the inequality

(12.5) $$\left| \langle \xi_0, x_2' - x_2 \rangle \right| \le \rho(1 + 2b_0)\delta + const\ \delta^2,$$

where the constant depends only on ρ and K. The inequality (12.5) shows that $\pi_2(I_k \cap \widetilde{R})$ is contained in a layer perpendicular to ξ_0 of the width $\rho(1 + 2b_0)\delta + const\ \delta^2$. Hence, there is $\bar{x}_2$ (in the 'center' of the layer) such that every $x = (x_1, x_2) \in I_k \cap \widetilde{R}$ must belong to the layer defined by the inequality

(12.6) $$|\langle \xi_0, x_2 - \bar{x}_2 \rangle| \le \rho(1 + 2b_0)\frac{\delta}{2} + const\ \delta^2.$$

We want to estimate the width of the layer where all the points from the core of the rectangle with 'short' unstable manifolds, cut by I_k, must lie. To that end let us take a point $y = (y_1, y_2)$ in the core of the rectangle $R(z; \delta)$ and such that $\|y_1 - \pi_1 z\| \le b_0 \frac{\delta}{2}$. If y satisfies the property (2_M) then, by Lemma 9.6, the projection $\pi_2 W_\delta^u(y)$ of the unstable manifold lies in the ball

$$\mathcal{B}(y_2; \rho\delta') \subset \mathcal{B}(y_2; \rho(1 + b_0)\frac{\delta}{2}).$$

Assuming that $W_\delta^u(y)$ is cut by I_k, there is $x = (x_1, x_2) \in I_k \cap \widetilde{R}$, for which

$$|\langle \xi_0, y_2 - x_2 \rangle| \le \rho(1 + b_0)\frac{\delta}{2}.$$

Hence, by (12.6), the point y must belong to the layer defined by the inequality

(12.7) $$|\langle \xi_0, y_2 - \bar{x}_2 \rangle| \le \rho(1 + b_0)\frac{\delta}{2} + \rho(1 + 2b_0)\frac{\delta}{2} + const\ \delta^2$$

The last step is to choose b_0 so small that this layer cannot cover all of the core. We prefer, for convenience, to fit a Cartesian product into the unstable core, and to prove that, if only b_0 is chosen sufficiently small, a fixed part of the measure of this set is not covered by the layer (12.7).

We choose such set to be

$$X(b_0) = \mathcal{B}(\pi_1 z; b_0 \frac{\delta}{2}) \times \mathcal{B}(\pi_2 z; s(b_0)\frac{\delta}{2})$$

where $s(b_0) = 1 - \rho - \rho b_0$. By the definition of a core, the set $X(b_0)$ is contained in the core of $R(z; \delta)$, and its measure is not less than a certain fixed part of the measure of the core, depending on b_0 (and the dimension d) but independent of δ.

If the layer (12.7) is sufficiently narrow, it cannot cover all of $X(b_0)$. The precise inequality, which guarantees that, is easily transformed into

(12.8) $$3\rho + const\ \delta < 1 - 4\rho b_0.$$

After a moment of reflection the reader will realize that if only $\rho < \frac{1}{3}$ we can choose b_0 so small that not only (12.8) is satisfied, but also a certain fixed part of $X(b_0)$ (depending on b_0 but independent of δ) is not covered by the layer (12.7). Thus, there is α, depending on ρ and b_0, such that more than 2α part of the measure of the core is free of points satisfying the property (2_M). Hence the rectangle R is **not** M-nonconnecting. ☐

If the reader finds it hard to follow the above argument, it is because we strived to use as little hyperbolicity as possible on our finite orbit. The amount of hyperbolicity is measured by the size ρ of the sector . We have managed to relax the condition on ρ up to $\rho < \frac{1}{3}$. It is not hard to see that, if the last condition is relaxed further, Proposition 12.2 will not hold in general.

13. 'Tail Bound'

We will be proving that, for every $\varepsilon > 0$, there is M such that the measure of points $z \in \mathcal{U}^1$ with the unstable manifold of size $\delta' < \delta$ cut by $\bigcup_{i \geq M+1} T^i \mathcal{S}^-$ does not exceed $\varepsilon\delta$. Comparing this set with the union of rectangles in $\mathcal{G}_\delta$ which are not α-connecting and not M-nonconnecting, we establish immediately that the measure of the union can be bigger by at most an absolute (=independent of δ) factor, made up of ρ, α and the overlap coefficient $k(c)$ (introduced prior to Definition 11.1). To arrive at this conclusion it is important that we consider only the rectangles from the covering $\mathcal{G}_\delta$ (and not all possible rectangles of size δ).

We start by exploring some of the consequences of the Sinai - Chernov Ansatz. No reference to the neighborhood $\mathcal{U}$ will be made at this stage. So we have assumed that almost all points in $\mathcal{S}^-$ (with respect to the measure $\mu_\mathcal{S}$) are strictly unbounded in the future. It follows from Theorem 6.8 that, for almost every point $p \in \mathcal{S}^-$,

$$\lim_{n \to +\infty} \inf_{0 \neq v \in \mathcal{C}(p)} \frac{\sqrt{\mathcal{Q}(D_p T^n v)}}{\|v\|} = +\infty.$$

For a linear monotone map, let us put

$$\sigma_*(L) = \inf_{0 \neq v \in \mathcal{C}(p)} \frac{\sqrt{\mathcal{Q}(Lv)}}{\|v\|}.$$

Consequently, for any (arbitrarily small) $h > 0$ and any (arbitrarily large) $t > 0$, there is $M = M(h, t)$ so large that the subset

$$\widetilde{E}_t = \{p \in \mathcal{S}^- \mid \sigma_*(D_p T^M) \leq t + 1\}$$

has measure

$$\mu_\mathcal{S}(\widetilde{E}_t) \leq h.$$

The map T^M is, in general, not even continuous in all of $\mathcal{S}^-$. The coefficient $\sigma_*(D_pT^M)$ is defined only for almost every point $p \in \mathcal{S}^-$. Hence, so far, the subset $\tilde{E}_t$ is defined modulo subsets of measure zero. We need a closed subset, since we plan to use Proposition 7.4.

The map T^M is discontinuous olny on $\mathcal{S}_M^+$, which was assumed to be a regular set. Using the proper alignment of singularity sets and the monotonicity of the system, we conclude that $\mathcal{S}_M^+$ is transversal to $\mathcal{S}^-$ (in the natural sense). It follows that the set $B_M = \left(\mathcal{S}_M^+ \cup \partial\mathcal{M}\right) \cap \mathcal{S}^-$ is a finite union of compact subsets of submanifolds of dimension $2d - 2$. Further, $\mathcal{S}^-$ is decomposed into a (possibly very large) finite number of pieces of submanifolds of dimension $2d - 1$ such that T^M is differentiable in the interior of every piece, and their boundaries are subsets of B_M. It follows that the coefficient $\sigma_*(D_pT^M)$ is continuous in the interior of every piece.

Let us choose ζ so small that the closure of the ζ-neighborhood of B_M in $\mathcal{S}^-$

$$B_M^\zeta = \{p \in \mathcal{S}^- \mid d(p, B_M) < \zeta\}$$

has small measure

$$\mu_{\mathcal{S}}\left(\overline{B_M^\zeta}\right) \leq h.$$

Now the set E_t defined by

$$E_t = \tilde{E}_t \setminus B_M^\zeta = \{p \in \mathcal{S}^- \setminus B_M^\zeta \mid \sigma_*(D_pT^M) \leq t + 1\}$$

is **closed**, and we have

$$\mu_{\mathcal{S}}\left(E_t \cup \overline{B_M^\zeta}\right) \leq 2h.$$

Let

$$\mathcal{S}_t = \{p \in \mathcal{S}^- \setminus B_M^\zeta \mid \sigma_*(D_pT^M) \geq t + 1\}.$$

$\mathcal{S}_t$ is a compact set and the coefficient $\sigma_*(D_pT^M)$ is continuous in a neighborhood of $\mathcal{S}_t$ in $\mathcal{M}$. Hence, there is $r > 0$ such that

$$\sigma_*(D_pT^M) > t,$$

for every point p in the r-neighborhood of $\mathcal{S}_t$ in $\mathcal{M}$, i.e., for every point in

$$\mathcal{S}_t^r = \{p \in \mathcal{M} \mid d(p, \mathcal{S}_t) < r\}.$$

Now we look at our neighborhood $\mathcal{U}$. Our goal is to estimate, for given δ, the measure of the set $Y(\delta, M)$ of points in $\mathcal{U}^1$ which have the unstable manifold of size $\delta' < \delta$ cut by $\bigcup_{i \geq M+1} T^i\mathcal{S}^-$. We will achieve this by splitting $Y(\delta, M)$ into convenient pieces and showing that their preimages must end up in extremely small neighborhoods of $\mathcal{S}^-$.

For $z \in Y(\delta, M)$ the unstable manifold $\mathcal{W}_{\delta'}^u(z)$ may be cut by several (possibly infinitely many) of the singularity sets $T^i\mathcal{S}^-$, $i = M + 1, \ldots$. Let $m(z)$ be the smallest $i \geq M + 1$ such that $\mathcal{W}_{\delta'}^u(z)$ is cut by $T^i\mathcal{S}^-$. Let further

$$k(z) = \#\{i \mid 1 \leq i \leq m(z) - M, \ T^{-i}z \in \mathcal{U}^1\}.$$

Roughly speaking $k(z)$ is the number of times the point z visits in $\mathcal{U}^1$ in the past in the time frame bounded by $m(z)$. We put for $k = 0, 1, \ldots$, $m = M + 1, \ldots$,

$$Y_m^k = \{z \in Y(\delta, M) \mid m(z) = m, k(z) = k\}.$$

We will now fix k and estimate the measure of

$$\bigcup_{m \geq M+1} Y_m^k.$$

Lemma 13.1. *For $m \neq m'$*

$$T^{-m} Y_m^k \cap T^{-m'} Y_{m'}^k = \emptyset.$$

Proof. Let $m < m'$. If $y \in T^{-m} Y_m^k \cap T^{-m'} Y_{m'}^k$ then for $z = T^m y$ and $z' = T^{m'} y$ we have

$$k(z') \geq k(z) + 1.$$

It contradicts the fact that $z \in Y_m^k$ and $z' \in Y_{m'}^k$. $\square$

By Lemma 13.1 we have

$$\mu\left(\bigcup_{m \geq M+1} Y_m^k \right) \leq \sum_{m \geq M+1} \mu(Y_m^k) = \sum_{m \geq M+1} \mu(T^{-m} Y_m^k) = \mu\left(\bigcup_{m \geq M+1} T^{-m} Y_m^k \right).$$

Let $z \in Y_m^k$ and $z' \in T^m \mathscr{S}^-$ be a point in the boundary of $W_\delta^u(z)$. We connect z and z' by the curve γ in $W_\delta^u(z)$ which projects under π_1 onto the linear segment from $\pi_1 z$ to $\pi_1 z'$. In the neighborhood $\mathcal{U}$ we have three ways of measuring the length of γ. We can use the quadratic form $\mathcal{Q}$, or the length of the projection onto the first component, or finally, we can use the Riemannian metric. All these metrics are equivalent in $\mathcal{U}$, and we will use the following coefficients defined by their ratios

$$\sup \left\{ \frac{\|v\|}{\|\xi\|} \mid 0 \neq v = (\xi, \eta) \in \mathscr{C}_\rho \right\} = \sqrt{1 + \rho^2}$$

and

$$q = \sup \left\{ \frac{\sqrt{\mathcal{Q}(v)}}{\|\xi\|} \mid 0 \neq v = (\xi, \eta) \in \mathscr{C}_\rho \right\},$$

where the last supremum is taken also over all of $\mathcal{U}$.

Our goal is to estimate the distance of $T^{-m} z$ and $T^{-m} z'$ in the Riemannian metric, such a distance clearly does not exceed the length of the curve $T^{-m} \gamma$. To that end, let $n \leq m - M$, be the time of the k-th visit in the past by z to $\mathcal{U}^1$, i.e., $T^{-n} z \in \mathcal{U}^1$. By Proposition 8.4, on every spaced return to $\mathcal{U}$ the projection of the preimage of γ is contracted by at least the factor ρ. In the k visits there must be at least $\frac{k}{N} - 1$ spaced returns. Hence, the projection of $T^{-n} \gamma$ has the length which, by (8.6u), does not exceed

$$c_1 \lambda^k \delta,$$

where

$$\lambda = \rho^{\frac{1}{N}} \quad \text{and} \quad c_1 = \frac{1}{\rho b} = \frac{1}{\rho \sqrt{1 - \rho^4}}.$$

It follows that the Riemannian length of $T^{-n} \gamma$ does not exceed

$$c_2 \lambda^k \delta,$$

where

$$c_2 = \frac{1}{\rho\sqrt{1-\rho^2}},$$

and its length in the metric $\mathfrak{Q}$ does not exceed

$$c_3 \lambda^k \delta,$$

where

$$c_3 = \frac{q}{\rho\sqrt{1-\rho^2}}.$$

Now we apply $T^{-(m-n)}$ to $T^{-n}\gamma$, and we use the fact that $m - n \geq M$. There are two different cases.

Case 1.

$$T^{-m}z' \in E_t \cup \overline{B_M^\zeta}$$

We use the noncontraction property. Under the action of $T^{-(m-n)}$ the Riemannian length of γ can expand at most by the factor $\frac{1}{a}$. We conclude that the Riemannian length of $T^{-m}\gamma$ does not exceed

$$\frac{c_2}{a}\lambda^k \delta.$$

Thus $T^{-m}z$ belongs to the neighborhood of $E_t \cup \overline{B_M^\zeta}$ in $\mathcal{M}$ of this radius. By Proposition 7.4 its measure does not exceed

(13.2) $$3h\frac{2c_2}{a}\lambda^k \delta,$$

if only δ is small enough ($\delta \leq \delta_0$ and δ_0 does not depend on k or m).

Case 2.

$$T^{-m}z' \in \mathcal{S}_t$$

We claim that, for sufficiently small δ, the length of $T^{-m}\gamma$ does not exceed

$$\frac{1}{t}c_3 \lambda^k \delta.$$

Indeed, it is so if $T^{-m}\gamma$ is contained in $\mathcal{S}_t^r$ (the r-neighborhood of $\mathcal{S}_t$ in $\mathcal{M}$). Since $m - n \geq M$, we have

$$\sigma_*(D_p T^{m-n}) > t,$$

for every point $p \in \mathcal{S}_t^r$ (if only the map T^{m-n} is differentiable at p). Hence, the length in the metric $\mathfrak{Q}$ of $T^{-n}\gamma$ is longer than the Riemannian length of $T^{-m}\gamma$ by at least the factor t. If $T^{-m}\gamma$ is not contained in $\mathcal{S}_t^r$, then there must be a segment of this curve in $\mathcal{S}_t^r$ which has at least length r. It follows that the image of this segment under T^{m-n} has the length in the metric $\mathfrak{Q}$ not less than tr, which is more than the total length in the metric $\mathfrak{Q}$ of $T^{-n}\gamma$, for sufficiently small δ. This contradiction shows that, for sufficiently small δ, $T^{-m}\gamma \subset \mathcal{S}_t^r$. We have proven our claim. It follows that $T^{-m}z$ belongs to the neighborhood of $\mathcal{S}^-$ of radius $\frac{1}{t}c_3\lambda^k\delta$. Using again Proposition 7.4, we can estimate the measure of this neighborhood by

$$(13.3) \qquad\qquad 2\mu_\mathcal{S}(\mathcal{S}^-)\frac{2c_3}{t}\lambda^k\delta,$$

if only δ is sufficiently small ($\delta \leq \delta_0$ and δ_0 does not depend on k or m).

Combining the estimates (13.2) and (13.3) we obtain that for any $k = 0, 1, \ldots$,

$$\mu\left(\bigcup_{m \geq M+1} T^{-m}Y_m^k\right) \leq \left(h\frac{6c_2}{a} + \frac{1}{t}2c_3\mu_\mathcal{S}(\mathcal{S}^-)\right)\lambda^k\,\delta.$$

It follows that

$$\mu(Y(\delta, M)) \leq \left(h\frac{6c_2}{a} + \frac{1}{t}4c_3\mu_\mathcal{S}(\mathcal{S}^-)\right)\frac{1}{1-\lambda}\,\delta.$$

The last inequality tells us how we should choose a small h and a large t at the beginning of our argument to guarantee that

$$\mu(Y(\delta, M)) \leq \varepsilon\delta.$$

The 'tail bound' is proven.

14. Applications

A. Billiard system in a convex scattering domains

We assume that the reader is familiar with billiard systems. If it is not the case, we recommend [W4] for a quick introduction into the subject. We will rely on the results of that paper.

Let us consider a domain in the plane bounded by a locally convex closed curve given by the natural equation $r = r(s), 0 \leq s \leq l$, describing the radius of curvature r as a function of the arc length s. We assume that the radius of curvature satisfies the condition

$$(14.1) \qquad\qquad \frac{d^2r}{ds^2} < 0, \quad \text{for all } s, \ 0 \leq s \leq l.$$

Curves satisfying this condition were called in [W4] strictly convex scattering.

Examples.

1. Perturbation of a circle.
2. Cardioid.

Such a domain cannot be convex, and there is a singular point in the boundary where the curve intersects itself. (If you do not like playing billiards on a table which is not convex, you may take the convex hull of our domain and everything below still applies.)

The following theorem is a fairly easy consequence of the Main Theorem.

Theorem 14.2. *The billiard system in a domain bounded by a strictly convex scattering curve (i.e., satisfying (14.1)) is ergodic.*

Let us consider the map T describing the first return map to the boundary. T is defined on the set $\mathcal{M}$ of unit tangent vectors pointing inwards. We parametrize $\mathcal{M}$ by the arc length parameter of the foot point $s, 0 \le s \le l$, and the angle $\varphi, 0 \le \varphi \le \pi$, which the unit vector makes with the boundary (oriented counterclockwise). In these coordinates $\mathcal{M}$ becomes the rectangle $[0, l] \times [0, \pi]$. The symplectic form (the invariant area element) is given by $\sin \varphi \, ds \wedge d\varphi$. After we derive the formula for the derivative of T, we will be able to check immediately that T preserves this area element.

The map T is discontinuous at those billiard orbits which hit the singular point of the boundary. They form a curve $\mathscr{S}^+$ in $\mathcal{M}$ which is a graph of a strictly decreasing function, decreasing curve for short. This curve divides the rectangle $\mathcal{M}$ into two curvilinear triangles, $\mathcal{M}_b^+$ with a side at the bottom and $\mathcal{M}_t^+$ with a side at the top.

To find the images of $\mathcal{M}_b^+$ and $\mathcal{M}_t^+$ we use the reversibility of our system. Namely, let $S : \mathcal{M} \to \mathcal{M}$ be defined by $S(s, \varphi) = (s, \pi - \varphi)$. We have

$$T \circ S = S \circ T^{-1}.$$

We can now claim that T^{-1} is continuous except on $\mathscr{S}^- = S\mathscr{S}^+$ which is an increasing curve (the graph of a strictly increasing function). $\mathscr{S}^+$ divides the rectangle $\mathcal{M}$ into two curvilinear triangles $\mathcal{M}_b^- = S\mathcal{M}_t^+$ and $\mathcal{M}_t^- = S\mathcal{M}_b^+$. We have constructed our symplectic boxes. T is a diffeomorphism on their interiors and a homeomorphism on the closure. The derivative of T does blow up at least at one point of the boundary $\mathscr{S}^+$ (different for $\mathcal{M}_b^+$ and for $\mathcal{M}_t^+$) corresponding to the two billiard orbits tangent to one of the branches of the boundary at the singular point. In the case of the cardioid the derivative blows up at any point of $\mathscr{S}^+$ and also at the vertical boundaries because the curvature at the cusp is infinite (see the formula for the derivative of T below). It is very handy that we did not have to require in Section 7 that our map is a diffeomorphism on the closed symplectic boxes.

The derivative of DT at (s_0, φ_0) has the form

(14.2)
$$\begin{pmatrix} \dfrac{\tau - d_0}{r_0 \sin \varphi_1} & \dfrac{\tau}{\sin \varphi_1} \\ \dfrac{\tau - d_0 - d_1}{r_0 r_1 \sin \varphi_1} & \dfrac{\tau - d_1}{r_1 \sin \varphi_1} \end{pmatrix}$$

where $T(s_0, \varphi_0) = (s_1, \varphi_1)$, τ is the time between consecutive hits (i.e., the length of the billiard orbit segment) and $d_i = r_i \sin \varphi_i$, $i = 1, 2$. This derivative can be obtained by straightforward implicit differentiation but we do not recommend it. There is a more geometric (and safer) way to obtain the derivative by resorting to the description of billiard orbit variations by Jacobi fields. In our two dimensional situation it amounts to introducing coordinates (J, J') in the tangent planes of $\mathcal{M}$

(14.3)
$$J = \sin \varphi ds,$$
$$J' = -\frac{1}{r} ds - d\varphi.$$

The evolution of (J, J') between collisions is given by the matrix

(14.4)
$$\begin{pmatrix} 1 & \tau \\ 0 & 1 \end{pmatrix}.$$

At the collision (J, J') is changed by

(14.5)
$$\begin{pmatrix} -1 & 0 \\ \frac{2}{d_1} & -1 \end{pmatrix}.$$

Now the derivative (14.2) is obtained by multiplying the matrices (14.4) and (14.5) and taking into account (14.3).

The geometric meaning of d_0, d_1, and the inequality

(14.6)
$$\tau > d_0 + d_1$$

is explained at length in [W4]. It was proven there that (14.6) holds for any billiard orbit segment, if the boundary curve is strictly convex scattering (actually these two properties are essentially equivalent). It follows from (14.6) that for a strictly convex scattering curve all elements in (14.2) are positive.

We choose as our family of sectors the constant sector between the horizontal line $\{d\varphi = 0\}$ and the vertical line $\{ds = 0\}$. We see immediately that the derivative DT is strictly monotone.

We are now ready to argue that the singularity sets $\mathscr{S}_n^- = \bigcup_{i=0}^{n} T^i \mathscr{S}^-$ are regular. We claim that $\mathscr{S}_n^-$ is a finite union of increasing curves which may intersect only at the endpoints. It can be proven by induction. Indeed $\mathscr{S}^-$ is an increasing curve and so it is also properly aligned. The singularity set $\mathscr{S}^+$ is a decreasing curve, and as such it may intersect each of the increasing curves of $\mathscr{S}_n^-$ in at most one point. Hence both $\mathcal{M}_b^+ \cap \mathscr{S}_n^-$ and $\mathcal{M}_t^+ \cap \mathscr{S}_n^-$ are finite unions of increasing curves with intersections only at the endpoints. Hence in view of the monotonicity of our system the images under T are also finite unions of increasing curves in $\mathcal{M}_b^-$ and $\mathcal{M}_t^-$ respectively. It is clear that we can safely add $\mathscr{S}^-$ to these images. We have thus checked that $\mathscr{S}_{n+1}^- = \mathscr{S}^- \cup T\mathscr{S}_n^-$ is also a finite union of increasing curves which intersect only at the endpoints. Note that the assumptions of Lemma 7.7 are too restrictive to allow its application in this case.

One can easily compute (and it was done explicitly in [W4]) that

(14.7)
$$\sigma(DT) = \sqrt{1 + \omega} + \sqrt{\omega}, \quad \text{where} \quad \omega = \frac{(\tau - d_0 - d_1)\tau}{d_0 d_1}.$$

It follows from (14.7) and from the supermultiplicativity of the coefficient of expansion σ that the only way in which an orbit can fail to be strictly unbounded is when the lengths of the segments of the orbit go to zero. It was shown by Halpern [Ha] that there are no such billiard orbits, if $r(s)$ is a C^1 function bounded away from zero. Hence, under such an assumption, which excludes the cardioid, all orbits for which an arbitrary power of T is differentiable are strictly unbounded. To include the cardioid, or more generally the curves with the radius of curvature $r(s)$ decreasing monotonously to zero at the endpoints of the interval, $0 \le s \le l$, (at the singular point), we shall argue that also for this class there is no accumulation of collisions at the singular point. Indeed, if an arc of the boundary between two consecutive hits by the billiard ball has monotone curvature, then the angle of incidence(reflection) is smaller where the curvature is bigger. Hence, as an orbit gets closer to the singularity point (the cusp for the cardioid), it is more and more perpendicular to the boundary, and so it cannot accumulate at the singularity.

This observation takes care of the Sinai - Chernov Ansatz. We are also guaranteed that the coefficient $\sigma(DT^n)$ can be made arbitrarily large by increasing n, except possibly for points which end up on the decreasing curve $\mathscr{S}^+$ in the future and the increasing curve $\mathscr{S}^-$ in the past. These are the points in $\mathscr{S}_n^+ \cap \mathscr{S}_m^-$, for some n and m, and so there are only countably many such points. (The orbit of such a point 'dies' both in the future and in the past, and it may fail to pick up enough hyperbolicity before then.) We can apply the Main Theorem to all other points, and they form a connected set. Hence, the local ergodicity obtained from the Main Theorem implies ergodicity.

It remains to check the noncontraction property. It was pointed out to us by Donnay [D1] that the derivative of T increases $|J'|^2$ on nonzero vectors from the sector. Indeed the interior of the sector is defined by

$$\frac{J'}{J} < -\frac{1}{d},$$

so that we have

$$\frac{|J'|}{|J|} > \frac{1}{d}.$$

If $DT(J_0, J'_0) = (J_1, J'_1)$, then we get from (14.4) and (14.5) that

$$J_1 = -J_0 - \tau J'_0.$$

It follows that

$$|J'_1| \geq \frac{1}{d_1}|J_1| = \frac{1}{d_1}|J_0 + \tau J'_0| \geq \frac{\tau}{d_1}|J'_0| - \frac{1}{d_1}|J_0| \geq \frac{\tau - d_0}{d_1}|J'_0|.$$

In view of (14.6) $\frac{\tau - d_0}{d_1} > 1$. So indeed $|J'|^2$ gets increased.

Moreover, for all vectors in the sector we have the following estimates

$$2(\frac{1}{r^2}ds^2 + d\varphi^2) \geq |J'|^2 = |\frac{1}{r}ds + d\varphi|^2 \geq \frac{1}{r^2}ds^2 + d\varphi^2.$$

The metric $\frac{1}{r^2}ds^2 + d\varphi^2$ is equivalent to the standard Riemannian metric in the (s, φ) coordinates $(ds^2 + d\varphi^2)$, if only r is bounded away from zero. Thus noncontraction is established under this additional assumption, which excludes the cardioid.

To cover the case of the cardioid, we observe that the noncontraction property is used only in the proof of the 'tail bound'. In that proof some subsets of the neighborhood $\mathcal{U}$ are transported back to the neighborhood of the singularity set $\mathcal{S}^-$. We need the property that vectors from the sector $\mathcal{C}$ are not contracted too much, along the orbits from the vicinity of the singularity set to the neighborhood $\mathcal{U}$, even if the orbit is very long. We obtain readily this property from the observation that although $|J'|^2$ is, in general, only bigger than the scaled standard Riemannian metric, it is clearly equivalent to one locally in the neighborhood $\mathcal{U}$.

The reader may be worried that the standard Riemannian metric in the (s, φ) coordinates does not generate the invariant area element. However, the Riemannian area is not smaller than the symplectic area. This is sufficient for the proof of Sinai Theorem. We could also handle this complication by introducing from the very beginning coordinates in $\mathcal{M}$ in which the symplectic form is standard.

We can conclude that T is ergodic, and so Theorem 14.2 is proven.

It follows from the results of Katok and Strelcyn [KS] that T is a Bernoulli system.

The framework of this paper allows to cover also the class of billiard systems in domains with more than one smooth piece in the boundary, which are not necessarily convex scattering. In the recent paper [D2] Donnay introduced a natural condition (focusing arc) on the convex pieces of the boundary of a billiard table. He proves that if two focusing arcs are connected by sufficiently long (extremely long may be required) straight segments, then the billiard system in such a (stadium like) domain has nonvanishing Lyapunov exponents. This work puts the original stadium of Bunimovich [B], which had arcs of circles in the boundary, into a large class of billiard systems with nonuniform hyperbolic behavior, larger than the class with convex scattering pieces introduced in [W4].

All the properties listed in Section 7 are satisfied for the billiards of Donnay in a straightforward fashion, with the notable exception of the noncontraction property. The problem is that the construction of the bundle of sectors depends heavily on the dynamics, and it is unlikely that there is a geometrically defined Lyapunov metric (like $|J'|^2$ for the convex scattering curves). Instead we use the following two ideas.

We have remarked in Section 7 that if the map T is differentiable up to and including the boundary of symplectic boxes, and DT is strictly monotone, then the noncontraction property holds automatically. In the billiards of Donnay the sectors are pushed strictly inside at the time of crossing from one convex piece to the other. Hence, we can use this observation on the compact part of the phase space made up of orbits which cross over from one convex piece to the other. We have the noncontraction property for the return map to this set, where we measure vectors in $\mathscr{C}$ using the form $\mathfrak{Q}$ defined by the bundle of sectors uniformly larger than $\mathscr{C}$. The construction of the bundle of sectors $\mathscr{C}$ by Donnay and his condition on the separation of convex pieces allows to introduce immediately these larger sectors with respect to which the derivative of the return map is monotone.

It remains to check the noncontraction property along "grazing" orbits which reflect many times in one convex piece. This is essentially done in [D2], where Lazutkin coordinates are used to put the map T in the vicinity of the boundary into a normal form.

These two observations, put together, give us the unconditional noncontraction property, and thus our Main Theorem applies.

B. Piecewise linear standard map

Let $T : \mathbb{T}^2 \to \mathbb{T}^2$ be defined by

$$T(x_1, x_2) = (x_1 + x_2 + Af(x_1), x_2 + Af(x_1))$$

where (x_1, x_2) are taken modulo 1, f is the periodic function

$$f(t) = |t| - \frac{1}{2}, \quad \text{for} \quad -\frac{1}{2} \leq t \leq \frac{1}{2},$$

and A is a real parameter. The mapping T preserves the Lebesgue measure. For $A = 1$ there is a simple invariant domain $\mathfrak{D}$ in the torus shown in Figure 9. It was proven in [W5] that the Lyapunov exponents are different from zero almost everywhere in $\mathfrak{D}$.

Theorem 14.8. T is ergodic in $\mathfrak{D}$.

As in the previous application, it follows that T is a Bernoulli system in $\mathfrak{D}$.

All the conditions of Section 7 are satisfied here in a very simple fashion. The reader can find all the necessary details in [W5] and [W6]. In this piecewise linear case one does not have to rely on the general results of Katok and Strelcyn. The existence of

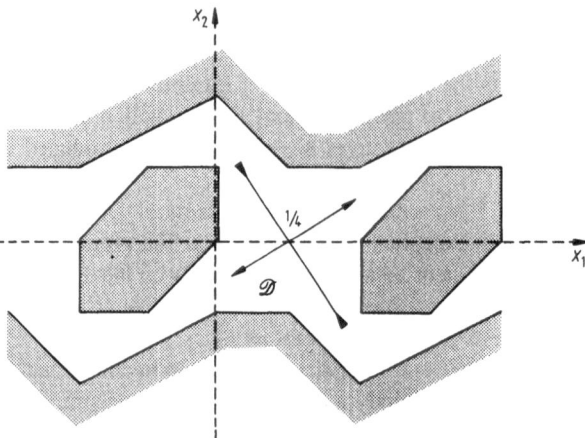

Figure 9. The domain $\mathcal{D}$

stable and unstable leaves can be obtained by the straightforward approach of Sections 1-3.

There are many other values of A for which nonvanishing of Lyapunov exponents was established for T in some domains in the torus, [W5],[W6]. The most interesting is the sequence of A's (roughly speaking) going to zero for which there is an invariant domain, with similar geometry as $\mathcal{D}$, where T has nonvanishing Lyapunov exponents. It is a piecewise linear model for the unstable layer containing the separatrices of the saddle fixed point $(0, \frac{1}{4})$. One can apply Main Theorem to all these special domains , so that in each case the map T is ergodic and hence Bernoulli. The reader should not have any difficulties in recovering the details based on the two papers cited above (incidentally even the noncontraction property was considered there).

C. The system of falling balls

One of the original motivations for our work was to prove ergodicity of the system of falling balls. This is a monotone system ([W7], [W8], [W3]), and all (semi-infinite) orbits are strictly unbounded. (The unboundedness of all orbits is obtained, under mild assumptions, by the application of Proposition 6.9) It follows that all Lyapunov exponents are different from zero, and it looks like a prime candidate for the application of the Main Theorem. It turns out, however, that in this example the singularity sets are not properly aligned, if the number of balls is greater than two. We will show this, and briefly discuss the case of two balls.

The system of falling balls is the system of point particles moving on a vertical line, which also interact by elastic collisions, and are subjected to a potential external field which forces the particles to fall down. To prevent the particles from falling into an abyss we introduce the hard floor, and assume that the bottom particle bounces back upon collision with it. The masses of the particles are in general different (the system of equal masses is completely integrable, since the elastic collision of equal masses in one dimension amounts to the exchanging of momenta).

The Hamiltonian of the system is

$$H = \sum_{i=1}^{N} \left(\frac{p_i^2}{2m_i} + m_i U(q_i) \right)$$

where q_i are the positions and $p_i = m_i v_i$ the momenta of the particles, $q_i, p_i \in \mathbb{R}, i = 1, \ldots, N$, and $U(q)$ is the potential of the external field . The differential equations of the system are

$$\dot{q}_i = \frac{p_i}{m_i}$$
$$\dot{p}_i = -m_i U'(q_i),$$

$i = 1, \ldots, N$.

The description of the dynamics is completed by the assumptions that the particles are impenetrable, and that they collide elastically with each other and with the floor $q = 0$.

We choose the following Lagrangian subspaces

$$V_1 = \{dp_1 = \cdots = dp_N = 0\} \quad \text{and} \quad V_2 = \{dh_1 = \cdots = dh_N = 0\},$$

where $h_i = \frac{p_i^2}{2m_i} + m_i U(q_i), i = 1, \ldots, N$, are individual energies of the particles.

We have

$$dh_i = \frac{p_i dp_i}{m_i} + m_i U'(q_i) dq_i,$$

$i = 1, \ldots, N$, so that V_1 and V_2 are indeed transversal if only $U' \neq 0$, i.e., if the external field is actually present.

The form $\mathfrak{Q}$ is equal to

$$\mathfrak{Q} = \sum_{i=1}^{N} \left(dq_i dp_i + \frac{p_i}{m_i^2 U'(q_i)} (dp_i)^2 \right).$$

It was shown in the papers cited above that the system is strictly monotone, provided that

$$U'(q) > 0 \quad \text{and} \quad U''(q) < 0,$$

and

$$m_1 > m_2 > \cdots > m_N.$$

The symplectic map T that naturally arises in this system is the map "from collision to collision". Our dynamical system is a suspension of the map. So that the system is ergodic if and only if the map T is ergodic. As usual, the actual computations are easier done in the full phase space of the flow.

Singularity set $\mathscr{S}^-$ corresponds to triple collisions: simultaneous collisions of three particles and the collision of two particles with the floor. Part of the first singularity set are not properly aligned. The second set is. So the methods of this paper apply only to the system of two particles.

Let us show that indeed the triple collision of three particles produces the singularity set which is not properly aligned. We consider the manifold

$$\{(q, p)|q_1 = q_2 = q_3\}.$$

Its tangent subspace is described by the equations

$$dq_1 = dq_2 = dq_3.$$

Its skew orthogonal complement is the two dimensional subspace given by equations

(14.9)
$$
\begin{aligned}
dq_i &= 0 \quad \text{for} \quad 1 \le i \le N, \\
dp_1 + dp_2 + dp_3 &= 0, \\
dp_i &= 0 \quad \text{for} \quad 4 \le i \le N.
\end{aligned}
$$

Restricting the form Ω to this plane we get

(14.10)
$$\sum_{i=1}^{3} \frac{p_i}{m_i^2 U'} (dp_i)^2.$$

We should assume that the particles emerge from collisions which means that

$$\frac{p_1}{m_1} < \frac{p_2}{m_2} < \frac{p_3}{m_3}.$$

But the momenta may, as well, be all negative which makes the quadratic form (14.10) negative definite. The actual characteristic line is obtained by intersecting the plane (14.9) by the tangent to the constant energy manifold. If all the momenta are negative, it is guaranteed to be outside of the sector. It is not hard to compute that the precise condition for the characteristic line to be contained in the sector is

$$\frac{v_1}{m_1} (v_2 - v_3)^2 + \frac{v_2}{m_2} (v_3 - v_1)^2 + \frac{v_3}{m_3} (v_1 - v_2)^2 \ge 0$$

where $v_i = \frac{p_i}{m_i}, i \ge 1$ are the velocities.

We close with the discussion of the system of two balls. For clarity, we restrict ourselves to the case of constant acceleration, $U(q) = q$. It was established in [W7], that also in this case all orbits are strictly monotone, if there are only two or three balls and their masses decrease. (For more than three balls technical problems arise, and it is an open problem to prove strict monotonicity almost everywhere.)

Let us fix the value of the total energy of the system, $H = \frac{1}{2}$. In this manifold we consider the two dimensional section $\mathcal{M}$ of the flow, corresponding to the bottom particle emerging from the collision with the floor; the surface $\mathcal{M}$ is given by $\{H = \frac{1}{2}, q_1 = 0, v_1 \ge 0\}$. The state of the system in M is completely described by the velocities of the particles (v_1, v_2); and we use the velocities as coordinates in $\mathcal{M}$. Hence, our phase space $\mathcal{M}$ is the domain bounded by the half-ellipse

$$m_1 v_1^2 + m_2 v_2^2 \le 1, \quad v_1 \ge 0.$$

Let us calculate the symplectic form in these coordinates. We have

$$\omega = dp_1 \wedge dq_1 + dp_2 \wedge dq_2.$$

On the surface of section $\mathcal{M}$

$$dq_1 \equiv 0 \quad \text{and} \quad dq_2 = -\frac{m_1}{m_2} v_1 dv_1 - v_2 dv_2.$$

Hence, we get

$$\omega = m_1 v_1 dv_1 \wedge dv_2.$$

The map $T : \mathcal{M} \to \mathcal{M}$ is defined by the first return of the flow to $\mathcal{M}$. Our symplectic box $\mathcal{M}$ is split into two symplectic boxes by $\mathscr{S}^+$, which is the arc of the ellipse $\{m_1 v_1^2 + m_2(v_2 - 2v_1)^2 = 1\}$ contained in $\mathcal{M}$. The symplectic box $\mathcal{M}_f^+$, above $\mathscr{S}^+$, contains all the initial states for which the bottom particle returns to the floor without colliding with the top particle. The map T in $\mathcal{M}_f^+$ is linear

$$T(v_1, v_2) = (v_1, v_2 - 2v_1).$$

The symplectic box $\mathcal{M}_c^+$, below $\mathscr{S}^+$, contains all the initial states for which there is a collision of the two particles before the bottom particle returns to the floor. The map T in $\mathcal{M}_c^+$ is nonlinear and is best described in the coordinate system (h, z), where

$$h = \frac{1}{2} m_1 v_1^2$$
$$z = v_2 - v_1.$$

The symplectic form $\omega = dh \wedge dz$. (This coordinate system is derived from the canonical system of coordinates in the full phase space furnished by the individual energies and velocities of the particles. The exceptional role of these coordinates is well documented in [W7], [CW].)

Note that both the energy of the bottom particle and the difference of velocities change only in collisions. Now $T = F_2 \circ F_1$, where

$$F_1(h, z) = (-h - az^2 + b, -z), \quad a = \frac{m_1 m_2 (m_1 - m_2)}{(m_1 + m_2)^2} \quad \text{and} \quad b = \frac{m_1}{m_1 + m_2},$$

describes the collision of the two particles, and

$$F_2(h, z) = (h, z + c\sqrt{h}), \quad c = \sqrt{\frac{8}{m_1}},$$

describes the collision of the bottom particle with the floor.

To find the image symplectic boxes $\mathcal{M}_f^-$ and $\mathcal{M}_c^-$ we can use the reversibility of our system. Namely, if we put $S(v_1, v_2) = (v_1, -v_2)$ then $T \circ S = S \circ T^{-1}$, and so $\mathcal{M}_f^- = S\mathcal{M}_f^+$, $\mathcal{M}_c^- = S\mathcal{M}_c^+$.

Our bundle of unstable sectors is constant in the coordinates (h, z) and equal to the positive (and negative) quadrant; the form $\mathfrak{A} = dhdz$. One can check immediately that $\mathscr{S}^+$ and $\mathscr{S}^- = S\mathscr{S}^+$ are properly aligned.

We can now check that T is monotone in $\mathcal{M}_f^+$ and strictly monotone in $\mathcal{M}_c^+$ (both F_1 and F_2 are monotone). Indeed, in the (h, z) coordinates we have

$$DF_1 = \begin{pmatrix} -1 & -2az \\ 0 & -1 \end{pmatrix} \quad \text{and} \quad DF_2 = \begin{pmatrix} 1 & 0 \\ \frac{c}{2\sqrt{h}} & 1 \end{pmatrix}.$$

Moreover the map T in $\mathcal{M}_f^+$ is equal in the coordinates (h, z) to F_2.

Since the collision of the two particles must eventually occur, we obtain strict mono-tonicity of all nondegenerate orbits. Unboundedness of all nondegenerate orbits follows from Proposition 6.9. So the Sinai-Chernov Ansatz holds.

To check the noncontraction property, we observe that the standard Riemannian metric in the coordinates (h, z) does not decrease on vectors from the sector, when we apply one of the above matrices.

Finally, we are guaranteed that the coefficient $\sigma(DT^n)$ can be made arbitrarily large by increasing n, except for points which end up on the singularity set $\mathcal{S}^+$ in the future and the singularity set $\mathcal{S}^-$ in the past. There are only countably many such points in view of the proper alignment of singularity sets, and the Main Theorem applies to all other points. It follows that T is ergodic and consequently, by the results of Katok and Strelcyn, it is a Bernoulli system.

The case of variable acceleration ($U'' < 0$) can be treated in a similar fashion. It is not possible in general to get explicit formulas for the return map T, but its derivative in the coordinates

$$\delta h = \frac{p_1}{m_1}\delta p_1$$
$$\delta z = \frac{1}{m_2 U'(q_2)}\delta p_2 - \frac{1}{m_1 U'(q_1)}\delta p_1,$$

was essentially calculated in [W8]. It is again a product of triangular matrices.

Afterword. This paper was greatly improved thanks to many insightful comments and corrections by the anonymous referees of the paper.

While we were writing this paper, several authors pursued similar goals. There are papers by Chernov [Ch1], [Ch2], the new version of his old preprint by Katok, in collaboration with Burns [K2], by Markarian [M], by Vaienti [Va], papers by Simányi [S1], [S2] and Szász [Sz].

References

[AW]   R.L. Adler, B. Weiss, *Entropy is a complete metric invariant for automorphisms of the torus*, Proc. Natl. Acad. Sci. USA **57** (1967), 1537–1576.

[AS]   D.V. Anosov, Ya.G. Sinai, *Certain smooth ergodic systems*, Russ. Math. Surv. **22** (1982), 103–167.

[B]   L.A. Bunimovich, *On the ergodic properties of nowhere dispersing billiards*, Commun.Math.Phys. **65** (1979), 295–312.

[BG]   K. Burns, M. Gerber, *Continuous invariant cone families and ergodicity of flows in dimension three*, Erg. Th. Dyn. Syst. **9** (1989), 19–25.

[CW]   J. Cheng, M.P. Wojtkowski, *Linear stability of a periodic orbit in the system of falling balls*, The Geometry of Hamiltonian Systems, Proceedings of a Workshop Held June 5-16,1989 MSRI Publications, Springer Verlag 1991 (ed. Tudor Ratiu), 53–71.

[Ch 1] N.I. Chernov, *The ergodicity of a Hamiltonian system of two particles in an external field*, Physica D **53** (1991), 233–239.

[Ch 2] N.I. Chernov, *On local ergodicity in hyperbolic systems with singularities*, Funct.-Anal.-Appl. **27** (1993), 51–59.

[CS] N.I. Chernov, Ya.G. Sinai, *Ergodic properties of some systems of 2-dimensional discs and 3-dimen- sional spheres*, Russ. Math. Surv. **42** (1987), 181–207.

[D1] V. Donnay, private communication (1988).

[D2] V. Donnay, *Using integrability to produce chaos: billiards with positive entropy*, Commun. Math. Phys. **141** (1991), 225 - 257.

[Ha] B. Halpern, *Strange Billiard Tables*, TAMS **232** (1977), 297–305.

[H] E. Hopf, *Statistik der Geodatischen Linien in Mannigfaltigkeiten Negativer Krummung*, Ber. Verh. Sächs. akad.wiss., Leipzig **91** (1939), 261–304.

[K1] A. Katok, *Invariant cone families and stochastic properties of smooth dynamical systems*, preprint (1988).

[K2] A. Katok in collaboration with K. Burns, *Infinitesimal Lyapunov functions, invariant cone families and stochastic properties of smooth dynamical systems*, to appear in Erg. Th. Dyn. Syst..

[KS] A. Katok, J.-M. Strelcyn with the collaboration of F. Ledrappier and F. Przytycki, *Invariant manifolds, entropy and billiards; smooth maps with singularities*, Lecture Notes in Math. 1222, Springer-Verlag, 1986.

[KSS] A. Krámli, N. Simányi, D. Szász, *A "Transversal" Fundamental Theorem for Semi-Dispersing Billiards*, (see also Erratum), Commun. Math. Phys. **129** (1990), 535–560.

[LW] C. Liverani, M.P. Wojtkowski, *Generalization of the Hilbert metric to the space of positive definite matrices*, to appear in Pac. J. Math..

[M] R. Markarian, *The Fundamental Theorem of Sinai–Chernov for dynamical systems with singularities*, Dynamical Systems (Santiago, 1990), Pitman Res. Notes Math. Ser., 285 Longman Sci. Tech., Harlow (1993), 131–158.

[O] V.I. Oseledets, *A Multiplicative Ergodic Theorem: Characteristic Lyapunov Exponents of Dynamical Systems*, Trans. Moscow Math. Soc. **19** (1968), 197–231.

[P] Ya.B. Pesin, *Lyapunov Characteristic Exponents and Smooth Ergodic Theory*, Russ. Math. Surveys **32,** 4 (1977), 55–114.

[Si1] N. Simányi, *The K-property of N billiard balls I*, Invent.-Math. **108** (1992), 521–548.

[Si] N. Simányi, *The K-property of N billiard balls II: Computation of neutral linear spaces*, Invent.-Math. **110** (1992), 151–172.

[S] Ya.G. Sinai, *Dynamical systems with elastic reflections*, Russ. Math. Surveys **25** (1970), 137–189.

[Sz] D. Szász, *On the K-property of some planar hyperbolic billiards*, Commun. Math. Phys. **145** (1992), 595–604.

[Va] S. Vaienti, *Ergodic properties of the discontinuous sawtooth map*, Jour. Stat. Phys. **67** (1992), 251–269.

[Ve] E. Vesentini, *Invariant metrics on convex cones*, Ann. Sc. Norm. Sup. Pisa ser. 4 **3** (1976), 671–696.

[W1] M.P. Wojtkowski, *Invariant families of cones and Lyapunov exponents*, Erg. Th. Dyn. Syst. **5** (1985), 145–161.

[W2] M.P. Wojtkowski, *Measure theoretic entropy of the system of hard spheres*, Erg. Th. Dyn. Syst. **8** (1988), 133–153.

[W3]   M.P. Wojtkowski, *Systems of classical interacting particles with nonvanishing Lyapunov exponents*, Lyapunov Exponents, Proceedings, Oberwolfach 1990, L. Arnold, H. Crauel, J.-P. Eckmann (Eds), Lecture Notes in Math. 1486, Springer-Verlag (1991), 243–262.

[W4]   M.P. Wojtkowski, *Principles for the design of billiards with nonvanishing Lyapunov exponents*, Commun. Math. Phys. **105** (1986), 391–414.

[W5]   M.P. Wojtkowski, *A model problem with the coexistence of stochastic and integrable behavior*, Commun. Math. Phys. **80** (1981), 453–464.

[W6]   M.P. Wojtkowski, *On the ergodic properties of piecewise linear perturbations of the twist map*, Ergodic Theory & Dynamical Systems **2** (1982), 525–542.

[W7]   M.P. Wojtkowski, *A system of one dimensional balls with gravity*, Commun. Math. Phys. **126** (1990), 507–533.

[W8]   M.P. Wojtkowski, *The system of one dimensional balls in an external field. II*, Commun. Math. Phys. **127** (1990), 425–432.

Linearization of Random Dynamical Systems

Thomas Wanner

Institut für Mathematik, Universität Augsburg, D–86135 Augsburg, Germany

1. Introduction

At the end of the last century the French mathematician Henri Poincaré laid the foundation for what we call nowadays the qualitative theory of ordinary differential equations. Roughly speaking, this theory is devoted to studying how the qualitative behavior (e.g. the asymptotic behavior) of solutions to certain initial value problems changes as the initial condition is varied. In order to make this more explicit, consider the autonomous differential equation

$$\dot{x} = f(x) \,, \tag{1.1}$$

where $f : \mathbb{R}^d \to \mathbb{R}^d$ is a C^1-mapping with $f(x_0) = 0$ for some $x_0 \in \mathbb{R}^d$. Obviously, the point x_0 is a constant solution of (1.1). But what can be said about the behavior of solutions of (1.1) starting in some small neighborhood of this point? Of course one is tempted to first study the linearized equation

$$\dot{x} = Df(x_0)x \tag{1.2}$$

near the origin, since it may be hoped that the nonlinear behavior of (1.1) near x_0 will be "basically" the same.

As to the behavior of the linear equation (1.2) it is well-known that the phase space $\mathbb{R}^d$ can be decomposed into a direct sum $E_- \oplus E_0 \oplus E_+$ of invariant subspaces, where E_-, E_0 or E_+ denote the sum of the generalized eigenspaces of $Df(x_0)$ corresponding to eigenvalues with negative, vanishing or positive real part. Furthermore, all solutions of equation (1.2) contained in E_- (or E_+) converge exponentially to the origin as $t \to \infty$ (or $t \to -\infty$) and the possible rates of convergence are determined by the real parts of the eigenvalues in the left (or right) half of the complex plane. Finally, the asymptotic behavior of solutions starting outside those three invariant subspaces is completely determined by the projections on E_-, E_0 and E_+. But, to what extend does this linear behavior determine the nonlinear behavior of (1.1) near x_0?

The answer to this question is provided by the linearization method which has been developed during the last decades. First of all, it can be shown that in a small neighborhood of the stationary point x_0 there exist five invariant manifolds for (1.1) corresponding to the invariant subspaces E_-, $E_- \oplus E_0$, E_0, $E_0 \oplus E_+$ and E_+ in the linearized case, which are called *stable, center-stable, center, center-unstable* and *unstable manifold*, respectively (cf. for example Amann [1], Hartman [16], Hirsch, Pugh, Shub [19] or Kelley [21]). Again, solutions starting on the stable (or unstable) manifold converge to the stationary point x_0 exponentially as $t \to \infty$ (or $t \to -\infty$) and the possible rates of convergence are equal to those observed in the linear case.

As to the behavior of solutions of (1.1) starting outside these five invariant manifolds Grobman [12, 13] and Hartman [14, 15] proved the following result known nowadays as the *classical Hartman-Grobman theorem*. Suppose that the Jacobian $Df(x_0)$ is hyperbolic, i.e. none of its eigenvalues lies on the imaginary axis. Then there is a neighborhood $U \subset \mathbf{R}^d$ of x_0 and a homeomorphism $h : U \to h(U) \subset \mathbf{R}^d$ with $h(x_0) = 0$ mapping solutions of (1.1) onto solutions of (1.2) as long as they are contained in U – and vice versa. In other words, the phase portrait of (1.1) near x_0 is nothing but the homeomorphic image of the phase portrait of the linearized equation (1.2) near the origin. For a non-hyperbolic Jacobian $Df(x_0)$ this result no longer remains true. But some twenty years ago Palmer [26] and Shoshitaishvili [28, 29] succeeded in proving a similar result for this case known as the *generalized Hartman-Grobman theorem* (cf. also Aulbach, Wanner [8], Hilger [17, 18] and Kirchgraber, Palmer [23]).

Parallel to this linearization theory for ordinary differential equations a linearization theory has been developed for difference equations as well, and both theories together form an important part of the theory of dynamical systems – recalling that certain autonomous differential or difference equations generate continuous-time or discrete-time dynamical systems through their general solutions. Yet, in recent times it has become more and more apparent that for modelling certain "real systems" dynamical systems are not always the appropriate tool. This is mainly due to the fact that real systems often are influenced by a rather complex environment, and the attempt to incorporate all these environmental influences into the definition of a suitable dynamical system with finite-dimensional phase space is condemned to run aground.

While one method to tackle this problem is the use of infinite-dimensional dynamical systems we will pursue another one. The basic idea behind this method is to consider the influences (which in the following will always be referred to as noise) and the dynamics as separate components of a new system where the noise acts on the dynamics in a very special way. The resulting object will be called *random dynamical system* and has been introduced during the last decade by Ludwig Arnold and his co-workers at the University of Bremen.

In order to present the exact definition of a random dynamical system we first have to introduce the mathematical model for the noise component.

Definition 1.1. *Let $(\Omega, \mathcal{F}, \mathbb{P})$ denote some probability space and let $\mathbb{T} = \mathbb{Z}$ or $\mathbb{T} = \mathbb{R}$. Then we call $(\Omega, \mathcal{F}, \mathbb{P}, (\theta_t)_{t \in \mathbb{T}})$ a metric dynamical system if the mappings $\theta_t : \Omega \to \Omega$ satisfy the following conditions:*

(a) The mapping $\mathbb{T} \times \Omega \ni (t, \omega) \mapsto \theta_t \omega \in \Omega$ is measurable.

(b) *The family* $(\theta_t)_{t \in \mathbf{T}}$ *forms a group, i.e. we have* $\theta_0 = \mathrm{id}_\Omega$ *as well as* $\theta_{t+s} = \theta_t \circ \theta_s$, *for arbitrary* $t, s \in \mathbf{T}$.

(c) *The mappings* θ_t *are* $\mathbb{P}$-*preserving, i.e. for arbitrary* $t \in \mathbf{T}$ *and* $F \in \mathcal{F}$ *the identity* $\mathbb{P}(\theta_t^{-1}(F)) = \mathbb{P}(F)$ *holds.*

A metric dynamical system $(\Omega, \mathcal{F}, \mathbb{P}, (\theta_t)_{t \in \mathbf{T}})$ *is called ergodic if every* θ_t-*invariant set has probability 0 or 1, i.e. if for all* $F \in \mathcal{F}$ *satisfying* $\theta_t^{-1}(F) = F$ *for every* $t \in \mathbf{T}$ *we have either* $\mathbb{P}(F) = 0$ *or* $\mathbb{P}(F) = 1$.

The following interpretations are to illuminate this definition. First of all, the set Ω consists of all possible states of the noise, and the group property of the family $(\theta_t)_{t \in \mathbf{T}}$ shows that the noise behaves like an ordinary deterministic dynamical system. However, the complex structure of this system makes it impossible to determine the exact state of the noise in practice, hence a probability $\mathbb{P}(F)$ is assigned to every set $F \in \mathcal{F}$ of states.

In most applications Ω will be an infinite-dimensional function space and $(\theta_t)_{t \in \mathbf{T}}$ a family of shift operators. In this case any function contained in Ω can be viewed as one possible temporal evolution of the noise.

Now let us introduce the dynamical component, together with the rules that control the impact of the noise on the dynamics.

Definition 1.2. *A measurable random dynamical system on* $\mathbb{R}^d$ *over some given metric dynamical system* $(\Omega, \mathcal{F}, \mathbb{P}, (\theta_t)_{t \in \mathbf{T}})$, *where* $\mathbf{T} = \mathbb{Z}$ *or* $\mathbf{T} = \mathbb{R}$, *is a measurable mapping* $\varphi : \mathbf{T} \times \Omega \times \mathbb{R}^d \to \mathbb{R}^d$ *forming a cocycle over* θ_t, *i.e. the mappings* $\varphi(t, \omega) := \varphi(t, \omega, \cdot)$ *satisfy*

$$\varphi(0, \omega) = \mathrm{id}_{\mathbb{R}^d}$$

as well as

$$\varphi(t + s, \omega) = \varphi(t, \theta_s \omega) \circ \varphi(s, \omega) ,$$

for arbitrary $t, s \in \mathbf{T}$ *and* $\omega \in \Omega$.

For $\omega \in \Omega$ *and* $\xi \in \mathbb{R}^d$ *the mapping* $\varphi(\cdot, \omega, \xi)$ *is called* ω-*orbit of* φ *through* ξ. *The random dynamical system* φ *is called* linear *if the mappings* $\varphi(t, \omega)$ *are linear, it is called* continuous *if the mappings* $\varphi(\cdot, \omega, \cdot)$ *are continuous, and it is called* smooth *of class* C^1 *if it is continuous, the mappings* $\varphi(t, \omega)$ *are of class* C^1 *and the derivative is continuous with respect to* (t, ξ), *for arbitrary* $t \in \mathbf{T}$ *and* $\omega \in \Omega$. *Finally, in the case* $\mathbf{T} = \mathbb{Z}$ *the random dynamical system is called* discrete-time *random dynamical system, whereas for* $\mathbf{T} = \mathbb{R}$ *it is called* continuous-time *random dynamical system.*

Obviously the above definition generalizes the well-known definition of a (measurable) dynamical system – one only has to consider the case that φ is independent of ω. But in the random case the temporal evolution of the system explicitly depends on the noise ω, and the cocycle property shows that the noise changes in time, too.

Nevertheless, there is a deterministic aspect in the concept of a random dynamical system. For this, consider the *extended phase space* $\Omega \times \mathbb{R}^d$ and define a family $(\Theta_t)_{t\in\mathbf{T}}$ of mappings $\Theta_t : \Omega \times \mathbb{R}^d \to \Omega \times \mathbb{R}^d$ by $\Theta_t(\omega,\xi) := (\theta_t\omega, \varphi(t,\omega,\xi))$. Then it can easily be verified that $(\Theta_t)_{t\in\mathbf{T}}$ forms a group, the so-called *skew-product flow* generated by φ. In other words, a random dynamical system induces an ordinary dynamical system on the space $\Omega \times \mathbb{R}^d$. The reader may find more about this and other aspects of random dynamical systems in Arnold, Crauel [4] and the forthcoming book of Arnold [2].

It will be the aim of this paper to present a linearization theory for random dynamical systems analogous to the one sketched at the beginning of this section for ordinary differential equations. The first and crucial step in that direction is the following result of Oseledets [24], his celebrated *multiplicative ergodic theorem*.

Theorem 1.1. *Let Φ denote a continuous linear random dynamical system on $\mathbb{R}^d$ over some ergodic metric dynamical system $(\Omega, \mathcal{F}, \mathbb{P}, (\theta_t)_{t\in\mathbf{T}})$ and assume that the integrability assumption*

$$\ln^+ \sup_{0\le t\le 1} \|\Phi(t,\omega)\| + \ln^+ \sup_{0\le t\le 1} \|\Phi(t,\omega)^{-1}\| \in L^1(\Omega, \mathcal{F}, \mathbb{P})$$

holds[1]. Then there exists a θ_t-invariant set $\tilde{\Omega} \in \mathcal{F}$ with $\mathbb{P}(\tilde{\Omega}) = 1$, as well as real numbers $\lambda_1 > \lambda_2 > \ldots > \lambda_p$ (the so-called Lyapunov exponents) with multiplicities d_i, $\sum_{i=1}^p d_i = d$, such that for every $\omega \in \tilde{\Omega}$ the following holds:

(a) *There is a splitting $\mathbb{R}^d = E_1(\omega) \oplus \ldots \oplus E_p(\omega)$, where the so-called Oseledets spaces $E_i(\omega)$ depend measurably on ω and satisfy $\dim E_i(\omega) = d_i$ as well as*

$$\Phi(t,\omega)E_i(\omega) = E_i(\theta_t\omega)$$

for arbitrary $t \in \mathbf{T}$ and $i = 1,\ldots,p$. In other words, the Oseledets spaces are invariant under the linear cocycle Φ.

(b) *For every $x \in \mathbb{R}^d$ the limits $\lim_{t\to\pm\infty} \frac{1}{t} \ln \|\Phi(t,\omega)x\|$ exist and*

$$\lim_{t\to\pm\infty} \frac{1}{t} \ln \|\Phi(t,\omega)x\| = \lambda_i \quad \Leftrightarrow \quad x \in E_i(\omega) \setminus \{0\}.$$

Proof. The proof can be found in Arnold [2, Section 3.3].   ◊

This theorem completely describes the asymptotic behavior of a linear random dynamical system, and it is clear that for $\mathbf{T} = \mathbb{R}$ the Lyapunov exponents are the random analogues to the real parts of the eigenvalues mentioned in our introductory discussion. Because of this it seems reasonable to expect that a vanishing Lyapunov exponent plays a special role in the theory to come. Therefore we make another definition.

Definition 1.3. *In the situation of Theorem 1.1 the linear random dynamical system Φ is called* hyperbolic *if none of its Lyapunov exponents vanishes.*

---

[1] As usual, the mapping $\ln^+$ is defined by $\ln^+ x := \max\{0, \ln x\}$.

In order to develop a linearization theory for random dynamical systems we will first restrict our attention to the discrete-time case and employ some rather deterministic way of thinking about random dynamical systems. The basic observation in this direction is that for a discrete-time random dynamical system φ the mapping $\varphi(\cdot, \omega, \xi)$ is the uniquely determined solution of the initial value problem

$$x_{k+1} = \varphi(1, \theta_k \omega, x_k) \quad , \quad x_0 = \xi \, ,$$

i.e. φ is nothing but the general solution of a special nonautonomous difference equation depending measurably on some parameter $\omega \in \Omega$.

Having this observation in mind we will derive in Section 2 a global linearization theory for general random difference equations, i.e. nonautonomous difference equations depending measurably on some parameter. In doing this we basically follow the lines of Aulbach, Wanner [8] where a deterministic version of the theory can be found. However, since we like to apply these results to random dynamical systems we have to modify the method presented in that monograph in order to allow norms depending on both time and chance.

With this linearization theory for random difference equations at hand we will now turn our attention to random dynamical systems. In Section 3 we derive (under rather restrictive conditions) a global linearization theory for both discrete-time and continuous-time random dynamical systems. This is accomplished by first considering the discrete-time case with the aid of the results of Section 2, and after that by reducing the continuous-time case to the discrete one.

Up to that point all results will be global ones, i.e. the conditions imposed on the systems under consideration are strong enough to guarantee global existence of the constructed objects (e.g. the invariant manifolds or the homeomorphism in the Hartman-Grobman theorem). Hence we indicate in Section 4 how local results for large classes of random dynamical systems can be proved by means of suitable cut-off techniques.

Finally, in the appendix we study the measurable dependence of fixed points on some parameter. These results will be needed in Section 2.

Acknowledgement. The results presented in this paper are taken from the authors doctoral thesis [30] which has been made while he held a scholarship at the *Graduiertenkolleg Mathematik – Analyse, Optimierung und Steuerung komplexer Systeme.* The author would like to thank the DFG and the Universität Augsburg for this support, as well as his advisors Ludwig Arnold and Bernd Aulbach. Finally, he would like to express his gratitude to Matthias Gundlach, Marco Holzmann and Stanislaus Maier-Paape for reading the manuscript and providing helpful comments.

2. Random Difference Equations

2.1 Preliminaries

As we already mentioned in the introduction this second section is devoted
to deriving a linearization theory for random difference equations. But before
we begin let us fix the needed notation and present some basic facts about
difference equations.

A subset $I \subset \mathbb{Z}$ is called *discrete interval* if it is the intersection of a real
interval with the set $\mathbb{Z}$, it is called *unbounded to the right* if either $I = \mathbb{Z}$ or $I =
\{k \in \mathbb{Z} : k \geq \kappa_0\}$ for some $\kappa_0 \in \mathbb{Z}$. Similarly, we may define discrete intervals
which are *unbounded to the left*. Since in the following discrete intervals with
a maximum cause some minor trouble we define – for a given discrete interval
I – a new discrete interval I^* by

$$I^* := \left\{ \begin{array}{cl} I \setminus \{\max I\} & \text{if the maximum exists} \\ I & \text{if } I \text{ is unbounded to the right} \end{array} \right.$$

Now let $I \subset \mathbb{Z}$ denote an arbitrary discrete interval with $I^* \neq \emptyset$, let $(\Omega, \mathcal{F})$
be a measurable space, and let P denote an arbitrary parameter set. Finally,
let $f : I^* \times \Omega \times \mathbb{R}^d \times P \to \mathbb{R}^d$ be an arbitrary mapping. Then an equation of
the form

$$x_{k+1} = f(k, \omega, x_k, p) \tag{2.1}$$

is called *(nonautonomous) parameter dependent difference equation (with pa-*
rameters ω and p). For fixed $\omega \in \Omega$ and $p \in P$ a mapping $\lambda : J \to \mathbb{R}^d$ defined
on some nonempty discrete interval $J \subset I$ is called ω, p-*solution* of *(2.1)* if
we have $\lambda(k+1) = f(k, \omega, \lambda(k), p)$ for all $k \in J^*$. Initial value problems may
be defined as in the case of ordinary differential equations. It is clear that for
given $\kappa_0 \in I$, $\xi_0 \in \mathbb{R}^d$, $\omega \in \Omega$ and $p \in P$ an ω, p-solution to the initial value
problem

$$x_{k+1} = f(k, \omega, x_k, p) \quad , \quad x_{\kappa_0} = \xi_0$$

exists uniquely on the discrete interval $\{k \in I : k \geq \kappa_0\}$ and is given recursively by

$$\lambda(k; \kappa_0, \omega, \xi_0, p) := \left\{ \begin{array}{cl} \xi_0 & \text{for } k = \kappa_0 \\ f(k-1, \omega, \lambda(k-1; \kappa_0, \omega, \xi_0, p), p) & \text{for } k > \kappa_0 \end{array} \right. . \tag{2.2}$$

If in addition the mapping $f(k, \omega, \cdot, p)$ is bijective for arbitrary $k \in I^*$, $\omega \in \Omega$
and $p \in P$, then an ω, p-solution to an arbitrary initial value problem exists
uniquely on the whole of I and is given recursively by

$$\lambda(k; \kappa_0, \omega, \xi_0, p) := \left\{ \begin{array}{cl} \xi_0 & \text{for } k = \kappa_0 \\ f(k-1, \omega, \lambda(k-1; \kappa_0, \omega, \xi_0, p), p) & \text{for } k > \kappa_0 \\ f^{-1}(k, \omega, \lambda(k+1; \kappa_0, \omega, \xi_0, p), p) & \text{for } k < \kappa_0 \end{array} \right. \tag{2.3}$$

where $f^{-1}(k,\omega,\cdot,p)$ denotes the inverse of $f(k,\omega,\cdot,p)$. In either of the above two cases, the mapping λ is called *general solution of the difference equation* (2.1) *(in the sense of (2.2), or (2.3))*. Moreover, the above definitions imply that the general solution satisfies

$$\lambda(k_2;\kappa,\omega,\xi,p) = \lambda(k_2;k_1,\omega,\lambda(k_1;\kappa,\omega,\xi,p),p) \qquad (2.4)$$

for all $\kappa, k_1, k_2 \in I$, $\omega \in \Omega$, $\xi \in \mathbb{R}^d$ and $p \in P$, where in the case of a non-invertible right-hand side we have to assume $\kappa \le k_1 \le k_2$.

So far, we did not impose any measurability or continuity conditions on f. Now suppose that P is a metric space equipped with the σ-algebra of Borel sets and that the mapping f is measurable. Then equation (2.1) is called *random difference equation (depending on the parameter p)*. It can readily be verified that the general solution λ in the sense of (2.2) is measurable if and only if the mapping f is. If in addition $f(k,\omega,\cdot,p)$ is bijective for all $k \in I^*$, $\omega \in \Omega$ and $p \in P$, then the general solution in the sense of (2.3) is measurable if and only if f, as well as the mapping $f^{-1} : I^* \times \Omega \times \mathbb{R}^d \times P \to \mathbb{R}^d$ defined as above are measurable.

Finally, the general solution in the sense of (2.2) is continuous with respect to the last two variables if and only if the mapping $f(k,\omega,\cdot,\cdot)$ is continuous for arbitrary $k \in I^*$ and $\omega \in \Omega$. If in addition $f(k,\omega,\cdot,p)$ is bijective for all $k \in I^*$, $\omega \in \Omega$ and $p \in P$, then the general solution in the sense of (2.3) is continuous with respect to the last two variables if and only if f and f^{-1} are.

In order to conclude this short introduction to difference equations let us add some remarks concerning the linear case. To that end, let $I \subset \mathbb{Z}$ again denote a discrete interval with $I^* \ne \emptyset$, let $(\Omega, \mathcal{F})$ be an arbitrary measurable space, and let $A : I^* \times \Omega \to GL(\mathbb{R}^d)$ and $b : I^* \times \Omega \to \mathbb{R}^d$ denote arbitrary mappings. Then the difference equation

$$x_{k+1} = A(k,\omega)x_k + b(k,\omega) \qquad (2.5)$$

is called *(inhomogeneous) nonautonomous linear difference equation depending on the parameter ω*, and the corresponding *homogeneous equation* is given by

$$x_{k+1} = A(k,\omega)x_k . \qquad (2.6)$$

As in the case of linear ordinary differential equations the general solution of (2.6) can be given in a nice way by means of the so-called *evolution operator* $\Phi(m,n,\omega)$. For this, we recursively define

$$\Phi(m,n,\omega) := \begin{cases} \mathrm{id}_{\mathbb{R}^d} & \text{for } m = n \\ A(m-1,\omega) \circ \ldots \circ A(n,\omega) & \text{for } m > n \\ A(m,\omega)^{-1} \circ \ldots \circ A(n-1,\omega)^{-1} & \text{for } m < n \end{cases}$$

for arbitrary $m, n \in I$ and $\omega \in \Omega$. It may readily be verified that then the general solution of (2.6) in the sense of (2.3) reads

$$\lambda(k;\kappa_0,\omega,\xi_0) = \Phi(k,\kappa_0,\omega)\xi_0 .$$

(Note that because of later applications we required the linear mapping $A(k,\omega)$ to be invertible, so all ω-solutions of (2.5) and (2.6) exist uniquely on the whole of I.) The following properties are immediate consequences of the above definition:

$$
\begin{aligned}
\Phi(k+1,m,\omega) &= A(k,\omega) \circ \Phi(k,m,\omega)\,, \\
\Phi(k,m,\omega) &= \Phi(k,n,\omega) \circ \Phi(n,m,\omega)\,, \\
\Phi(k,m,\omega)^{-1} &= \Phi(m,k,\omega)\,,
\end{aligned}
$$

where $m,n,k \in I$ and $\omega \in \Omega$ are arbitrary, and in the first equation we additionally have to require $k \in I^*$. As in the deterministic case it is possible to obtain a formula for the general solution of (2.5). This "variation of constants" formula reads as follows:

$$
\lambda(k;\kappa_0,\omega,\xi_0) = \begin{cases}
\xi_0 & \text{for} \quad k = \kappa_0 \\
\Phi(k,\kappa_0,\omega)\xi_0 + \sum\limits_{i=\kappa_0}^{k-1} \Phi(k,i+1,\omega)b(i,\omega) & \text{for} \quad k > \kappa_0 \\
\Phi(k,\kappa_0,\omega)\xi_0 - \sum\limits_{i=k}^{\kappa_0-1} \Phi(k,i+1,\omega)b(i,\omega) & \text{for} \quad k < \kappa_0
\end{cases}
$$

$$(2.7)$$

Finally, assume that the mappings A and b are measurable. Then (2.5) and (2.6) are called *linear random difference equations*, and it can easily be verified that the evolution operator $\Phi : I \times I \times \Omega \to \mathrm{GL}(\mathbb{R}^d)$ is measurable.

2.2 Quasiboundedness and Its Consequences

On our way to a linearization theory for random difference equations this subsection will be the most important, yet most technical part. It will turn out soon that it is crucial to study special solutions of random difference equations, so-called *quasibounded solutions*, which are characterized by their asymptotic growth rate. In view of later applications we are forced to use families of norms depending on both time and chance, rather than one single norm as in the deterministic case treated in Aulbach, Wanner [8].

Definition 2.1. *Let $(\Omega,\mathcal{F})$ denote an arbitrary measurable space and $\|\cdot\|_{k,\omega}$, $k \in \mathbb{Z}$, $\omega \in \Omega$, a family of norms on $\mathbb{R}^d$ such that $\mathbb{Z} \times \Omega \times \mathbb{R}^d \ni (k,\omega,x) \mapsto \|x\|_{k,\omega} \in \mathbb{R}_0^+$ is measurable. Moreover, let $I \subset \mathbb{Z}$ be a discrete interval, $\mu : I \to \mathbb{R}^d$ an arbitrary mapping and $\gamma \in \mathbb{R}^+$ a positive constant.*

The mapping μ is called γ^+-quasibounded (with respect to $\omega \in \Omega$) if I is unbounded to the right and there exists some $\kappa \in I$ and $C \in \mathbb{R}_0^+$ such that

$$\|\mu(k)\|_{k,\omega} \le C\gamma^k \quad \text{for all} \quad k \ge \kappa\,.$$

In this case, we denote the least possible constant C for which the above estimate holds by $\|\mu\|_{\kappa,\omega,\gamma}^+$, i.e. we set $\|\mu\|_{\kappa,\omega,\gamma}^+ := \sup\{\gamma^{-k}\|\mu(k)\|_{k,\omega} : k \ge \kappa, k \in I\}$.

Similarly, γ^--quasiboundedness (with respect to $\omega \in \Omega$) may be defined if I is unbounded to the left and the above estimate holds for arbitrary $k \leq \kappa$, and we abbreviate the least possible constant C to $\|\mu\|^-_{\kappa,\omega,\gamma}$.

Finally, in the case $I = \mathbb{Z}$ the mapping μ is called γ-quasibounded (with respect to $\omega \in \Omega$) if there exists some $C \in \mathbb{R}^+_0$ with

$$\|\mu(k)\|_{k,\omega} \leq C\gamma^k \quad \text{for all} \quad k \in \mathbb{Z},$$

and the least possible constant C is denoted by $\|\mu\|_{\omega,\gamma}$.

The following two lemmas investigate to some extent questions of existence and uniqueness of quasibounded solutions for certain nonlinear difference equations. More precisely, we like to show that corresponding properties of linear equations carry over to a (sufficiently small) nonlinear perturbation.

To begin with, we consider equations having exactly one γ^--quasibounded solution for certain values of γ. While for the unperturbed linear equation this will be enforced by a condition on the evolution operator, the conditions on the nonlinear perturbation are mainly of Lipschitz type.

Lemma 2.1. *Consider the difference equation*

$$x_{k+1} = A^-(k,\omega)x_k + f(k,\omega,x_k,p) + f_0(k,\omega,p) \tag{2.8}$$

depending on two parameters $\omega \in \Omega$ and $p \in P$, where P is an arbitrary metric space. Let the discrete interval $I \subset \mathbb{Z}$ be unbounded to the left, and let $A^- : I^* \times \Omega \to \mathrm{GL}(\mathbb{R}^{d^-})$, $f : I^* \times \Omega \times \mathbb{R}^{d^-} \times P \to \mathbb{R}^{d^-}$ and $f_0 : I^* \times \Omega \times P \to \mathbb{R}^{d^-}$ denote arbitrary mappings. Finally, let $\Phi^-(m,n,\omega)$ denote the evolution operator of the homogeneous linear difference equation $x_{k+1} = A^-(k,\omega)x_k$.

Now suppose that for some (fixed) parameters $\omega \in \Omega$ and $p \in P$ the following three conditions hold for arbitrary $m, n \in I$, $k \in I^*$ and $x, \bar{x} \in \mathbb{R}^{d^-}$:

$$\|\Phi^-(m,n,\omega)\|_{n,m,\omega} \leq K\alpha_-^{m-n} \quad \text{for} \quad m \geq n, \tag{2.9}$$
$$f(k,\omega,0,p) = 0,$$
$$\|f(k,\omega,x,p) - f(k,\omega,\bar{x},p)\|_{k+1,\omega} \leq L\|x - \bar{x}\|_{k,\omega},$$

with real constants $\alpha_- > 0$, $K \geq 1$ and $L \geq 0$[2]. Then for every $\gamma > \alpha_- + KL$ and $\kappa \in I$ we have: If the number

$$M_{\omega,p} := \sup\{\gamma^{-k}\|f_0(k,\omega,p)\|_{k+1,\omega} : k < \kappa, k \in I\} \tag{2.10}$$

is finite, then there is a uniquely determined ω, p-solution $\mu(\cdot,\omega,p) : I \to \mathbb{R}^{d^-}$ of (2.8) which is γ^--quasibounded with respect to ω, and the estimate

$$\|\mu(\cdot,\omega,p)\|^-_{\kappa,\omega,\gamma} \leq \frac{KM_{\omega,p}}{\gamma - \alpha_- - KL} \tag{2.11}$$

---

[2] In (2.9) we use the definition $\|\Phi^-(m,n,\omega)\|_{n,m,\omega} := \sup\{\|\Phi^-(m,n,\omega)x\|_{m,\omega} : x \in \mathbb{R}^{d^-}, \|x\|_{n,\omega} \leq 1\}$.

holds. If in the case $I = \mathbb{Z}$ the set $\{\gamma^{-k}\|f_0(k,\omega,p)\|_{k+1,\omega} : k \in \mathbb{Z}\}$ is bounded, then $\mu(\cdot,\omega,p)$ is even γ-quasibounded with respect to ω and

$$\|\mu(\cdot,\omega,p)\|_{\omega,\gamma} \leq \frac{K}{\gamma - \alpha_- - KL} \sup\{\gamma^{-k}\|f_0(k,\omega,p)\|_{k+1,\omega} : k \in \mathbb{Z}\}. \quad (2.12)$$

Now suppose that the mappings A^-, f and f_0 are measurable and assume that the above conditions hold for all $\omega \in \Omega$ and $p \in P$. Then we have:

(a) *The above-defined mapping $\mu : I \times \Omega \times P \to \mathbb{R}^{d^-}$ is measurable.*
(b) *If the mappings $f(k,\omega,\cdot,\cdot)$ and $f_0(k,\omega,\cdot)$ are continuous for arbitrary $k \in I^*$ and $\omega \in \Omega$, and if the set $\{M_{\omega,p} : p \in P\}$ is bounded for all $\omega \in \Omega$, then the mapping $\mu(k,\omega,\cdot) : P \to \mathbb{R}^{d^-}$ is continuous for all $k \in I$ and $\omega \in \Omega$.*

Proof. The estimate contained in (2.12) follows easily from (2.11) observing that $\|\mu(\cdot,\omega,p)\|_{\omega,\gamma} = \lim_{\kappa \to \infty} \|\mu(\cdot,\omega,p)\|^-_{\kappa,\omega,\gamma}$. As for the remaining assertions of the lemma we proceed in several steps.

(I) To begin with we consider the special case $I = (-\infty,\kappa] \cap \mathbb{Z}$, $f_0(k,\omega,p) \equiv 0$ on $I^* \times \Omega \times P$ and $L = 0$. Let $\mu : I \to \mathbb{R}^{d^-}$ denote an ω-solution of the homogeneous linear equation

$$x_{k+1} = A^-(k,\omega)x_k \quad (2.13)$$

which is γ^--quasibounded with respect to $\omega \in \Omega$, and let $\kappa_0 \in I$ be arbitrary, but fixed. Then the identity $\mu(\kappa_0) = \Phi^-(\kappa_0,k,\omega)\mu(k)$ furnishes the estimate

$$\|\mu(\kappa_0)\|_{\kappa_0,\omega} \leq \|\Phi^-(\kappa_0,k,\omega)\|_{k,\kappa_0,\omega}\|\mu(k)\|_{k,\omega} \leq K\alpha_-^{\kappa_0}\left(\frac{\gamma}{\alpha_-}\right)^k \|\mu\|^-_{\kappa_0,\omega,\gamma}$$

for arbitrary $k \leq \kappa_0$, and because of $\gamma > \alpha_- + KL = \alpha_-$ the right-hand side of this inequality converges to 0 as $k \to -\infty$. Hence we have $\mu(\kappa_0) = 0$ for all $\kappa_0 \in I$, i.e. the trivial solution is the only ω-solution of (2.13) which is γ^--quasibounded with respect to ω – and defining $\mu(k,\omega,p) := 0$ for $k \in I$ proves the lemma in this case.

(II) Again let $I = (-\infty,\kappa] \cap \mathbb{Z}$ and $L = 0$, but this time let the mapping f_0 be arbitrary with $M_{\omega,p} < \infty$. Because of *(I)* there is at most one ω,p-solution of the inhomogeneous linear difference equation

$$x_{k+1} = A^-(k,\omega)x_k + f_0(k,\omega,p) \quad (2.14)$$

which is γ^--quasibounded with respect to ω, since the difference of two such solutions is a γ^--quasibounded ω-solution of (2.13) and therefore vanishes identically. As for the existence we define

$$\mu(k,\omega,p) := \sum_{i=-\infty}^{k-1} \Phi^-(k,i+1,\omega)f_0(i,\omega,p)$$

for arbitrary $k \in I$ and leave it to the reader to verify that this sum actually converges, yields an ω, p-solution of (2.14) and satisfies everything claimed in the lemma.

(III) In this third part of the proof we consider the special case $I = (-\infty, \kappa] \cap \mathbb{Z}$ once again, but now for arbitrary $L \geq 0$ and arbitrary f_0 with $M_{\omega, p} < \infty$. Since we make use of two fixed point results that will be proved in the appendix, let X_ω denote the Banach space of all mappings $\nu : I \to \mathbb{R}^{d^-}$ which are γ^--quasibounded with respect to ω, equipped with the norm $\| \cdot \|_{\kappa, \omega, \gamma}^-$. Then it is routine to check that the evaluation mappings $X_\omega \ni \nu \mapsto \nu(k) \in \mathbb{R}^{d^-}$ satisfy the assumptions of Lemma 5.2 and Lemma 5.3 for arbitrary $k \in I$ and $\omega \in \Omega$. Now let $\omega \in \Omega$ and $p \in P$ be chosen as in the formulation of Lemma 2.1. Furthermore, for the construction of a contraction $T_{\omega, p}$ on X_ω let $\nu_{\omega, p} \in X_\omega$ be arbitrary. Then because of

$$\gamma^{-k} \| f(k, \omega, \nu_{\omega, p}(k), p) + f_0(k, \omega, p) \|_{k+1, \omega} \leq L \| \nu_{\omega, p} \|_{\kappa, \omega, \gamma}^- + M_{\omega, p} \quad (2.15)$$

for all $k < \kappa$ we may apply part *(II)* to the inhomogeneous linear difference equation

$$x_{k+1} = A^-(k, \omega) x_k + f(k, \omega, \nu_{\omega, p}(k), p) + f_0(k, \omega, p) \,,$$

i.e. this equation has a uniquely determined ω, p-solution $\nu_{\omega, p}^* : I \to \mathbb{R}^{d^-}$ which is γ^--quasibounded with respect to ω, and the definition $T_{\omega, p} \nu_{\omega, p} := \nu_{\omega, p}^*$ yields an operator $T_{\omega, p} : X_\omega \to X_\omega$. Moreover, *(II)* and (2.15) furnish

$$\| T_{\omega, p} \nu_{\omega, p} \|_{\kappa, \omega, \gamma}^- \leq \frac{K}{\gamma - \alpha_-} (L \| \nu_{\omega, p} \|_{\kappa, \omega, \gamma}^- + M_{\omega, p}) \,, \quad (2.16)$$

and the reader may verify that the operator family $T_{\omega, p}$, $(\omega, p) \in \Omega \times P$, satisfies everything required in Lemma 5.2*(b)* and that for every $\omega \in \Omega$ Lemma 5.3*(b)* holds for the family $T_{\omega, p}$, $p \in P$.

Next we have to show that the above-defined operator $T_{\omega, p}$ indeed is a contraction on X_ω. To that end, let $\nu_1, \nu_2 \in X_\omega$ be arbitrary. Then the mapping $T_{\omega, p} \nu_1 - T_{\omega, p} \nu_2$ is an ω, p-solution of

$$x_{k+1} = A^-(k, \omega) x_k + f(k, \omega, \nu_1(k), p) - f(k, \omega, \nu_2(k), p)$$

which is even γ^--quasibounded with respect to ω. Applying *(II)* to this equation yields the estimate

$$\| T_{\omega, p} \nu_1 - T_{\omega, p} \nu_2 \|_{\kappa, \omega, \gamma}^- \leq \frac{KL}{\gamma - \alpha_-} \| \nu_1 - \nu_2 \|_{\kappa, \omega, \gamma}^- \,,$$

i.e. $T_{\omega, p}$ is a contraction on X_ω which has a unique fixed point $\mu(\cdot, \omega, p) \in X_\omega$. Since a mapping $\nu \in X_\omega$ is a fixed point of $T_{\omega, p}$ if and only if it is a γ^--quasibounded (with respect to ω) ω, p-solution of (2.8) the first claim of Lemma 2.1 follows. Finally, the estimate (2.11) is an immediate consequence of (2.16) and $T_{\omega,p}\mu(\cdot,\omega,p) = \mu(\cdot,\omega,p)$, and the assertions contained in *(a)* and *(b)* follow from Lemma 5.2 and Lemma 5.3, respectively.

(IV) As for concluding the proof of Lemma 2.1 let $I \subset \mathbb{Z}$ denote an arbitrary discrete interval which is unbounded to the left, and let $\kappa \in I$ be arbitrary, but fixed. Furthermore, let $\tilde{\mu}(\cdot,\omega,p) : (-\infty,\kappa] \cap \mathbb{Z} \to \mathbb{R}^{d^-}$ be the mapping guaranteed by *(III)* if we restrict (2.8) to the discrete interval $(-\infty,\kappa] \cap \mathbb{Z} \subset I$. Then it may easily be verified that the mapping $\mu(\cdot,\omega,p) : I \to \mathbb{R}^{d^-}$ defined as $\mu(k,\omega,p) := \lambda(k;\kappa,\omega,\tilde{\mu}(\kappa,\omega,p),p)$ for arbitrary $k > \kappa$ and $\mu(k,\omega,p) := \tilde{\mu}(k,\omega,p)$ for $k \leq \kappa$ satisfies everything claimed in the above lemma[3].   ◊

The second lemma provides conditions implying that every ω-solution of a certain nonlinear difference equation depending on a parameter $\omega \in \Omega$ is γ^--quasibounded with respect to ω.

Lemma 2.2. *Consider the difference equation*

$$x_{k+1} = A^+(k,\omega)x_k + f(k,\omega,x_k) + f_0(k,\omega) \qquad (2.17)$$

depending on the parameter $\omega \in \Omega$. Let the discrete interval $I \subset \mathbb{Z}$ be unbounded to the left, and let $A^+ : I^* \times \Omega \to \mathrm{GL}(\mathbb{R}^{d^+})$, $f : I^* \times \Omega \times \mathbb{R}^{d^+} \to \mathbb{R}^{d^+}$ and $f_0 : I^* \times \Omega \to \mathbb{R}^{d^+}$ denote arbitrary mappings. Finally, let $\Phi^+(m,n,\omega)$ denote the evolution operator of the homogeneous linear difference equation $x_{k+1} = A^+(k,\omega)x_k$.

Now suppose that for some (fixed) parameter $\omega \in \Omega$ the following three conditions hold for arbitrary $m,n \in I$, $k \in I^*$ and $x,\bar{x} \in \mathbb{R}^{d^+}$:

$$\|\Phi^+(m,n,\omega)\|_{n,m,\omega} \leq K\alpha_+^{m-n} \quad \text{for} \quad m \leq n \,,$$
$$f(k,\omega,0) = 0 \,,$$
$$\|f(k,\omega,x) - f(k,\omega,\bar{x})\|_{k+1,\omega} \leq L\|x - \bar{x}\|_{k,\omega} \,,$$

with real constants $K \geq 1$, $L \geq 0$ and $\alpha_+ > KL$. Then for every $k \in I^*$ the mapping $A^+(k,\omega) + f(k,\omega,\cdot) + f_0(k,\omega)$ is a bijection on $\mathbb{R}^{d^+}$, i.e. all ω-solutions of (2.17) exist uniquely on the whole of I. Furthermore, for arbitrary $0 < \gamma < \alpha_+ - KL$ and $\kappa \in I$ we have: If the number

$$M_\omega := \sup\{\gamma^{-k}\|f_0(k,\omega)\|_{k+1,\omega} : k < \kappa, k \in I\}$$

is finite, then every ω-solution $\mu : I \to \mathbb{R}^{d^+}$ of (2.17) is γ^--quasibounded with respect to ω and the estimate

$$\|\mu\|_{\kappa,\omega,\gamma}^- \leq K\gamma^{-\kappa}\|\mu(\kappa)\|_{\kappa,\omega} + \frac{KM_\omega}{\alpha_+ - \gamma - KL}$$

holds.

---

[3] λ denotes the general solution of (2.8) in the sense of (2.2).

Proof. As to the invertibility of the right-hand side of (2.17) for fixed $k \in I^*$ let $y \in \mathbb{R}^{d^+}$ be arbitrary and define an operator $T_{k,\omega,y} : \mathbb{R}^{d^+} \to \mathbb{R}^{d^+}$ via

$$T_{k,\omega,y}(x) := A^+(k,\omega)^{-1}y - A^+(k,\omega)^{-1}f(k,\omega,x) - A^+(k,\omega)^{-1}f_0(k,\omega) .$$

Due to $A^+(k,\omega)^{-1} = \Phi^+(k,k+1,\omega)$ we have $\|T_{k,\omega,y}(x) - T_{k,\omega,y}(\bar{x})\|_{k,\omega} \leq \frac{KL}{\alpha_+}\|x - \bar{x}\|_{k,\omega}$ for all $x, \bar{x} \in \mathbb{R}^{d^+}$, so $T_{k,\omega,y}$ has a unique fixed point in $\mathbb{R}^{d^+}$. But then the claimed invertibility is an easy consequence of the fact that x is a fixed point of the mapping $T_{k,\omega,y}$ if and only if $y = A^+(k,\omega)x + f(k,\omega,x) + f_0(k,\omega)$.

Now let $\mu : I \to \mathbb{R}^{d^+}$ be an arbitrary ω-solution of (2.17) and $\kappa \in I$ be arbitrary, but fixed. Then the variation of constants formula (2.7) furnishes for every $k \leq \kappa$ the identity

$$\mu(k) = \Phi^+(k,\kappa,\omega)\mu(\kappa) - \sum_{i=k}^{\kappa-1} \Phi^+(k,i+1,\omega)\left(f(i,\omega,\mu(i)) + f_0(i,\omega)\right) .$$

¿From this we deduce the estimate

$$\alpha_+^{-k}\|\mu(k)\|_{k,\omega} \leq K\alpha_+^{-\kappa}\|\mu(\kappa)\|_{\kappa,\omega} + \frac{KM_\omega}{\alpha_+}\sum_{i=k}^{\kappa-1}\left(\frac{\gamma}{\alpha_+}\right)^i + \frac{KL}{\alpha_+}\sum_{i=k}^{\kappa-1}\alpha_+^{-i}\|\mu(i)\|_{i,\omega}$$

and an application of a discrete version of Gronwall's lemma (cf. Aulbach, Wanner [8]) yields after some straightforward calculations

$$\alpha_+^{-k}\|\mu(k)\|_{k,\omega} \leq \left(\frac{\gamma}{\alpha_+}\right)^k \left(K\gamma^{-\kappa}\|\mu(\kappa)\|_{\kappa,\omega} + \frac{KM_\omega}{\alpha_+ - \gamma - KL}\right)$$

for arbitrary $k \leq \kappa$, which completes the proof of Lemma 2.2. ◇

Now that we have enough information on quasibounded solutions of random difference equations at our disposal we can turn our attention to the central result of this subsection. For the sake of motivation let us first have a look at the linear random difference equation

$$x_{k+1} = \begin{pmatrix} A^+(k,\omega) & 0 \\ 0 & A^-(k,\omega) \end{pmatrix} x_k , \tag{2.18}$$

where the mappings $A^\pm : \mathbb{Z} \times \Omega \to GL(\mathbb{R}^{d^\pm})$ are measurable. Obviously, the two subspace bundles

$$\mathcal{U}^+(\omega) := \{(k,x^+,x^-) \in \mathbb{Z} \times \mathbb{R}^{d^+} \times \mathbb{R}^{d^-} : x^- = 0\} \text{ and}$$
$$\mathcal{U}^-(\omega) := \{(k,x^+,x^-) \in \mathbb{Z} \times \mathbb{R}^{d^+} \times \mathbb{R}^{d^-} : x^+ = 0\}$$

are invariant with respect to (2.18), i.e. every solution starting on $\mathcal{U}^\pm(\omega)$ remains on $\mathcal{U}^\pm(\omega)$ for all times. But what happens if we add a sufficiently small nonlinear perturbation to the equation (2.18)? Of course in general

$\mathcal{U}^{\pm}(\omega)$ will not be invariant for the perturbed equation anymore. Nevertheless one may hope that in the nonlinear case there are two invariant fiber bundles where the fibers are manifolds in $\mathbb{R}^d$ rather than subspaces (cf. Figure 2.1) – and the following proposition shows that this hope actually is a fact.

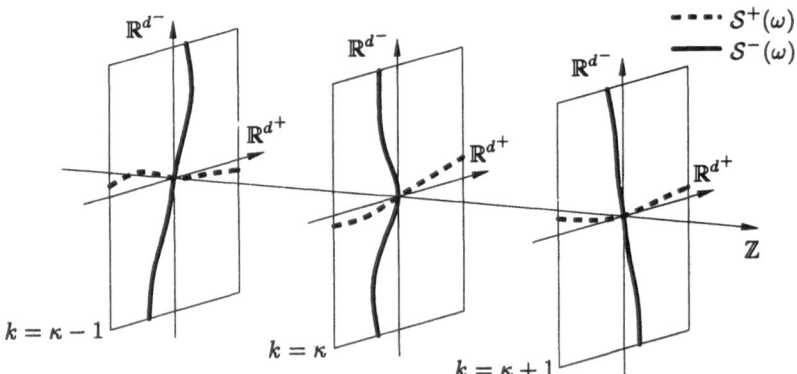

Fig. 2.1. Random invariant fiber bundles

Proposition 2.1. *Consider the following parameter dependent random difference equation*

$$x_{k+1} = \begin{pmatrix} A^+(k,\omega) & 0 \\ 0 & A^-(k,\omega) \end{pmatrix} x_k + \begin{pmatrix} F^+(k,\omega,x_k,p) \\ F^-(k,\omega,x_k,p) \end{pmatrix} \qquad (2.19)$$

where P is an arbitrary metric space, $\mathbb{R}^d = \mathbb{R}^{d^+} \times \mathbb{R}^{d^-}$, the mappings $A^{\pm} : \mathbb{Z} \times \Omega \to \mathrm{GL}(\mathbb{R}^{d^{\pm}})$ and $F^{\pm} : \mathbb{Z} \times \Omega \times \mathbb{R}^d \times P \to \mathbb{R}^{d^{\pm}}$ are measurable, and for arbitrary $k \in \mathbb{Z}$, $\omega \in \Omega$ and $x \in \mathbb{R}^d$ the mappings $F^{\pm}(k,\omega,x,\cdot)$ are continuous. Moreover, suppose that the following holds:

(1) The evolution operators $\Phi^{\pm}(m,n,\omega)$ of the homogeneous linear random difference equations $x_{k+1}^{\pm} = A^{\pm}(k,\omega)x_k^{\pm}$ satisfy

$$\|\Phi^+(m,n,\omega)\|_{n,m,\omega} \le K\alpha_+^{m-n} \quad \text{for all} \quad m \le n\,,$$
$$\|\Phi^-(m,n,\omega)\|_{n,m,\omega} \le K\alpha_-^{m-n} \quad \text{for all} \quad m \ge n$$

with $m,n \in \mathbb{Z}$, $\omega \in \Omega$, as well as real constants $\alpha_+ > \alpha_- > 0$ and $K \ge 1$. Here $\|\cdot\|_{k,\omega}$ denotes a norm on $\mathbb{R}^{d^+}$ or $\mathbb{R}^{d^-}$ according to Definition 2.1. Finally, choose an arbitrary, but fixed, constant $0 < \delta < \frac{\alpha_+ - \alpha_-}{2}$.

(2) For every $k \in \mathbb{Z}$, $\omega \in \Omega$, $x, \bar{x} \in \mathbb{R}^d$ and $p \in P$ we have $F^{\pm}(k,\omega,0,p) = 0$ as well as

$$\|F^{\pm}(k,\omega,x,p) - F^{\pm}(k,\omega,\bar{x},p)\|_{k+1,\omega} \le L\|x - \bar{x}\|_{k,\omega}\,,$$

with $0 \le L < \frac{\delta}{2K}$ and $\|x\|_{k,\omega} := \|x^+\|_{k,\omega} + \|x^-\|_{k,\omega}$ for arbitrary $x = (x^+, x^-) \in \mathbb{R}^{d^+} \times \mathbb{R}^{d^-} = \mathbb{R}^d$.

Finally, assume that for every $\omega \in \Omega$ and $p \in P$ all ω, p-solutions of (2.19) exist uniquely on $\mathbb{Z}$ and that the general solution $\lambda = (\lambda^+, \lambda^-) : \mathbb{Z} \times \mathbb{Z} \times \Omega \times \mathbb{R}^d \times P \to \mathbb{R}^{d^+} \times \mathbb{R}^{d^-}$ is measurable, and continuous with respect to the last two variables. Then the following holds:

(a) *There are uniquely determined mappings $s^\pm : \mathbb{Z} \times \Omega \times \mathbb{R}^{d^\pm} \times P \to \mathbb{R}^{d^\mp}$ such that a point $(\kappa, \xi) \in \mathbb{Z} \times \mathbb{R}^d$ is contained in the random (parameter dependent) fiber bundle*

$$S^+(\omega,p) := \{(k, x^+, s^+(k, \omega, x^+, p)) : k \in \mathbb{Z}, x^+ \in \mathbb{R}^{d^+}\} \text{ or}$$
$$S^-(\omega,p) := \{(k, s^-(k, \omega, x^-, p), x^-) : k \in \mathbb{Z}, x^- \in \mathbb{R}^{d^-}\}$$

if and only if the solution $\lambda(\cdot; \kappa, \omega, \xi, p)$ is γ^-- or γ^+-quasibounded with respect to ω, respectively, for every choice of $\gamma \in [\alpha_- + \delta, \alpha_+ - \delta]$.

(b) *The mappings $s^\pm$ are measurable, the mappings $s^\pm(\kappa, \omega, \cdot, \cdot)$ are continuous, and both the estimate*

$$\|s^\pm(\kappa, \omega, \xi_1^\pm, p) - s^\pm(\kappa, \omega, \xi_2^\pm, p)\|_{\kappa,\omega} \leq \frac{K^2 L(\delta - KL)}{\delta(\delta - 2KL)} \|\xi_1^\pm - \xi_2^\pm\|_{\kappa,\omega}$$

and the identity $s^\pm(\kappa, \omega, 0, p) = 0$ hold for arbitrary $\kappa \in \mathbb{Z}$, $\omega \in \Omega$, $\xi_1^\pm, \xi_2^\pm \in \mathbb{R}^{d^\pm}$ and $p \in P$.

(c) *The graphs $S^\pm(\omega, p)$ are random invariant fiber bundles for (2.19) in the following sense: An inclusion of the form $(\kappa, \xi) \in S^\pm(\omega, p)$ implies $(k, \lambda(k; \kappa, \omega, \xi, p)) \in S^\pm(\omega, p)$ for arbitrary $k \in \mathbb{Z}$.*

Proof. We only prove the assertions concerning s^+. As for the existence and the corresponding properties of s^- all we have to do is to replace the sub- or superscripts "+" by "−", and vice versa, reverse the direction of time, and use the following dual versions of Lemmas 2.1 and 2.2: In Lemma 2.1 consider an interval I which is unbounded to the right, change the $m \geq n$ in (2.9) to $m \leq n$, and then show that for every $0 < \gamma < \alpha_- - KL$ there is a unique γ^+-quasibounded solution of the respective equation provided the constant $M_{\omega,p}$ in (2.10) is finite and in the definition of this constant $k < \kappa$ is changed to $k \geq \kappa$. Similarly, the dual version of Lemma 2.2 shows that after reversing the conditions as above all solutions of the respective equation are γ^+-quasibounded.

In order to prove the assertions of Proposition 2.1 concerning s^+ we will first verify the following claim:

(C) *For every choice of $\kappa \in \mathbb{Z}$, $\omega \in \Omega$, $\xi^+ \in \mathbb{R}^{d^+}$, $p \in P$ and $\gamma \in [\alpha_- + \delta, \alpha_+ - \delta]$ there is a unique point $s_\gamma^+(\kappa, \omega, \xi^+, p) \in \mathbb{R}^{d^-}$ such that $\lambda^-(\cdot; \kappa, \omega, \xi^+, s_\gamma^+(\kappa, \omega, \xi^+, p), p)$ is γ^--quasibounded with respect to ω. Moreover, the mapping s_γ^+ is measurable and $s_\gamma^+(\kappa, \omega, \xi^+, \cdot)$ is continuous for arbitrary $\kappa \in \mathbb{Z}$, $\omega \in \Omega$ and $\xi^+ \in \mathbb{R}^{d^+}$.*

The proof of this claim makes use of Lemmas 5.2 and 5.3 from the appendix. First of all, let $\kappa \in \mathbf{Z}$ and $\gamma \in [\alpha_- + \delta, \alpha_+ - \delta]$ be arbitrary, but fixed, define $I := (-\infty, \kappa] \cap \mathbf{Z}$ and let X_ω denote the Banach space of all mappings $\nu : I \to \mathbf{R}^{d^-}$ which are γ^--quasibounded with respect to ω, equipped with the norm $\| \cdot \|_{\kappa,\omega,\gamma}^-$.

Now let $\omega \in \Omega$, $\xi^+ \in \mathbf{R}^{d^+}$ and $p \in P$ be arbitrary, but fixed. In order to construct a contraction $T_{\omega,\xi^+,p}$ on X_ω let $\nu_{\omega,\xi^+,p} \in X_\omega$ be arbitrary, and consider the initial value problem

$$x_{k+1}^+ = A^+(k,\omega)x_k^+ + F^+(k,\omega,x_k^+,\nu_{\omega,\xi^+,p}(k),p) \quad , \quad k \in I \quad , \quad x_\kappa^+ = \xi^+ .$$

According to Lemma 2.2 it has a unique ω,p-solution $\mu_{\omega,\xi^+,p} : I \to \mathbf{R}^{d^+}$, and this solution is even γ^--quasibounded with respect to ω. Now let us have a look at the random difference equation

$$x_{k+1}^- = A^-(k,\omega)x_k^- + F^-(k,\omega,\mu_{\omega,\xi^+,p}(k),\nu_{\omega,\xi^+,p}(k),p) \quad , \quad k \in I .$$

Because of $\gamma^{-k}\|F^-(k,\omega,\mu_{\omega,\xi^+,p}(k),\nu_{\omega,\xi^+,p}(k),p)\|_{k+1,\omega} \le L\|\nu_{\omega,\xi^+,p}\|_{\kappa,\omega,\gamma}^- + L\|\mu_{\omega,\xi^+,p}\|_{\kappa,\omega,\gamma}^-$ for every $k < \kappa$, Lemma 2.1 guarantees a uniquely determined ω,p-solution $\nu_{\omega,\xi^+,p}^* : I \to \mathbf{R}^{d^-}$ of the above equation which is γ^--quasibounded with respect to ω. Hence the definition $T_{\omega,\xi^+,p}\nu_{\omega,\xi^+,p} := \nu_{\omega,\xi^+,p}^*$ generates an operator $T_{\omega,\xi^+,p} : X_\omega \to X_\omega$, and the reader may verify with the help of Lemmas 2.1 and 5.1 that the family $T_{\omega,\xi^+,p}$, $(\omega,\xi^+,p) \in \Omega \times \mathbf{R}^{d^+} \times P$, satisfies (b) of Lemma 5.2, and for every $\omega \in \Omega$ and $\xi^+ \in \mathbf{R}^{d^+}$ the family $T_{\omega,\xi^+,p}$, $p \in P$, satisfies the assumptions contained in (b) of Lemma 5.3.

But before we are able to apply these two fixed point results we have to derive some estimates. For this let $\omega \in \Omega$, $\xi_1^+, \xi_2^+ \in \mathbf{R}^{d^+}$, $\nu_1, \nu_2 \in X_\omega$ and $p \in P$ be arbitrary, and let μ_i denote the solution of the initial value problem

$$x_{k+1}^+ = A^+(k,\omega)x_k^+ + F^+(k,\omega,x_k^+,\nu_i(k),p) \quad , \quad k \in I \quad , \quad x_\kappa^+ = \xi_i^+ .$$

Then the mapping $\mu := \mu_1 - \mu_2$ is an ω,p-solution of

$$\begin{aligned} x_{k+1}^+ &= A^+(k,\omega)x_k^+ + F^+(k,\omega,x_k^+ + \mu_2(k),\nu_1(k),p) - \\ &\quad - F^+(k,\omega,\mu_2(k),\nu_2(k),p) , \, k \in I , \end{aligned}$$

and since we may apply Lemma 2.2 to this equation we deduce

$$\|\mu_1 - \mu_2\|_{\kappa,\omega,\gamma}^- \le K\gamma^{-\kappa}\|\xi_1^+ - \xi_2^+\|_{\kappa,\omega} + \frac{KL}{\alpha_+ - \gamma - KL}\|\nu_1 - \nu_2\|_{\kappa,\omega,\gamma}^- .$$

Now let $\nu_i^* := T_{\omega,\xi_i^+,p}\nu_i$. Then the mapping $\nu^* := \nu_1^* - \nu_2^*$ is an ω,p-solution of

$$\begin{aligned} x_{k+1}^- &= A^-(k,\omega)x_k^- + F^-(k,\omega,\mu_1(k),\nu_1(k),p) - \\ &\quad - F^-(k,\omega,\mu_2(k),\nu_2(k),p) \quad , \quad k \in I , \end{aligned}$$

and with the help of the last estimate we get

$$\gamma^{-k}\|F^-(k,\omega,\mu_1(k),\nu_1(k),p) - F^-(k,\omega,\mu_2(k),\nu_2(k),p)\|_{k+1,\omega} \leq$$
$$\leq KL\gamma^{-\kappa}\|\xi_1^+ - \xi_2^+\|_{\kappa,\omega} + \frac{L(\alpha_+ - \gamma)}{\alpha_+ - \gamma - KL}\|\nu_1 - \nu_2\|_{\kappa,\omega,\gamma}^-$$

for every $k < \kappa$. Due to the γ^--quasiboundedness (with respect to ω) of ν^* according to our above construction an application of Lemma 2.1 immediately furnishes

$$\|\nu^*\|_{\kappa,\omega,\gamma}^- \leq \frac{K^2 L}{\gamma - \alpha_-}\gamma^{-\kappa}\|\xi_1^+ - \xi_2^+\|_{\kappa,\omega} +$$
$$+ \frac{KL(\alpha_+ - \gamma)}{(\alpha_+ - \gamma - KL)(\gamma_- - \alpha_-)}\|\nu_1 - \nu_2\|_{\kappa,\omega,\gamma}^- ,$$

and $T_{\omega,\xi_i^+,p}\nu_i = \nu_i^*$ and $\gamma \in [\alpha_- + \delta, \alpha_+ - \delta]$ eventually yield

$$\|T_{\omega,\xi_1^+,p}\nu_1 - T_{\omega,\xi_2^+,p}\nu_2\|_{\kappa,\omega,\gamma}^- \leq \frac{K^2 L}{\delta}\gamma^{-\kappa}\|\xi_1^+ - \xi_2^+\|_{\kappa,\omega} + \frac{KL}{\delta - KL}\|\nu_1 - \nu_2\|_{\kappa,\omega,\gamma}^- .$$
$$(2.20)$$

Furthermore, this estimate implies for arbitrary $\omega \in \Omega$, $\xi^+ \in \mathbb{R}^{d^+}$ and $p \in P$

$$\|T_{\omega,\xi^+,p}\nu_1 - T_{\omega,\xi^+,p}\nu_2\|_{\kappa,\omega,\gamma}^- \leq \frac{KL}{\delta - KL}\|\nu_1 - \nu_2\|_{\kappa,\omega,\gamma}^- .$$

According to the assumptions of the proposition we have $KL < \delta - KL$, so $T_{\omega,\xi^+,p}$ indeed is a contraction on X_ω and has a unique fixed point $\nu_{\omega,\xi^+,p} \in X_\omega$. If we finally define $s_\gamma^+(\kappa,\omega,\xi^+,p) := \nu_{\omega,\xi^+,p}(\kappa)$, then it can easily be verified that $s_\gamma^+(\kappa,\omega,\xi^+,p)$ is the unique point in $\mathbb{R}^{d^-}$ for which the mapping $\lambda^-(\cdot;\kappa,\omega,\xi^+,s_\gamma^+(\kappa,\omega,\xi^+,p),p)$ is γ^--quasibounded with respect to ω. In addition, Lemma 5.2 furnishes the claimed measurability of s_γ^+ and Lemma 5.3 shows that $s_\gamma^+(\kappa,\omega,\xi^+,\cdot)$ is continuous for every choice of $\kappa \in \mathbb{Z}$, $\omega \in \Omega$ and $\xi^+ \in \mathbb{R}^{d^+}$. This completes the proof of the above claim *(C)*. In order to complete the proof of Proposition 2.1 we only have to verify *(a)*, *(b)* and *(c)*.

Proof of (a): First of all, let $\gamma_1, \gamma_2 \in [\alpha_- + \delta, \alpha_+ - \delta]$ be arbitrary with $\gamma_1 > \gamma_2$. Since then γ_1^--quasiboundedness implies γ_2^--quasiboundedness, *(C)* yields $s_{\gamma_1}^+ = s_{\gamma_2}^+$, i.e. s_γ^+ actually is independent of the special choice of $\gamma \in [\alpha_- + \delta, \alpha_+ - \delta]$. Hence the index γ may be omitted.

As for the dynamical characterization of $\mathcal{S}^+(\omega,p)$ for every $\gamma \in [\alpha_- + \delta, \alpha_+ - \delta]$ let $\lambda(\cdot;\kappa,\omega,\xi,p) = \lambda(\cdot;\kappa,\omega,\xi^+,\xi^-,p)$ be a γ^--quasibounded mapping with respect to ω. Then $\lambda^-(\cdot;\kappa,\omega,\xi^+,\xi^-,p)$ obviously is γ^--quasibounded, too – and *(C)* yields $\xi^- = s^+(\kappa,\omega,\xi^+,p)$ or $(\kappa,\xi) = (\kappa,\xi^+,\xi^-) \in \mathcal{S}^+(\omega,p)$.

Conversely, let $(\kappa, \xi) \in \mathcal{S}^+(\omega, p)$ be arbitrary. Then the claim *(C)* implies the γ^--quasiboundedness of $\lambda^-(\cdot; \kappa, \omega, \xi, p)$ with respect to ω. If we now apply Lemma 2.2 to the difference equation

$$x_{k+1}^+ = A^+(k, \omega)x_k^+ + F^+(k, \omega, x_k^+, \lambda^-(k; \kappa, \omega, \xi, p), p)$$

we get (observing that $\alpha_+ - \gamma \geq \delta$)

$$\|\lambda^+(\cdot; \kappa, \omega, \xi, p)\|_{\kappa, \omega, \gamma}^- \leq K\gamma^{-\kappa}\|\xi^+\|_{\kappa, \omega} + \frac{KL}{\delta - KL}\|\lambda^-(\cdot; \kappa, \omega, \xi, p)\|_{\kappa, \omega, \gamma}^-,$$
(2.21)

i.e. $\lambda^+(\cdot; \kappa, \omega, \xi, p)$ is γ^--quasibounded with respect to ω – and therefore the same holds for $\lambda(\cdot; \kappa, \omega, \xi, p)$, too.

Proof of (b): The measurability of s^+ is already clear, and since the trivial mapping is an ω, p-solution of (2.19) for arbitrary $\omega \in \Omega$ and $p \in P$ which is γ^--quasibounded with respect to ω, the identity $s^+(\kappa, \omega, 0, p) = 0$ follows from the uniqueness statement contained in claim *(C)*.

Now let $\kappa \in \mathbb{Z}$ be arbitrary, but fixed, and let $\nu_{\omega, \xi_1^+, p}$ and $\nu_{\omega, \xi_2^+, p}$ denote the fixed point of $T_{\omega, \xi_1^+, p}$ and $T_{\omega, \xi_2^+, p}$, respectively. Then (2.20) yields after some straightforward calculations

$$\|\nu_{\omega, \xi_1^+, p} - \nu_{\omega, \xi_2^+, p}\|_{\kappa, \omega, \gamma}^- \leq \frac{K^2 L(\delta - KL)}{\delta(\delta - 2KL)}\gamma^{-\kappa}\|\xi_1^+ - \xi_2^+\|_{\kappa, \omega},$$
(2.22)

and finally

$$\|s^+(\kappa, \omega, \xi_1^+, p) - s^+(\kappa, \omega, \xi_2^+, p)\|_{\kappa, \omega} = \|(\nu_{\omega, \xi_1^+, p} - \nu_{\omega, \xi_2^+, p})(\kappa)\|_{\kappa, \omega} \leq$$

$$\leq \gamma^\kappa \|\nu_{\omega, \xi_1^+, p} - \nu_{\omega, \xi_2^+, p}\|_{\kappa, \omega, \gamma}^- \leq \frac{K^2 L(\delta - KL)}{\delta(\delta - 2KL)}\|\xi_1^+ - \xi_2^+\|_{\kappa, \omega}.$$

Hence, only the continuity assertion remains to be verified. Let $\kappa \in \mathbb{Z}$ and $\omega \in \Omega$ be arbitrary. Then the last estimate furnishes for all $\xi^+, \xi_0^+ \in \mathbb{R}^{d^+}$ and $p, p_0 \in P$ the inequality

$$\|s^+(\kappa, \omega, \xi^+, p) - s^+(\kappa, \omega, \xi_0^+, p_0)\|_{\kappa, \omega} \leq$$

$$\leq \frac{K^2 L(\delta - KL)}{\delta(\delta - 2KL)}\|\xi^+ - \xi_0^+\|_{\kappa, \omega} + \|s^+(\kappa, \omega, \xi_0^+, p) - s^+(\kappa, \omega, \xi_0^+, p_0)\|_{\kappa, \omega},$$

which – together with the already proved continuity of $s^+(\kappa, \omega, \xi_0^+, \cdot)$ – completes the proof of *(b)*.

Proof of (c): Let $(\kappa, \xi) \in \mathcal{S}^+(\omega, p)$ and $k \in \mathbb{Z}$ be arbitrary. Then the mapping $\lambda(\cdot; \kappa, \omega, \xi, p)$ is γ^--quasibounded with respect to ω. Now (2.4) yields the identity

$$\lambda(\cdot; k, \omega, \lambda(k; \kappa, \omega, \xi, p), p) = \lambda(\cdot; \kappa, \omega, \xi, p),$$

which in turn implies the γ^--quasiboundedness of $\lambda(\cdot; k, \omega, \lambda(k; \kappa, \omega, \xi, p), p)$ with respect to ω, i.e. we have $(k, \lambda(k; \kappa, \omega, \xi, p)) \in \mathcal{S}^+(\omega, p)$ as claimed. $\Diamond$

With Proposition 2.1 we now have the crucial tool at hand for developing a linearization theory for random difference equations. It will play a central role in the proofs of all major results to come.

But before we end this technical subsection, let us add some remarks on the asymptotic behavior of ω, p-solutions which are contained in the invariant fiber bundles of the proposition. For this, let μ denote a solution contained in $S^+(\omega, p)$. Then the reader may verify that for every $0 < \gamma \leq \alpha_+ - \delta$ the estimates (2.21) and (2.22) imply after some straightforward calculations

$$\|\mu(k)\|_{k,\omega} \leq \frac{K(\delta - KL)}{\delta - 2KL}\gamma^{k-\kappa}\|\mu(\kappa)\|_{\kappa,\omega} \quad \text{for all} \quad k \leq \kappa.$$

With the aid of $\mu(\kappa) = \lambda(\kappa; k, \omega, \lambda(k; \kappa, \omega, \mu(\kappa)))$ we further deduce

$$\|\mu(k)\|_{k,\omega} \geq \frac{\delta - 2KL}{K(\delta - KL)}\gamma^{k-\kappa}\|\mu(\kappa)\|_{\kappa,\omega} \quad \text{for all} \quad k \geq \kappa.$$

Similarly, for every ω, p-solution ν of (2.19) which is contained in $S^-(\omega, p)$ and every choice of $\gamma \geq \alpha_- + \delta$ we have

$$\|\nu(k)\|_{k,\omega} \leq \frac{K(\delta - KL)}{\delta - 2KL}\gamma^{k-\kappa}\|\nu(\kappa)\|_{\kappa,\omega} \quad \text{for all} \quad k \geq \kappa,$$

$$\|\nu(k)\|_{k,\omega} \geq \frac{\delta - 2KL}{K(\delta - KL)}\gamma^{k-\kappa}\|\nu(\kappa)\|_{\kappa,\omega} \quad \text{for all} \quad k \leq \kappa.$$

One first, and easy, consequence of these estimates is the fact that only the trivial solution of (2.19) is contained in both $S^+(\omega, p)$ and $S^-(\omega, p)$.

2.3 Random Invariant Fiber Bundles

Beginning with this subsection we present the results that constitute our linearization theory for nonautonomous random difference equations. But before going into that let us try to explain the assumptions on the linear part contained in Proposition 2.1(1), since this will probably help to clarify the assumptions to come.

Suppose that the linear part of (2.19) is independent of time and chance, i.e. consider the autonomous deterministic special case $A = \text{diag}(A^+, A^-)$. Then it can easily be shown that the assumptions on the evolution operators $\Phi^\pm$ are satisfied with respect to the Euclidean norm if all eigenvalues of A^+ or A^- have modulus strictly greater than α_+ or strictly less than α_-, respectively (cf. Aulbach, Wanner [8]). In other words, the spectrum $\sigma(A)$ of A has to have empty intersection with the ring $\{z \in \mathbb{C} : \alpha_- \leq |z| \leq \alpha_+\}$ as depicted in Figure 2.2(a). Yet in many situations we will have more information on the spectrum of A at our disposal. For instance, it might occur that the eigenvalues of A – in generalization of the above situation – can be separated by $p - 1$ rings (as shown in Figure 2.2(b) for the case

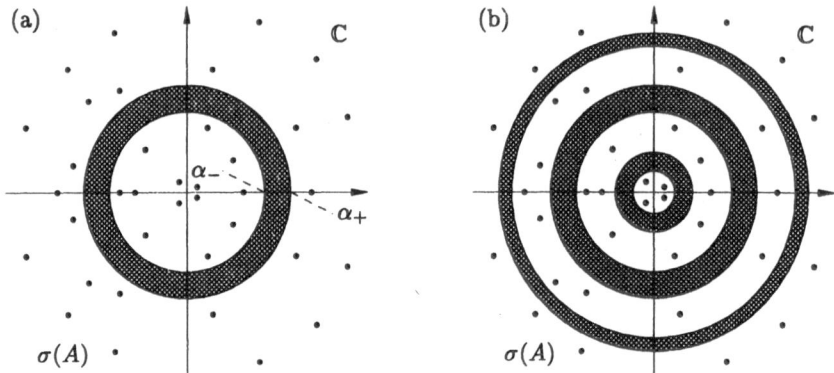

Fig. 2.2. On the assumptions concerning the linear part

$p = 4$), and hopefully this will enable us to deduce even more information on the nonlinear perturbation of our equation.

In what follows, we will show that this indeed is the case, and this subsection will be devoted to the existence of numerous random invariant fiber bundles (to be precise, we prove the existence of $\frac{(p-1)(p+2)}{2}$ invariant fiber bundles).

To begin with, let us state once and for all the assumptions we impose on the nonautonomous random difference equation

$$x_{k+1} = A(k,\omega)x_k + F(k,\omega,x_k) \tag{2.23}$$

on $\mathbb{R}^d$ which will be studied from now on. As for the linear part of (2.23) we require the following conditions reflecting the above-discussed situation.

(A1) *There are measurable mappings $A_i : \mathbb{Z} \times \Omega \to \mathrm{GL}(\mathbb{R}^{d_i})$, $i = 1,\ldots,p$, such that for arbitrary $k \in \mathbb{Z}$ and $\omega \in \Omega$ we have $A(k,\omega) = diag(A_1(k,\omega), A_2(k,\omega),\ldots,A_p(k,\omega))$, i.e. $A(k,\omega)$ is in block-diagonal form with $p \geq 2$ blocks. On every $\mathbb{R}^{d_i}$ there is a norm as in Definition 2.1 depending on both k and ω^4, and for $x = (x^1, x^2,\ldots, x^p) \in \mathbb{R}^{d_1} \times \mathbb{R}^{d_2} \times \ldots \times \mathbb{R}^{d_p} = \mathbb{R}^d$ we define $\|x\|_{k,\omega} := \|x^1\|_{k,\omega} + \|x^2\|_{k,\omega} + \ldots + \|x^p\|_{k,\omega}$. Moreover, let $\alpha_{i,+}$, $\alpha_{i,-}$ denote real constants with*

$$\infty > \alpha_{1,-} > \alpha_{1,+} > \alpha_{2,-} > \ldots > \alpha_{p-1,+} > \alpha_{p,-} > \alpha_{p,+} > 0\,,$$

and assume that the evolution operators $\Phi_i(m,n,\omega)$ of the homogeneous linear difference equations $x^i_{k+1} = A_i(k,\omega)x^i_k$ satisfy

$$\|\Phi_i(m,n,\omega)\|_{n,m,\omega} \leq K\alpha_{i,+}^{m-n} \quad \text{for all} \quad m \leq n\,,$$

$$\|\Phi_i(m,n,\omega)\|_{n,m,\omega} \leq K\alpha_{i,-}^{m-n} \quad \text{for all} \quad m \geq n$$

---

[4] By abuse of notation all these (different!) norms are denoted by $\|\cdot\|_{k,\omega}$.

with $m, n \in \mathbb{Z}$, $\omega \in \Omega$, $i = 1, \ldots, p$, as well as some constant $K \geq 1$. Finally, let

$$0 < \delta < \min \left\{ \frac{\alpha_{1,+} - \alpha_{2,-}}{2}, \ldots, \frac{\alpha_{p-1,+} - \alpha_{p,-}}{2}, \alpha_{p,+} \right\}$$

be arbitrary, but fixed.

As for the (in general) nonlinear perturbation F we require the following.

(A2) *The mapping $F : \mathbb{Z} \times \Omega \times \mathbb{R}^d \to \mathbb{R}^d$ is measurable with $F(k, \omega, 0) = 0$ for arbitrary $k \in \mathbb{Z}$ and $\omega \in \Omega$. Furthermore, we have*

$$\|F_i(k, \omega, x) - F_i(k, \omega, \bar{x})\|_{k+1,\omega} \leq L \|x - \bar{x}\|_{k,\omega}$$

for every $i = 1, \ldots, p$, $k \in \mathbb{Z}$, $\omega \in \Omega$ and $x, \bar{x} \in \mathbb{R}^d$, with some constant $L \geq 0$ and

$$F(k, \omega, x) =: (F_1(k, \omega, x), \ldots, F_p(k, \omega, x)) \in \mathbb{R}^{d_1} \times \ldots \times \mathbb{R}^{d_p}.$$

Of course, the Lipschitz constant L has to be sufficiently small. More precise statements on the right choice of L will be given explicitly in the results to come. Besides the above two assumptions the general solution of (2.23) has to satisfy the following condition.

(A3) *For arbitrary $\omega \in \Omega$ all ω-solutions of (2.23) exist uniquely on the whole of $\mathbb{Z}$, and the general solution $\lambda = (\lambda_1, \ldots, \lambda_p) : \mathbb{Z} \times \mathbb{Z} \times \Omega \times \mathbb{R}^d \to \mathbb{R}^{d_1} \times \ldots \times \mathbb{R}^{d_p} = \mathbb{R}^d$ is measurable, and continuous with respect to the last variable.*

While assumptions (A1), (A2) and (A3) suffice for all results contained in Subsections 2.3, 2.4 and 2.5, we will require an additional assumption in Subsection 2.6.

(A4) *For arbitrary $i = 1, \ldots, p$, $k \in \mathbb{Z}$, $\omega \in \Omega$ and $x \in \mathbb{R}^d$ the estimate*

$$\|F_i(k, \omega, x)\|_{k+1,\omega} \leq M$$

holds with some constant $M \geq 0$.

Finally, let us introduce some notation. For every $x = (x^1, \ldots, x^p) \in \mathbb{R}^{d_1} \times \ldots \times \mathbb{R}^{d_p} = \mathbb{R}^d$ and $i, j \in \{1, \ldots, p\}$ with $i \leq j$ define $\lambda_{\leq i} := (\lambda_1, \ldots, \lambda_i)$, $\lambda_{>i} := (\lambda_{i+1}, \ldots, \lambda_p)$,

$$
\begin{aligned}
x^{\leq i} &:= (x^1, \ldots, x^i) &\in& \quad \mathbb{R}^{d_1} \times \ldots \times \mathbb{R}^{d_i}, \\
x^{>i} &:= (x^{i+1}, \ldots, x^p) &\in& \quad \mathbb{R}^{d_{i+1}} \times \ldots \times \mathbb{R}^{d_p}, \\
x^{i \leq j} &:= (x^i, \ldots, x^j) &\in& \quad \mathbb{R}^{d_i} \times \ldots \times \mathbb{R}^{d_j},
\end{aligned}
$$

and so on, and for every subset $\{i_1, \ldots, i_m\} \subset \{1, \ldots, p\}$ let

$$\|(x^{i_1}, \ldots, x^{i_m})\|_{k,\omega} := \|x^{i_1}\|_{k,\omega} + \ldots + \|x^{i_m}\|_{k,\omega}.$$

In other words, in the following subvectors of vectors will be given by super-scripts, whereas for mappings we will use subscripts to abbreviate combinations of the respective component mappings.

Now we have gathered everything in order to state and prove the main result of this subsection, the existence of many random invariant fiber bundles for (2.23).

Theorem 2.1. *Consider the random difference equation*

$$x_{k+1} = A(k,\omega)x_k + F(k,\omega,x_k) \tag{2.24}$$

satisfying (A1), (A2) and (A3) from above with $0 \le L < \frac{\delta}{2K^2p}(K+2-\sqrt{K^2+4})$. Then for every choice of $1 \le i \le j \le p$ with $(i,j) \neq (1,p)$ there is a measurable mapping $s^{i,j} : \mathbb{Z} \times \Omega \times \mathbb{R}^{d_i} \times \ldots \times \mathbb{R}^{d_j} \to \mathbb{R}^{d_1} \times \ldots \times \mathbb{R}^{d_{i-1}} \times \mathbb{R}^{d_{j+1}} \times \ldots \times \mathbb{R}^{d_p}$ such that the set

$$\mathcal{S}^{i,j}(\omega) := \{(k,x) \in \mathbb{Z} \times \mathbb{R}^d : (x^{<i}, x^{>j}) = s^{i,j}(k,\omega,x^{i \le j})\}$$

is a random invariant fiber bundle for (2.24). For $1 \le i < p$ we have the dynamical characterizations

$$
\begin{aligned}
\mathcal{S}^{1,i}(\omega) &= \{(\kappa,\xi) \in \mathbb{Z} \times \mathbb{R}^d : \lambda(\cdot;\kappa,\omega,\xi) \text{ is } \gamma^- \text{-quasibounded} \\
&\qquad \text{with respect to } \omega\} = \\
&= \{(\kappa,\xi) \in \mathbb{Z} \times \mathbb{R}^d : \lambda(\cdot;\kappa,\omega,\xi) \text{ is } \gamma^- \text{- and } \gamma_1^+ \text{-quasibounded} \\
&\qquad \text{with respect to } \omega\}, \\
\mathcal{S}^{i+1,p}(\omega) &= \{(\kappa,\xi) \in \mathbb{Z} \times \mathbb{R}^d : \lambda(\cdot;\kappa,\omega,\xi) \text{ is } \gamma^+ \text{-quasibounded} \\
&\qquad \text{with respect to } \omega\} = \\
&= \{(\kappa,\xi) \in \mathbb{Z} \times \mathbb{R}^d : \lambda(\cdot;\kappa,\omega,\xi) \text{ is } \gamma^+ \text{- and } \gamma_2^- \text{-quasibounded} \\
&\qquad \text{with respect to } \omega\},
\end{aligned}
$$

for arbitrary $\gamma \in [\alpha_{i+1,-} + \delta, \alpha_{i,+} - \delta]$, $\gamma_1 \ge \alpha_{1,-} + \delta$ and $0 < \gamma_2 \le \alpha_{p,+} - \delta$. Furthermore, for $1 < i \le j < p$ we get the dynamical characterization

$$
\begin{aligned}
\mathcal{S}^{i,j}(\omega) &= \{(\kappa,\xi) \in \mathbb{Z} \times \mathbb{R}^d : \lambda(\cdot;\kappa,\omega,\xi) \text{ is } \gamma_1^+ \text{- and } \gamma_2^- \text{-quasibounded} \\
&\qquad \text{with respect to } \omega\},
\end{aligned}
$$

for every choice of $\gamma_1 \in [\alpha_{i,-} + \delta, \alpha_{i-1,+} - \delta]$ and $\gamma_2 \in [\alpha_{j+1,-} + \delta, \alpha_{j,+} - \delta]$. Finally, for arbitrary $\kappa \in \mathbb{Z}$ and $\omega \in \Omega$ the identity $s^{i,j}(\kappa,\omega,0) = 0$ holds and the mapping $s^{i,j}(\kappa,\omega,\cdot)$ is globally Lipschitz continuous with respect to the norm $\|\cdot\|_{\kappa,\omega}$ with Lipschitz constant $\frac{2C(L)}{1-C(L)}$, where $C(L) = C(L,K,\delta,p) = \frac{K^2Lp(\delta-KLp)}{\delta(\delta-2KLp)}$.

Proof. The reader might have guessed already that in proving the above theorem we shall make extensive use of Proposition 2.1. To begin with, let $i \in \{1, \ldots, p-1\}$ be arbitrary, but fixed. Since we first like to apply the proposition to prove the existence of $\mathcal{S}^{1,i}(\omega)$ and $\mathcal{S}^{i+1,p}(\omega)$ we define $\mathbb{R}^{d^+} := \mathbb{R}^{d_1} \times \ldots \times \mathbb{R}^{d_i}$, $\mathbb{R}^{d^-} := \mathbb{R}^{d_{i+1}} \times \ldots \times \mathbb{R}^{d_p}$, and for every $x \in \mathbb{R}^d$ we write $x = (x^+, x^-) \in \mathbb{R}^{d^+} \times \mathbb{R}^{d^-}$ and set $\|x^+\|_{k,\omega} := \|x^{\le i}\|_{k,\omega}$, $\|x^-\|_{k,\omega} := \|x^{>i}\|_{k,\omega}$. Then it may readily be verified that the evolution operators corresponding to $A^+(k,\omega) := \mathrm{diag}(A_1(k,\omega), \ldots, A_i(k,\omega))$ and $A^-(k,\omega) := \mathrm{diag}(A_{i+1}(k,\omega), \ldots, A_p(k,\omega))$ satisfy all the assumptions of Proposition 2.1*(1)* with $\alpha_+ := \alpha_{i,+}$ and $\alpha_- := \alpha_{i+1,-}$. Furthermore, with $F^+ := (F_1, \ldots, F_i)$ and $F^- := (F_{i+1}, \ldots, F_p)$ we deduce the estimates

$$\|F^+(k,\omega,x) - F^+(k,\omega,\bar{x})\|_{k+1,\omega} \le iL\|x - \bar{x}\|_{k,\omega} \le$$
$$\le pL\|x - \bar{x}\|_{k,\omega} \, ,$$
$$\|F^-(k,\omega,x) - F^-(k,\omega,\bar{x})\|_{k+1,\omega} \le (p-i)L\|x - \bar{x}\|_{k,\omega} \le$$
$$\le pL\|x - \bar{x}\|_{k,\omega} \, ,$$

and because of $0 \le L < \frac{\delta}{2K^2 p}(K + 2 - \sqrt{K^2 + 4}) \le \frac{\delta}{2Kp}$ an application of Proposition 2.1 now furnishes the assertions concerning $\mathcal{S}^{1,i}(\omega)$ and $\mathcal{S}^{i+1,p}(\omega)$ if we note that the Lipschitz constant of $s^{1,i}$ and $s^{i+1,p}$ even is $C(L) \le \frac{2C(L)}{1 - C(L)}$ and that the alternative dynamical characterizations are consequences of an application of Lemma 2.2 and its dual version.

As for the remaining fiber bundles $\mathcal{S}^{i,j}(\omega)$ the dynamical characterization contained in Theorem 2.1 already indicates that they are nothing but the intersection of two suitable fiber bundles from above. Hence, let $1 < i \le j < p$ be arbitrary, but fixed, and define $\mathcal{X}_1 := \mathbb{R}^{d_1} \times \ldots \times \mathbb{R}^{d_{i-1}}$, $\mathcal{X}_2 := \mathbb{R}^{d_i} \times \ldots \times \mathbb{R}^{d_j}$, as well as $\mathcal{X}_3 := \mathbb{R}^{d_{j+1}} \times \ldots \times \mathbb{R}^{d_p}$. We know by now that the invariant fiber bundles $\mathcal{S}^{1,j}(\omega)$ and $\mathcal{S}^{i,p}(\omega)$ are generated by measurable mappings $s^{1,j}$ and $s^{i,p}$. If we define a family of mappings $T_{\kappa,\omega,\xi_2} : \mathcal{X}_1 \times \mathcal{X}_3 \to \mathcal{X}_1 \times \mathcal{X}_3$ for $\kappa \in \mathbb{Z}$, $\omega \in \Omega$ and $\xi_2 \in \mathcal{X}_2$ by

$$T_{\kappa,\omega,\xi_2}(\xi_1, \xi_3) := (s^{i,p}(\kappa, \omega, \xi_2, \xi_3), s^{1,j}(\kappa, \omega, \xi_1, \xi_2)) \, ,$$

then the mapping $T_{\cdot,\cdot,\cdot}(\cdot, \cdot)$ is measurable and for arbitrary $\kappa \in \mathbb{Z}$, $\omega \in \Omega$, $\xi_2 \in \mathcal{X}_2$, $\xi_1, \bar{\xi}_1 \in \mathcal{X}_1$ and $\xi_3, \bar{\xi}_3 \in \mathcal{X}_3$ the global Lipschitz condition of $s^{1,j}$ and $s^{i,p}$ yields the estimate

$$\|T_{\kappa,\omega,\xi_2}(\xi_1, \xi_3) - T_{\kappa,\omega,\xi_2}(\bar{\xi}_1, \bar{\xi}_3)\|_{\kappa,\omega} =$$
$$= \|s^{i,p}(\kappa, \omega, \xi_2, \xi_3) - s^{i,p}(\kappa, \omega, \xi_2, \bar{\xi}_3)\|_{\kappa,\omega} +$$
$$+ \|s^{1,j}(\kappa, \omega, \xi_1, \xi_2) - s^{1,j}(\kappa, \omega, \bar{\xi}_1, \xi_2)\|_{\kappa,\omega} \le$$
$$\le C(L)\|(\xi_1, \xi_3) - (\bar{\xi}_1, \bar{\xi}_3)\|_{\kappa,\omega} \, .$$

Now observe that the condition concerning L contained in Theorem 2.1 has been tailor-made for assuring $C(L) < 1$, so T_{κ,ω,ξ_2} is a contraction on $\mathcal{X}_1 \times \mathcal{X}_3$ with respect to the norm $\|(\xi_1, \xi_3)\|_{\kappa,\omega} = \|\xi_1\|_{\kappa,\omega} + \|\xi_3\|_{\kappa,\omega}$. If we denote the

226 Thomas Wanner

unique fixed point by $s^{i,j}(\kappa,\omega,\xi_2)$, then an application of Lemma 5.1 furnishes the measurability of $s^{i,j}$. Furthermore, a point (ξ_1,ξ_3) is a fixed point of T_{κ,ω,ξ_2} if and only if $(\kappa,\xi_1,\xi_2,\xi_3)\in\mathcal{S}^{1,j}(\omega)\cap\mathcal{S}^{i,p}(\omega)$, hence the set $\mathcal{S}^{i,j}(\omega)$ defined in the theorem indeed is a random invariant fiber bundle which contains the trivial solution and admits the required dynamical characterization.

In order to conclude the proof of the theorem we only have to verify the global Lipschitz condition. For this, fix $\kappa\in\mathbb{Z}$ and $\omega\in\Omega$, and let $s^{i,j}(\kappa,\omega,\xi_2)=:(s_1^{i,j}(\kappa,\omega,\xi_2),s_3^{i,j}(\kappa,\omega,\xi_2))\in\mathcal{X}_1\times\mathcal{X}_3$. Since the point $(\kappa,s_1^{i,j}(\kappa,\omega,\xi_2),\xi_2,s_3^{i,j}(\kappa,\omega,\xi_2))$ is contained in both $\mathcal{S}^{1,j}(\omega)$ and $\mathcal{S}^{i,p}(\omega)$ for every $\xi_2\in\mathcal{X}_2$ according to our above construction, we deduce for arbitrary $\xi_2,\bar{\xi}_2\in\mathcal{X}_2$ the estimate

$$\|s_3^{i,j}(\kappa,\omega,\xi_2)-s_3^{i,j}(\kappa,\omega,\bar{\xi}_2)\|_{\kappa,\omega}=$$
$$=\ \|s^{1,j}(\kappa,\omega,s_1^{i,j}(\kappa,\omega,\xi_2),\xi_2)-s^{1,j}(\kappa,\omega,s_1^{i,j}(\kappa,\omega,\bar{\xi}_2),\bar{\xi}_2)\|_{\kappa,\omega}\le$$
$$\le\ C(L)\|s_1^{i,j}(\kappa,\omega,\xi_2)-s_1^{i,j}(\kappa,\omega,\bar{\xi}_2)\|_{\kappa,\omega}+C(L)\|\xi_2-\bar{\xi}_2\|_{\kappa,\omega}\ ,$$

as well as an analogous estimate for $\|s_1^{i,j}(\kappa,\omega,\xi_2)-s_1^{i,j}(\kappa,\omega,\bar{\xi}_2)\|_{\kappa,\omega}$. Adding those two inequalities eventually proves the desired Lipschitz condition of $s^{i,j}$. ◊

Remark 2.1. In view of later applications let us introduce an abbreviating notation. For arbitrary $1\le i\le j\le p$ with $(i,j)\ne(1,p)$, $\kappa\in\mathbb{Z}$, $\omega\in\Omega$ and $\xi\in\mathbb{R}^{d_i}\times\ldots\times\mathbb{R}^{d_j}$ let

$$(\kappa,\pi^{i,j}(\kappa,\omega,\xi))\in\mathbb{Z}\times\mathbb{R}^d$$

denote the unique point on $\mathcal{S}^{i,j}(\omega)$ which is determined by κ and ξ, i.e. using the abbreviations of the last proof we have $\pi^{i,j}(\kappa,\omega,\xi)=(\xi_1,\xi_2,\xi_3)\in\mathcal{X}_1\times\mathcal{X}_2\times\mathcal{X}_3$, where $\xi_2:=\xi$ and $(\xi_1,\xi_3):=s^{i,j}(\kappa,\omega,\xi_2)$. Now the random invariant fiber bundle $\mathcal{S}^{i,j}(\omega)$ can be written alternatively as $\mathcal{S}^{i,j}(\omega)=\{(\kappa,\pi^{i,j}(\kappa,\omega,\xi)):\kappa\in\mathbb{Z},\xi\in\mathbb{R}^{d_i}\times\ldots\times\mathbb{R}^{d_j}\}$.

Furthermore, in the following we will often write $\mathcal{S}^{\le i}(\omega)$ or $\mathcal{S}^{>i}(\omega)$ instead of $\mathcal{S}^{1,i}(\omega)$ or $\mathcal{S}^{i+1,p}(\omega)$, respectively. □

With Theorem 2.1 we have the first major result of our linearization theory for nonautonomous random difference equations at hand. We managed to show that – as in the classical case discussed at the beginning of the introduction – many invariant subspace bundles of a given linear equation survive a sufficiently small nonlinear perturbation as random invariant fiber bundles, whose fibers are Lipschitzian manifolds.

Yet the alert reader might have noticed the following: For the linear part of (2.23) there are not only $\frac{(p-1)(p+2)}{2}$ invariant subspace bundles, but rather 2^p-2 ones! So it is only natural to ask whether or not they survive the nonlinear perturbation, too.

The answer to this question cannot be given now – unfortunately we will have to wait until the end of Subsection 2.5. However, we like to note already that the nonlinear analogues (provided they exist) cannot have fibers which are Lipschitzian manifolds. An example proving this assertion can be found in Aulbach [7] or Aulbach, Wanner [8]. In other words, the best we can expect in general are random invariant fiber bundles whose fibers are ordinary manifolds.

2.4 Asymptotic Phases

Whereas in the last subsection we constructed numerous random invariant fiber bundles for the random difference equation

$$x_{k+1} = A(k,\omega)x_k + F(k,\omega,x_k) \qquad (2.25)$$

in this subsection we like to study the asymptotic behavior of solutions of (2.25) which are not contained in any of these invariant fiber bundles. Our first step towards this aim is the following lemma, which shows that for every given ω-solution μ of (2.25) there are invariant fiber bundles having exactly this solution in common. Obviously, this result is to generalize the results of the last subsection, where we considered the special case $\mu = 0$.

Lemma 2.3. *Consider the random difference equation*

$$x_{k+1} = A(k,\omega)x_k + F(k,\omega,x_k) \qquad (2.26)$$

satisfying (A1), (A2) and (A3) with $0 \leq L < \frac{\delta}{2Kp}$. Then for every choice of $i \in \{1,\ldots,p-1\}$ there are two families of mappings

$$
\begin{aligned}
s^{\leq i}_{\kappa_0,\xi_0} &: \quad \mathbb{Z} \times \Omega \times \mathbb{R}^{d_1} \times \ldots \times \mathbb{R}^{d_i} \quad \to \quad \mathbb{R}^{d_{i+1}} \times \ldots \times \mathbb{R}^{d_p}, \\
s^{>i}_{\kappa_0,\xi_0} &: \quad \mathbb{Z} \times \Omega \times \mathbb{R}^{d_{i+1}} \times \ldots \times \mathbb{R}^{d_p} \quad \to \quad \mathbb{R}^{d_1} \times \ldots \times \mathbb{R}^{d_i},
\end{aligned}
$$

with $\kappa_0 \in \mathbb{Z}$ and $\xi_0 \in \mathbb{R}^d$ such that the mappings $s^{\leq i}_{\cdot,\cdot}(\cdot,\cdot,\cdot)$ and $s^{>i}_{\cdot,\cdot}(\cdot,\cdot,\cdot)$ are measurable, and for arbitrary $\kappa_0 \in \mathbb{Z}$ and $\xi_0 \in \mathbb{R}^d$ the sets

$$
\begin{aligned}
S^{\leq i}_{\kappa_0,\xi_0}(\omega) &:= \quad \{(k,x) \in \mathbb{Z} \times \mathbb{R}^d : x^{>i} = s^{\leq i}_{\kappa_0,\xi_0}(k,\omega,x^{\leq i})\}, \\
S^{>i}_{\kappa_0,\xi_0}(\omega) &:= \quad \{(k,x) \in \mathbb{Z} \times \mathbb{R}^d : x^{\leq i} = s^{>i}_{\kappa_0,\xi_0}(k,\omega,x^{>i})\}
\end{aligned}
$$

are random invariant fiber bundles for (2.26), the so-called hierarchy bundles through (κ_0,ξ_0) (or through the solution $\lambda(\cdot;\kappa_0,\omega,\xi_0)$). These fiber bundles can be characterized dynamically as

$$
\begin{aligned}
S^{\leq i}_{\kappa_0,\xi_0}(\omega) &= \{(\kappa,\xi) : \lambda(\cdot;\kappa,\omega,\xi) - \lambda(\cdot;\kappa_0,\omega,\xi_0) \text{ is } \gamma^-\text{-quasibounded} \\
&\qquad\qquad \text{with respect to } \omega\}, \\
S^{>i}_{\kappa_0,\xi_0}(\omega) &= \{(\kappa,\xi) : \lambda(\cdot;\kappa,\omega,\xi) - \lambda(\cdot;\kappa_0,\omega,\xi_0) \text{ is } \gamma^+\text{-quasibounded} \\
&\qquad\qquad \text{with respect to } \omega\},
\end{aligned}
$$

for every choice of $\gamma \in [\alpha_{i+1,-} + \delta, \alpha_{i,+} - \delta]$. Finally, the mappings $s^{\leq i}_{\cdot,\cdot}(\cdot,\omega,\cdot)$ and $s^{>i}_{\cdot,\cdot}(\cdot,\omega,\cdot)$ are continuous for $\omega \in \Omega$, and the mappings $s^{\leq i}_{\kappa_0,\xi_0}(\kappa,\omega,\cdot)$

and $s^{>i}_{\kappa_0,\xi_0}(\kappa,\omega,\cdot)$ *are Lipschitz continuous with respect to the norm* $\|\cdot\|_{\kappa,\omega}$ *with Lipschitz constant* $C(L)$ *(which has been defined in Theorem 2.1) for all* $\kappa, \kappa_0 \in \mathbb{Z}$ *and* $\omega \in \Omega$.

Proof. Let $i \in \{1,\dots,p-1\}$ be arbitrary, but fixed. Because of symmetry we only prove the assertions concerning $S^{\leq i}_{\kappa_0,\xi_0}(\omega)$. To that end, consider the random difference equation

$$x_{k+1} = A(k,\omega)x_k + \tilde{F}(k,\omega,x_k,\kappa_0,\xi_0)\,, \tag{2.27}$$

depending continuously on the parameter $(\kappa_0,\xi_0) \in \mathbb{Z} \times \mathbb{R}^d$, where

$$\tilde{F}(k,\omega,x,\kappa_0,\xi_0) := F(k,\omega,x + \lambda(k;\kappa_0,\omega,\xi_0)) - F(k,\omega,\lambda(k;\kappa_0,\omega,\xi_0))\,.$$

If we define $\mathbb{R}^{d\pm}$ as in the first part of the proof of Theorem 2.1, it can easily be shown that this *random difference equation of perturbed motion* (cf. Aulbach, Wanner [8]) satisfies all the assumptions of Proposition 2.1. Let $s^{\leq i}$ denote the (parameter dependent!) mapping guaranteed by this proposition. Then the definition

$$s^{\leq i}_{\kappa_0,\xi_0}(\kappa,\omega,\xi^{\leq i}) := s^{\leq i}(\kappa,\omega,\xi^{\leq i} - \lambda_{\leq i}(\kappa;\kappa_0,\omega,\xi_0),\kappa_0,\xi_0) + \lambda_{>i}(\kappa;\kappa_0,\omega,\xi_0)$$

implies the measurability of $s^{\leq i}_{\cdot,\cdot}(\cdot,\cdot,\cdot)$, as well as the continuity of $s^{\leq i}_{\cdot,\cdot}(\cdot,\omega,\cdot)$ for arbitrary $\omega \in \Omega$. In addition, $s^{\leq i}_{\kappa_0,\xi_0}(\kappa,\omega,\cdot)$ satisfies the required global Lipschitz condition.

As for the dynamical characterization we only have to mention that a mapping ν is an ω,κ_0,ξ_0-solution of (2.27) if and only if the mapping $\nu + \lambda(\cdot;\kappa_0,\omega,\xi_0)$ is an ω-solution of (2.26) – so the dynamical characterization follows immediately from the corresponding characterization contained in Proposition 2.1.

Finally, let us have a look at the desired invariance of $S^{\leq i}_{\kappa_0,\xi_0}(\omega)$. Let $(\kappa,\xi) \in S^{\leq i}_{\kappa_0,\xi_0}(\omega)$ and $k \in \mathbb{Z}$ be arbitrary. Then the above-proved dynamical characterization implies that the difference $\lambda(\cdot;\kappa,\omega,\xi) - \lambda(\cdot;\kappa_0,\omega,\xi_0)$ is γ^--quasibounded with respect to ω, for every $\gamma \in [\alpha_{i+1,-} + \delta, \alpha_{i,-} - \delta]$. Since (2.4) furnishes the identity $\lambda(\cdot;k,\omega,\lambda(k;\kappa,\omega,\xi)) = \lambda(\cdot;\kappa,\omega,\xi)$, the difference $\lambda(\cdot;k,\omega,\lambda(k;\kappa,\omega,\xi)) - \lambda(\cdot;\kappa_0,\omega,\xi_0)$ is γ^--quasibounded with respect to ω, too. Applying the dynamical characterization once again implies $(k,\lambda(k;\kappa,\omega,\xi)) \in S^{\leq i}_{\kappa_0,\xi_0}(\omega)$ – and the proof of Lemma 2.3 is complete. $\Diamond$

Remark 2.2. The above proof shows that under the stronger condition on L required in Theorem 2.1 it would be possible to construct even more random invariant fiber bundles through an arbitrarily given solution. However, since these fiber bundles will play hardly any role in the following we omitted the explicit construction.   □

It is worthwhile to have a second look at the last proof. Basically, we did the following: In order to construct random invariant fiber bundles through a given solution μ we transformed the given equation (2.26) into a new equation (2.27) whose behavior around the trivial solution reflected exactly the behavior of the original equation around μ. Now we applied Proposition 2.1 to (2.27) and after transforming things back again we arrived at the desired random invariant fiber bundles through μ. So the idea behind Lemma 2.3 actually is very simple – yet it forces us to consider nonautonomous equations, since even if the original equation (2.26) is autonomous or of the special form needed for generating discrete-time random dynamical systems, the transformed equation (2.27) will no longer be of the respective type.

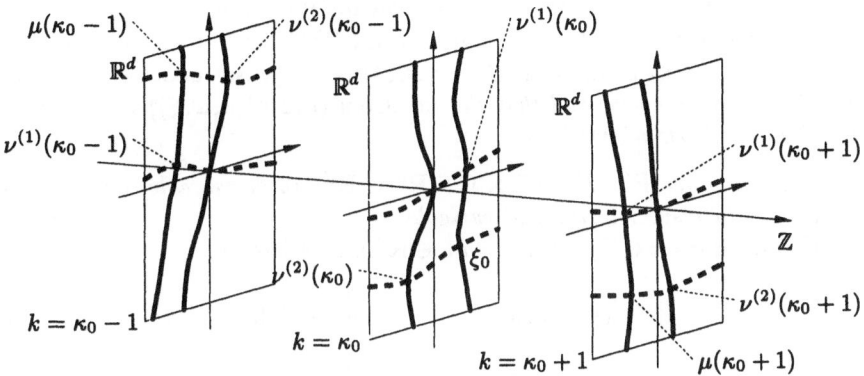

Fig. 2.3. The asymptotic phases

But although the above lemma is so simple, it plays a crucial role in deriving the next results of our linearization theory. First of all, let us return to the problem mentioned in the beginning of this subsection, i.e. the study of the asymptotic behavior of solutions of (2.25) which are not contained in any of the random invariant fiber bundles guaranteed by Theorem 2.1. In Figure 2.3 we have depicted two invariant fiber bundles through the trivial solution, as well as two invariant fiber bundles through a nontrivial ω-solution μ with $\mu(\kappa_0) = \xi_0$. Then it is almost obvious that it should be possible to uniquely assign two ω-solutions $\nu^{(1)}$ and $\nu^{(2)}$ to μ in such a way that $\nu^{(1)}$ (or $\nu^{(2)}$) is contained in $S^{\leq i}(\omega)$ (or $S^{>i}(\omega)$) and that $\mu - \nu^{(1)}$ (or $\mu - \nu^{(2)}$) is γ^+-quasibounded (or γ^--quasibounded) with respect to ω, for every choice of $\gamma \in [\alpha_{i+1,-} + \delta, \alpha_{i,+} - \delta]$. Especially in the case $\gamma < 1$ there would be exactly one ω-solution in $S^{\leq i}(\omega)$ which approaches our given solution μ exponentially as $k \to \infty$[5].

---

[5] Note however that for measuring the distance between the two solutions we use the time-dependent norms $\| \cdot \|_{k,\omega}$, so it might happen that the Euclidean distance of these solutions is unbounded as $k \to \infty$! Fortunately, this will not

The following theorem is devoted to formulating the above assertions exactly and to proving them. Yet the idea for the proof can be given already now by having another look at Figure 2.3: All we have to do is to intersect the fiber bundles $\mathcal{S}^{\leq i}_{\kappa_0,\xi_0}(\omega)$ and $\mathcal{S}^{>i}(\omega)$, or $\mathcal{S}^{>i}_{\kappa_0,\xi_0}(\omega)$ and $\mathcal{S}^{\leq i}(\omega)$.

Theorem 2.2. *Consider the random difference equation*

$$x_{k+1} = A(k,\omega)x_k + F(k,\omega,x_k) \tag{2.28}$$

satisfying assumptions (A1), (A2) and (A3) with $0 \leq L < \frac{\delta}{2K^2p}(K + 2 - \sqrt{K^2+4})$. Then for every $i \in \{1,\dots,p-1\}$, $\kappa \in \mathbb{Z}$, $\omega \in \Omega$ and $\xi \in \mathbb{R}^d$ there is a uniquely determined point $P^{\leq i}(\kappa,\omega,\xi) \in \mathbb{R}^d$ such that the difference $\lambda(\cdot;\kappa,\omega,\xi) - \lambda(\cdot;\kappa,\omega,P^{\leq i}(\kappa,\omega,\xi))$ is γ^+-quasibounded with respect to ω for arbitrary $\gamma \in [\alpha_{i+1,-} + \delta, \alpha_{i,+} - \delta]$ and that the inclusion $(\kappa, P^{\leq i}(\kappa,\omega,\xi)) \in \mathcal{S}^{\leq i}(\omega)$ holds. Likewise, there is a unique point $P^{>i}(\kappa,\omega,\xi) \in \mathbb{R}^d$ such that the difference $\lambda(\cdot;\kappa,\omega,\xi) - \lambda(\cdot;\kappa,\omega,P^{>i}(\kappa,\omega,\xi))$ is γ^--quasibounded with respect to ω for all γ as above and that the inclusion $(\kappa, P^{>i}(\kappa,\omega,\xi)) \in \mathcal{S}^{>i}(\omega)$ is satisfied. Furthermore we have:

(a) *The mappings $P^{\leq i}, P^{>i} : \mathbb{Z} \times \Omega \times \mathbb{R}^d \to \mathbb{R}^d$ are measurable, and continuous with respect to the last variable.*

(b) *For arbitrary $\kappa \in \mathbb{Z}$, $\omega \in \Omega$ and $\xi \in \mathbb{R}^d$ the estimates*

$$\|P^{\leq i}(\kappa,\omega,\xi)\|_{\kappa,\omega} \leq \frac{1}{1-C(L)}\|\xi\|_{\kappa,\omega} \quad \text{and}$$

$$\|P^{>i}(\kappa,\omega,\xi)\|_{\kappa,\omega} \leq \frac{1}{1-C(L)}\|\xi\|_{\kappa,\omega}$$

hold with $C(L) = C(L,K,\delta,p) = \frac{K^2Lp(\delta-KLp)}{\delta(\delta-2KLp)} < 1$ (cf. Theorem 2.1).

(c) *If μ is an arbitrary ω-solution of (2.28), then both $P^{\leq i}(\cdot,\omega,\mu(\cdot))$ and $P^{>i}(\cdot,\omega,\mu(\cdot))$ are ω-solutions of (2.28), too, which are contained in $\mathcal{S}^{\leq i}(\omega)$ or $\mathcal{S}^{>i}(\omega)$, respectively.*

Proof. Because of symmetry we only prove the assertions concerning $P^{\leq i}$. For this let $\mathcal{X} := \mathbb{R}^{d_1} \times \dots \times \mathbb{R}^{d_i}$ and define a family of mappings

$$T_{\kappa,\omega,\xi} : \mathcal{X} \ni \eta \mapsto s^{>i}_{\kappa,\xi}(\kappa,\omega,s^{\leq i}(\kappa,\omega,\eta)) \in \mathcal{X}$$

for $\kappa \in \mathbb{Z}$, $\omega \in \Omega$ and $\xi \in \mathbb{R}^d$. Then obviously $T_{\cdot,\cdot,\cdot}(\cdot)$ is measurable, and $T_{\kappa,\omega,\cdot}(\cdot)$ is continuous for arbitrary $\kappa \in \mathbb{Z}$ and $\omega \in \Omega$. Moreover, Lemma 2.3 furnishes the estimate

$$\begin{aligned}
\|T_{\kappa,\omega,\xi}(\eta_1) - T_{\kappa,\omega,\xi}(\eta_2)\|_{\kappa,\omega} &\leq C(L)\|s^{\leq i}(\kappa,\omega,\eta_1) - s^{\leq i}(\kappa,\omega,\eta_2)\|_{\kappa,\omega} \leq \\
&\leq C(L)^2\|\eta_1 - \eta_2\|_{\kappa,\omega},
\end{aligned}$$

---

be the case in our applications on random dynamical systems as we shall see in the next section.

for arbitrary $\kappa \in \mathbb{Z}$, $\omega \in \Omega$, $\xi \in \mathbb{R}^d$ and $\eta_1, \eta_2 \in \mathcal{X}$. Hence – having in mind that the condition on L implies $C(L) < 1$ – the mapping $T_{\kappa,\omega,\xi}$ is a contraction on $\mathcal{X}$ and has a unique fixed point $\tilde{P}(\kappa,\omega,\xi) \in \mathcal{X}$. According to Lemma 5.1 the mapping $\tilde{P}$ is measurable, and a standard application of the uniform contraction principle yields the continuity of $\tilde{P}(\kappa,\omega,\cdot)$ for every $\kappa \in \mathbb{Z}$ and $\omega \in \Omega$. Furthermore, a point $\eta \in \mathcal{X}$ is a fixed point of the mapping $T_{\kappa,\omega,\xi}$ if and only if we have

$$\eta = s_{\kappa,\xi}^{>i}(\kappa,\omega, s^{\leq i}(\kappa,\omega,\eta)) \quad \Leftrightarrow \quad (\kappa, \eta, s^{\leq i}(\kappa,\omega,\eta)) \in S_{\kappa,\xi}^{>i}(\omega) \Leftrightarrow$$
$$\Leftrightarrow \quad (\kappa, \pi^{1,i}(\kappa,\omega,\eta)) \in S_{\kappa,\xi}^{>i}(\omega),$$

where the mapping $\pi^{1,i}$ has been introduced in Remark 2.1. Now the definition

$$P^{\leq i}(\kappa,\omega,\xi) := \pi^{1,i}(\kappa,\omega,\tilde{P}(\kappa,\omega,\xi))$$

eventually furnishes the uniquely determined point in $\mathbb{R}^d$ having the properties stated in the above theorem, and the assertions of (a) are satisfied, too.

As for the proof of (b) let $\kappa \in \mathbb{Z}$, $\omega \in \Omega$ and $\xi \in \mathbb{R}^d$ be arbitrary. Then the estimate

$$\|T_{\kappa,\omega,\xi}(0)\|_{\kappa,\omega} \leq$$
$$\leq \|s_{\kappa,\xi}^{>i}(\kappa,\omega,0) - s_{\kappa,\xi}^{>i}(\kappa,\omega,\xi^{>i})\|_{\kappa,\omega} + \|\underbrace{s_{\kappa,\xi}^{>i}(\kappa,\omega,\xi^{>i})}_{=\xi^{\leq i}}\|_{\kappa,\omega} \leq$$
$$\leq C(L)\|\xi^{>i}\|_{\kappa,\omega} + \|\xi^{\leq i}\|_{\kappa,\omega} \leq \|\xi^{>i}\|_{\kappa,\omega} + \|\xi^{\leq i}\|_{\kappa,\omega} = \|\xi\|_{\kappa,\omega}$$

implies

$$\|\tilde{P}(\kappa,\omega,\xi)\|_{\kappa,\omega} \leq \|T_{\kappa,\omega,\xi}(\tilde{P}(\kappa,\omega,\xi)) - T_{\kappa,\omega,\xi}(0)\|_{\kappa,\omega} + \|T_{\kappa,\omega,\xi}(0)\|_{\kappa,\omega} \leq$$
$$\leq C(L)^2\|\tilde{P}(\kappa,\omega,\xi)\|_{\kappa,\omega} + \|\xi\|_{\kappa,\omega},$$

and eventually we deduce

$$\|P^{\leq i}(\kappa,\omega,\xi)\|_{\kappa,\omega} = \|\tilde{P}(\kappa,\omega,\xi)\|_{\kappa,\omega} + \|s^{\leq i}(\kappa,\omega,\tilde{P}(\kappa,\omega,\xi))\|_{\kappa,\omega} \leq$$
$$\leq (1+C(L))\|\tilde{P}(\kappa,\omega,\xi)\|_{\kappa,\omega} \leq \frac{1}{1-C(L)}\|\xi\|_{\kappa,\omega},$$

which is the desired inequality.

Finally let us turn our attention to the proof of (c). Let μ denote an arbitrary ω-solution of (2.28). Furthermore, let $\kappa \in \mathbb{Z}$ be arbitrary, but fixed, $\xi := \mu(\kappa)$ and $\nu := \lambda(\cdot;\kappa,\omega,P^{\leq i}(\kappa,\omega,\xi))$. Then ν obviously is an ω-solution of (2.28) which is contained in $S^{\leq i}(\omega)$ due to $(\kappa,P^{\leq i}(\kappa,\omega,\xi)) \in S^{\leq i}(\omega)$. In addition to that we know already that the difference $\mu-\nu$ is γ^+-quasibounded with respect to ω for every $\gamma \in [\alpha_{i+1,-} + \delta, \alpha_{i,+} - \delta]$.

If we now let $k \in \mathbb{Z}$ be arbitrary, then $\xi^* := P^{\leq i}(k,\omega,\mu(k))$ is the uniquely determined point in $\mathbb{R}^d$ such that the difference $\lambda(\cdot;k,\omega,\mu(k)) - \lambda(\cdot;k,\omega,\xi^*)$

is γ^+-quasibounded with respect to ω and that the inclusion $(k, \xi^*) \in S^{\leq i}(\omega)$ holds. Using the γ^+-quasiboundedness of $\mu - \nu = \lambda(\cdot; k, \omega, \mu(k)) - \lambda(\cdot; k, \omega, \nu(k))$ and the inclusion $(k, \nu(k)) \in S^{\leq i}(\omega)$ (note that ν is contained in $S^{\leq i}(\omega)$!) we finally get $P^{\leq i}(k, \omega, \mu(k)) = \xi^* = \nu(k)$. Since $k \in \mathbb{Z}$ has been chosen arbitrarily, this completes the proof of the theorem. $\qquad\qquad\Diamond$

Right now we could present a first corollary of Theorem 2.2, namely a *reduction principle for nonautonomous random difference equations*. This principle – which in the case of autonomous differential equations dates from Kelley [22] and Pliss [27], and in the nonautonomous case from Aulbach [6] – allows to determine the stability of the trivial solution of (2.28) with the help of a lower-dimensional new equation. However, due to lack of space we refrain from presenting this result here, and proceed with developing our linearization theory.

2.5 Topological Decoupling

In the last subsection we demonstrated how to assign to a given ω-solution μ of the random difference equation

$$x_{k+1} = A(k, \omega)x_k + F(k, \omega, x_k) \qquad\qquad (2.29)$$

a new ω-solution in a canonical way which is contained in the random invariant fiber bundle $S^{\leq i}(\omega)$. According to Theorem 2.2 this solution is given by $P^{\leq i}(\cdot, \omega, \mu(\cdot))$. But, recalling that points $(\kappa, \xi) \in S^{\leq i}(\omega)$ are already determined exactly by the values of $\kappa \in \mathbb{Z}$ and $\xi^{\leq i} \in \mathbb{R}^{d_1} \times \ldots \times \mathbb{R}^{d_i}$ we see that actually only the first i components $P^{\leq i}_{\leq i}(\cdot, \omega, \mu(\cdot))$ of this solution are of interest – and these components satisfy a random difference equation again: The fact that $P^{\leq i}(\cdot, \omega, \mu(\cdot))$ is an ω-solution of (2.29) lying in $S^{\leq i}(\omega)$ immediately implies that $P^{\leq i}_{\leq i}(\cdot, \omega, \mu(\cdot))$ is an ω-solution of the $(d_1 + \ldots + d_i)$-dimensional equation

$$x^{\leq i}_{k+1} = A_{\leq i}(k, \omega)x^{\leq i}_k + F_{\leq i}(k, \omega, x^{\leq i}_k, s^{\leq i}(k, \omega, x^{\leq i}_k)) \ .$$

Analogously, the last $p - i$ components of the mapping $P^{>i}(\cdot, \omega, \mu(\cdot))$ are an ω-solution of the $(d_{i+1} + \ldots + d_p)$-dimensional equation

$$x^{>i}_{k+1} = A_{>i}(k, \omega)x^{>i}_k + F_{>i}(k, \omega, s^{>i}(k, \omega, x^{>i}_k), x^{>i}_k) \ .$$

If we now define a mapping $D^i : \mathbb{Z} \times \Omega \times \mathbb{R}^d \to \mathbb{R}^d$ by

$$D^i(\kappa, \omega, \xi) := (P^{\leq i}_{\leq i}(\kappa, \omega, \xi), P^{>i}_{>i}(\kappa, \omega, \xi)) \ ,$$

then for every ω-solution μ of (2.29) the mapping $D^i(\cdot, \omega, \mu(\cdot))$ is an ω-solution of the d-dimensional equation

$$\begin{aligned} x^{\leq i}_{k+1} &= A_{\leq i}(k, \omega)x^{\leq i}_k + F_{\leq i}(k, \omega, x^{\leq i}_k, s^{\leq i}(k, \omega, x^{\leq i}_k)) \\ x^{>i}_{k+1} &= A_{>i}(k, \omega)x^{>i}_k + F_{>i}(k, \omega, s^{>i}(k, \omega, x^{>i}_k), x^{>i}_k) \end{aligned} \qquad (2.30)$$

which is simpler than the original equation, since it is decoupled. Even more, one can show that for every $\kappa \in \mathbb{Z}$ and $\omega \in \Omega$ the mapping $D^i(\kappa, \omega, \cdot)$ is bijective, and if we denote the corresponding inverse by $\tilde{D}^i(\kappa, \omega, \cdot)$, then for every ω-solution ν of (2.30) the mapping $\tilde{D}^i(\cdot, \omega, \nu(\cdot))$ is an ω-solution of (2.29).

This observation should remind the reader of the Hartman-Grobman theorem mentioned at the beginning of the introduction, since there solutions of a complicated equation where mapped (locally) onto solutions of a linear equation, and vice versa, too. Of course we cannot expect in general that the above equation (2.30) already is linear, yet the passage from the coupled equation (2.29) to a decoupled equation is the first step in deriving a Hartman-Grobman theorem for nonautonomous random difference equations. But rather than considering (2.30) we shall prove in the following proposition that our original equation can be reduced in the above sense to a decoupled one having exactly p subequations – which of course reflects assumption (A1) concerning the linear part.

Proposition 2.2. *Consider the random difference equation*

$$x_{k+1} = A(k, \omega)x_k + F(k, \omega, x_k) \qquad (2.31)$$

satisfying the assumptions (A1), (A2) and (A3) from the beginning of Subsection 2.3. Then there is a constant $0 < L^* \leq \frac{\delta}{2K^2 p}(K + 2 - \sqrt{K^2 + 4})$ such that for all $0 \leq L < L^*$ the following holds: The decoupled equation

$$x_{k+1} = A(k, \omega)x_k + \tilde{F}(k, \omega, x_k) \qquad (2.32)$$

with $\tilde{F}_i(k, \omega, x) := F_i(k, \omega, \pi^{i,i}(k, \omega, x^i))$ for $i = 1, \ldots, p$ satisfies all the assumptions of Theorem 2.1, and the random invariant fiber bundles guaranteed by this theorem are actually subspace bundles, where the generating mappings $\tilde{s}^{i,j}$ vanish identically. Furthermore, there are measurable mappings $D, \tilde{D} : \mathbb{Z} \times \Omega \times \mathbb{R}^d \to \mathbb{R}^d$ with the following properties:

(a) For arbitrary $\kappa \in \mathbb{Z}$ and $\omega \in \Omega$ the mappings $D(\kappa, \omega, \cdot)$ and $\tilde{D}(\kappa, \omega, \cdot)$ are continuous and inverse to each other, i.e. they are homeomorphisms on $\mathbb{R}^d$.

(b) For every ω-solution μ of (2.31) the mapping $D(\cdot, \omega, \mu(\cdot))$ is an ω-solution of (2.32). Similarly, if ν denotes an arbitrary ω-solution of (2.32), then $\tilde{D}(\cdot, \omega, \nu(\cdot))$ is an ω-solution of (2.31).

(c) For arbitrary $\kappa \in \mathbb{Z}$, $\omega \in \Omega$ and $\xi \in \mathbb{R}^d$ the estimates

$$\|D(\kappa, \omega, \xi)\|_{\kappa, \omega} \leq B\|\xi\|_{\kappa, \omega} \quad \text{and} \quad \|\tilde{D}(\kappa, \omega, \xi)\|_{\kappa, \omega} \leq B\|\xi\|_{\kappa, \omega}$$

hold, with some constant $B = B(L, K, \delta, p) \geq 1$. In particular, quasi-boundedness is preserved under D and $\tilde{D}$, so the random invariant fiber bundles of (2.31) according to Theorem 2.1 are mapped onto the corresponding invariant subspace bundles of the decoupled equation (2.32), and vice versa.

In other words, equations (2.31) and (2.32) are topologically equivalent via D and $\tilde{D}$.

Proof. To begin with, let us prove the assertions concerning the invariant fiber bundles of equation (2.32). For arbitrary $0 \leq L < \frac{\delta}{2K^2p}(K + 2 - \sqrt{K^2 + 4})$ Theorem 2.1 furnishes after some straightforward calculations the estimate

$$\|F_i(k, \omega, \pi^{i,i}(k, \omega, x^i)) - F_i(k, \omega, \pi^{i,i}(k, \omega, \bar{x}^i))\|_{k+1,\omega} \leq$$

$$\leq L\left(\frac{2C(L)}{1 - C(L)} + 1\right)\|x^i - \bar{x}^i\|_{k,\omega},$$

for arbitrary $i = 1, \ldots, p$, $k \in \mathbb{Z}$, $\omega \in \Omega$ and $x^i, \bar{x}^i \in \mathbb{R}^{d_i}$. Hence, $\lim_{L \to 0} C(L) = 0$ implies the existence of a constant $0 < L_1^* \leq \frac{\delta}{2K^2p}(K + 2 - \sqrt{K^2 + 4})$ such that for all $0 \leq L < L_1^*$ both the estimates

$$\|F_i(k, \omega, \pi^{i,i}(k, \omega, x^i)) - F_i(k, \omega, \pi^{i,i}(k, \omega, \bar{x}^i))\|_{k+1,\omega} \leq$$

$$\leq 2L\|x^i - \bar{x}^i\|_{k,\omega} \leq 2L\|x - \bar{x}\|_{k,\omega}$$

and $2L < \frac{\delta}{2K^2p}(K + 2 - \sqrt{K^2 + 4}) < \frac{\delta}{K}$ hold. But then for every $0 \leq L < L_1^*$ we may apply Lemmas 2.1 and 2.2, as well as their dual versions, to the equation

$$x_{k+1}^i = A_i(k, \omega)x_k^i + F_i(k, \omega, \pi^{i,i}(k, \omega, x_k^i)) \tag{2.33}$$

and this immediately yields the following:

- All ω-solutions of (2.33) exist uniquely on the whole of $\mathbb{Z}$, and the corresponding general solution is measurable, as well as continuous with respect to the last variable (here we use the uniform contraction principle once again). Together with the above two estimates we further deduce that Theorem 2.1 may be applied to (2.32), i.e. there are random invariant fiber bundles $\tilde{S}^{i,j}(\omega)$ which are generated by measurable mappings $\tilde{s}^{i,j}$.
- The trivial solution is the only ω-solution of equation (2.33) which is γ^--quasibounded with respect to ω for every $\gamma \geq \alpha_{i,-} + \delta$, and for these values of γ all ω-solutions of (2.33) are γ^+-quasibounded with respect to ω.
- The trivial solution is the only ω-solution of equation (2.33) which is γ^+-quasibounded with respect to ω for every $0 < \gamma \leq \alpha_{i,+} - \delta$, and for these values of γ all ω-solutions of equation (2.33) are γ^--quasibounded with respect to ω.

With the help of these facts it can easily be verified that the mappings $\tilde{s}^{i,j}$ have to vanish identically.

As for the proof of the remaining assertions of Proposition 2.2 we proceed by induction on p. First of all, consider the case $p = 2$. If we define the mapping D as

$$D(\kappa, \omega, \xi) := (P_1^{\leq 1}(\kappa, \omega, \xi), P_2^{>1}(\kappa, \omega, \xi))$$

(where $P^{\leq 1}$ and $P^{>1}$ denote the mappings guaranteed by Theorem 2.2), then D is measurable, and continuous with respect to the last variable. Moreover, in view of the discussion from the beginning of this subsection the assertions in *(b)* and *(c)* concerning D are immediate consequences of Theorem 2.2*(c)* and *(b)*, respectively. Now let us construct the mapping $\tilde{D}$. Looking back at the proof of Theorem 2.2 and the above definition of D it is sensible to intersect the fiber bundles $\mathcal{S}^{\leq 1}_{\kappa, \pi^{2,2}(\kappa, \omega, \xi^2)}(\omega)$ and $\mathcal{S}^{>1}_{\kappa, \pi^{1,1}(\kappa, \omega, \xi^1)}(\omega)$ in the fiber $k = \kappa$ and call the (as we would expect) unique intersection point $\tilde{D}(\kappa, \omega, \xi)$. To be more rigorous, define a family of mappings $T_{\kappa, \omega, \xi} : \mathbb{R}^{d_1} \times \mathbb{R}^{d_2} \to \mathbb{R}^{d_1} \times \mathbb{R}^{d_2}$ by

$$T_{\kappa, \omega, \xi}(\eta_1, \eta_2) := \left(s^{>1}_{\kappa, \pi^{1,1}(\kappa, \omega, \xi^1)}(\kappa, \omega, \eta_2), s^{\leq 1}_{\kappa, \pi^{2,2}(\kappa, \omega, \xi^2)}(\kappa, \omega, \eta_1) \right),$$

for $\kappa \in \mathbb{Z}$, $\omega \in \Omega$ and $\xi \in \mathbb{R}^d$. Again, the mapping $T_{.,.,.}(\cdot, \cdot)$ is measurable, and for arbitrary $\kappa \in \mathbb{Z}$ and $\omega \in \Omega$ the mapping $T_{\kappa, \omega, .}(\cdot, \cdot)$ is continuous. In addition, for all $\kappa \in \mathbb{Z}$, $\omega \in \Omega$, $\xi \in \mathbb{R}^d$, $\eta_1, \bar{\eta}_1 \in \mathbb{R}^{d_1}$ and $\eta_2, \bar{\eta}_2 \in \mathbb{R}^{d_2}$ the estimate

$$\|T_{\kappa, \omega, \xi}(\eta_1, \eta_2) - T_{\kappa, \omega, \xi}(\bar{\eta}_1, \bar{\eta}_2)\|_{\kappa, \omega} \leq C(L) \|(\eta_1, \eta_2) - (\bar{\eta}_1, \bar{\eta}_2)\|_{\kappa, \omega}$$

holds, so due to $C(L) < 1$ the mapping $T_{\kappa, \omega, \xi}$ is a contraction on $\mathbb{R}^{d_1} \times \mathbb{R}^{d_2} = \mathbb{R}^d$ with a unique fixed point $\tilde{D}(\kappa, \omega, \xi)$. Lemma 5.1 and the uniform contraction principle now imply the measurability of $\tilde{D}$, as well as the continuity with respect to the last variable. Finally, since a point (η_1, η_2) is a fixed point of $T_{\kappa, \omega, \xi}$ if and only if we have

$$\eta_1 = s^{>1}_{\kappa, \pi^{1,1}(\kappa, \omega, \xi^1)}(\kappa, \omega, \eta_2) \quad \text{and} \quad \eta_2 = s^{\leq 1}_{\kappa, \pi^{2,2}(\kappa, \omega, \xi^2)}(\kappa, \omega, \eta_1)$$
$$\Leftrightarrow \quad (\kappa, \eta_1, \eta_2) \in \mathcal{S}^{>1}_{\kappa, \pi^{1,1}(\kappa, \omega, \xi^1)}(\omega) \quad \text{and} \quad (\kappa, \eta_1, \eta_2) \in \mathcal{S}^{\leq 1}_{\kappa, \pi^{2,2}(\kappa, \omega, \xi^2)}(\omega)$$

the reader may check that the mappings $D(\kappa, \omega, \cdot)$ and $\tilde{D}(\kappa, \omega, \cdot)$ indeed are inverse to each other, for arbitrary $\kappa \in \mathbb{Z}$ and $\omega \in \Omega$, and that the assertion in *(b)* concerning $\tilde{D}$ is an immediate consequence of the corresponding assertion for D. In order to conclude the proof of the case $p = 2$ we only have to verify the estimate for $\tilde{D}$ contained in *(c)*. Yet this can be done analogously to the proof of the estimate for $P^{\leq i}$ in Theorem 2.2.

As for the proof of the case $p \geq 3$ let us begin with a short outline of our further proceeding. First of all, we shall apply the results proved so far in order to show that (2.31) is topologically equivalent to a partially decoupled equation consisting of a d_1- and a $(d_2 + \ldots + d_p)-$dimensional subequation. After that the inductive hypothesis implies the topological equivalence of the $(d_2 + \ldots + d_p)-$dimensional subequation to a completely decoupled equation consisting of $p-1$ subequations, and putting things together in the right way this already furnishes the topological equivalence of (2.31) and (2.32).

More precisely, according to the above-proved special case there is a constant $0 < L_2^* < L_1^*$ such that for all $0 \leq L < L_2^*$ the original equation (2.31) is topologically equivalent to the partially decoupled equation

$$
\begin{aligned}
x_{k+1}^1 &= A_1(k,\omega)x_k^1 + F_1(k,\omega,x_k^1,s^{\le 1}(k,\omega,x_k^1)) \\
x_{k+1}^2 &= A_2(k,\omega)x_k^2 + F_2(k,\omega,s^{>1}(k,\omega,x_k^2,\ldots,x_k^p),x_k^2,\ldots,x_k^p) \\
&\vdots \\
x_{k+1}^p &= A_p(k,\omega)x_k^p + F_p(k,\omega,s^{>1}(k,\omega,x_k^2,\ldots,x_k^p),x_k^2,\ldots,x_k^p)
\end{aligned}
\tag{2.34}
$$

via two mappings $D^{(1)}$ and $\tilde{D}^{(1)}$. Furthermore, it can easily be verified that all ω-solutions of (2.34) exist uniquely on the whole of $\mathbb{Z}$ and that the corresponding general solution is measurable, as well as continuous with respect to the last variable[6]. In addition, similar assertions are true for the subequation

$$
\begin{aligned}
x_{k+1}^2 &= A_2(k,\omega)x_k^2 + F_2(k,\omega,s^{>1}(k,\omega,x_k^2,\ldots,x_k^p),x_k^2,\ldots,x_k^p) \\
&\vdots \\
x_{k+1}^p &= A_p(k,\omega)x_k^p + F_p(k,\omega,s^{>1}(k,\omega,x_k^2,\ldots,x_k^p),x_k^2,\ldots,x_k^p)
\end{aligned}
\tag{2.35}
$$

since (2.34) is decoupled. As in the beginning of this proof it can be shown that the nonlinearities in (2.35) satisfy a global Lipschitz condition, where the Lipschitz constant converges to 0 as $L \to 0$. Hence there is a $0 < L^* \le L_2^*$ such that for all $0 \le L < L^*$ the inductive hypothesis may be applied to (2.35) – furnishing the topological equivalence of this equation and the completely decoupled equation

$$
\begin{aligned}
x_{k+1}^2 &= A_2(k,\omega)x_k^2 + F_2(k,\omega,s^{>1}(k,\omega,\hat{\pi}^{2,2}(k,\omega,x_k^2)),\hat{\pi}^{2,2}(k,\omega,x_k^2)) \\
&\vdots \\
x_{k+1}^p &= A_p(k,\omega)x_k^p + F_p(k,\omega,s^{>1}(k,\omega,\hat{\pi}^{p,p}(k,\omega,x_k^p)),\hat{\pi}^{p,p}(k,\omega,x_k^p))
\end{aligned}
$$

via two mappings $D^{(2)}$ and $\tilde{D}^{(2)}$. Here the random invariant fiber bundles for (2.35) guaranteed by Theorem 2.1 are generated by the mappings $\hat{\pi}^{i,j}(\omega)$ from Remark 2.1. If we now define for arbitrary $\kappa \in \mathbb{Z}$, $\omega \in \Omega$ and $\xi \in \mathbb{R}^d$ two mappings D and $\tilde{D}$ by $D(\kappa,\omega,\xi) := (D_1^{(1)}(\kappa,\omega,\xi), D^{(2)}(\kappa,\omega,D_{>1}^{(1)}(\kappa,\omega,\xi)))$ and $\tilde{D}(\kappa,\omega,\xi) := \tilde{D}^{(1)}(\kappa,\omega,\xi^1,\tilde{D}^{(2)}(\kappa,\omega,\xi^{>1}))$, then equation (2.31) is topologically equivalent to the decoupled equation

$$
\begin{aligned}
x_{k+1}^1 &= A_1(k,\omega)x_k^1 + F_1(k,\omega,x_k^1,s^{\le 1}(k,\omega,x_k^1)) \\
x_{k+1}^2 &= A_2(k,\omega)x_k^2 + F_2(k,\omega,s^{>1}(k,\omega,\hat{\pi}^{2,2}(k,\omega,x_k^2)),\hat{\pi}^{2,2}(k,\omega,x_k^2)) \\
&\vdots \\
x_{k+1}^p &= A_p(k,\omega)x_k^p + F_p(k,\omega,s^{>1}(k,\omega,\hat{\pi}^{p,p}(k,\omega,x_k^p)),\hat{\pi}^{p,p}(k,\omega,x_k^p))
\end{aligned}
\tag{2.36}
$$

via D and $\tilde{D}$. Furthermore, the assertions contained in *(a)*, *(b)* and *(c)* can easily be verified, yet with equation (2.32) replaced by (2.36).

So for completing the proof of Proposition 2.2 we only have to show that the random difference equations (2.32) and (2.36) actually are identical. First of all, Theorem 2.1 implies for arbitrary $\kappa \in \mathbb{Z}$, $\omega \in \Omega$ and $\xi^1 \in \mathbb{R}^{d_1}$

---

[6] We only have to note that this is true for (2.31), and then use the fact that (2.31) and (2.34) are topologically equivalent.

the identity $(\xi^1, s^{\leq 1}(\kappa, \omega, \xi^1)) = \pi^{1,1}(\kappa, \omega, \xi^1)$. Now let $i \in \{2, \ldots, p\}$, $\kappa \in \mathbb{Z}$, $\omega \in \Omega$ and $\xi^i \in \mathbb{R}^{d_i}$ be arbitrary, but fixed. Moreover, let μ denote the unique ω-solution of (2.35) satisfying the initial condition $x_{\kappa}^{\geq 1} = \hat{\pi}^{i,i}(\kappa, \omega, \xi^i)$. Then because of Theorem 2.1 and Remark 2.1 the mapping μ is γ_1^+- and γ_2^--quasibounded with respect to ω, for every $\gamma_1 \in [\alpha_{i,-} + \delta, \alpha_{i-1,+} - \delta]$ and $\gamma_2 \in [\alpha_{i+1,-} + \delta, \alpha_{i,+} - \delta]$. Furthermore, the mapping $\nu := (s^{>1}(\cdot, \omega, \mu(\cdot)), \mu(\cdot))$ is an ω-solution of (2.31). But now the estimate

$$\|\nu(k)\|_{k,\omega} = \|s^{>1}(k, \omega, \mu(k))\|_{k,\omega} + \|\mu(k)\|_{k,\omega} \leq (C(L)+1)\|\mu(k)\|_{k,\omega}, k \in \mathbb{Z},$$

yields the γ_1^+- and γ_2^--quasiboundedness (with respect to ω) of ν, with γ_1 and γ_2 as above. Applying Theorem 2.1 and Remark 2.1 once again – yet this time to the original random difference equation (2.31) – implies $(\kappa, \nu(\kappa)) \in \mathcal{S}^{i,i}(\omega)$ or $\nu(\kappa) = \pi^{i,i}(\kappa, \omega, \nu_i(\kappa))$, and $\nu_i(\kappa) = \mu_i(\kappa) = \xi^i$ eventually furnishes

$$\pi^{i,i}(\kappa, \omega, \xi_i) = \nu(\kappa) = (s^{>1}(\kappa, \omega, \hat{\pi}^{i,i}(\kappa, \omega, \xi^i)), \hat{\pi}^{i,i}(\kappa, \omega, \xi^i)),$$

i.e. equations (2.32) and (2.36) are identical indeed.   ◊

Now let us return to the question posed at the end of Subsection 2.3. There we realized that Theorem 2.1 only guaranteed $\frac{(p-1)(p+2)}{2}$ random invariant fiber bundles, whereas the corresponding linear equation had $2^p - 2$ invariant subspace bundles. Even more, we indicated that within the Lipschitz category we cannot expect more in general. But if we leave this category there are nonlinear analogues to all of the above $2^p - 2$ invariant subspace bundles of the linear equation. For this we only have to note that the decoupled equation (2.32) in Proposition 2.2 has exactly $2^p - 2$ invariant subspace bundles, and by applying the mapping $\tilde{D}$ this immediately furnishes $2^p - 2$ random invariant fiber bundles for (2.31). However, let us stress once again that in general only the fiber bundles of Theorem 2.1 have Lipschitzian fibers.

2.6 Topological Linearization

By now we managed to carry over almost all linearization results mentioned in the introduction to the case of nonautonomous random difference equations – with only one exception, namely the Hartman-Grobman theorems. They will be the subject of this last subsection of Section 2.

In view of Proposition 2.2 we still have to show that almost all subequations of equation (2.32) are topologically equivalent to the corresponding linear equations, and this will be achieved with the help of the following preparatory results which in the deterministic case date from Palmer [25].

Lemma 2.4. *Consider the random difference equations*

$$\begin{aligned} x_{k+1} &= A^-(k, \omega)x_k + f_1(k, \omega, x_k), & (2.37) \\ x_{k+1} &= A^-(k, \omega)x_k + f_2(k, \omega, x_k) & (2.38) \end{aligned}$$

with measurable mappings $A^- : \mathbb{Z} \times \Omega \rightarrow \mathrm{GL}(\mathbb{R}^{d^-})$ and $f_1, f_2 : \mathbb{Z} \times \Omega \times \mathbb{R}^{d^-} \rightarrow \mathbb{R}^{d^-}$. As usual, let $\Phi^-(m, n, \omega)$ denote the evolution operator of the homogeneous linear difference equation $x_{k+1} = A^-(k, \omega)x_k$, and suppose that for arbitrary $m, n \in \mathbb{Z}$, $\omega \in \Omega$, $x, \bar{x} \in \mathbb{R}^{d^-}$ and $i = 1, 2$ the following holds:

$$\|\Phi^-(m, n, \omega)\|_{n, m, \omega} \leq K\alpha_-^{m-n} \quad \text{for} \quad m \geq n \, ,$$

$$\|f_i(k, \omega, x)\|_{k+1, \omega} \leq M \quad \text{and} \quad \|f_i(k, \omega, x) - f_i(k, \omega, \bar{x})\|_{k+1, \omega} \leq L\|x - \bar{x}\|_{k, \omega} \, ,$$

with real constants $0 < \alpha_- < 1$, $K \geq 1$, $M \geq 0$ and $0 \leq L < \frac{1-\alpha_-}{K}$. Finally assume that all ω-solutions of both (2.37) and (2.38) exist uniquely on the whole of $\mathbb{Z}$, and that the corresponding general solutions $\lambda^{(1)}, \lambda^{(2)} : \mathbb{Z} \times \mathbb{Z} \times \Omega \times \mathbb{R}^{d^-} \rightarrow \mathbb{R}^{d^-}$ are measurable, as well as continuous with respect to the last variable.

Then for every choice of $\kappa \in \mathbb{Z}$, $\omega \in \Omega$ and $\xi \in \mathbb{R}^{d^-}$ there is a uniquely determined point $H^-(\kappa, \omega, \xi) \in \mathbb{R}^{d^-}$ such that the difference $\lambda^{(2)}(\cdot; \kappa, \omega, H^-(\kappa, \omega, \xi)) - \lambda^{(1)}(\cdot; \kappa, \omega, \xi)$ is 1-quasibounded with respect to ω. The so-defined mapping $H^- : \mathbb{Z} \times \Omega \times \mathbb{R}^{d^-} \rightarrow \mathbb{R}^{d^-}$ is measurable, continuous with respect to the last variable, and for arbitrary $\kappa \in \mathbb{Z}$, $\omega \in \Omega$ and $\xi \in \mathbb{R}^{d^-}$ the estimate

$$\|H^-(\kappa, \omega, \xi) - \xi\|_{\kappa, \omega} \leq \frac{2KM}{1 - \alpha_- - KL}$$

holds. Furthermore, if ν is an arbitrary ω-solution of equation (2.37), then $H^-(\cdot, \omega, \nu(\cdot))$ is an ω-solution of (2.38).

Proof. Consider the random difference equation

$$x_{k+1} = A^-(k, \omega)x_k + f_2(k, \omega, x_k + \lambda^{(1)}(k; \kappa_0, \omega, \xi_0)) - f_1(k, \omega, \lambda^{(1)}(k; \kappa_0, \omega, \xi_0)) \tag{2.39}$$

depending on the parameter $(\kappa_0, \xi_0) \in \mathbb{Z} \times \mathbb{R}^{d^-}$. Then it may easily be checked that all ω, κ_0, ξ_0-solutions of (2.39) exist uniquely on the whole of $\mathbb{Z}$, and that the corresponding general solution $\lambda^{(3)} : \mathbb{Z} \times \mathbb{Z} \times \Omega \times \mathbb{R}^{d^-} \times \mathbb{Z} \times \mathbb{R}^{d^-} \rightarrow \mathbb{R}^{d^-}$ is measurable, as well as continuous with respect to the last three variables. Furthermore, the difference equation (2.39) satisfies all the assumptions of Lemma 2.1 with $\gamma := 1$ and $f_0(k, \omega, \kappa_0, \xi_0) := f_2(k, \omega, \lambda^{(1)}(k; \kappa_0, \omega, \xi_0)) - f_1(k, \omega, \lambda^{(1)}(k; \kappa_0, \omega, \xi_0))$. Hence there is a uniquely determined mapping $\mu : \mathbb{Z} \times \Omega \times \mathbb{Z} \times \mathbb{R}^{d^-} \rightarrow \mathbb{R}^{d^-}$ such that for all $\kappa, \kappa_0 \in \mathbb{Z}$, $\omega \in \Omega$ and $\xi_0 \in \mathbb{R}^{d^-}$ the mapping $\lambda^{(3)}(\cdot; \kappa, \omega, \mu(\kappa, \omega, \kappa_0, \xi_0), \kappa_0, \xi_0)$ is 1-quasibounded with respect to ω. In addition, μ is measurable, continuous with respect to the last two variables, and for arbitrary $\kappa, \kappa_0 \in \mathbb{Z}$, $\omega \in \Omega$ and $\xi_0 \in \mathbb{R}^{d^-}$ the estimate

$$\|\mu(\kappa, \omega, \kappa_0, \xi_0)\|_{\kappa, \omega} \leq \frac{2KM}{1 - \alpha_- - KL}$$

holds. Noting that a mapping $\nu : \mathbb{Z} \rightarrow \mathbb{R}^{d^-}$ is an ω, κ_0, ξ_0-solution of (2.39) if and only if the mapping $\nu + \lambda^{(1)}(\cdot; \kappa_0, \omega, \xi_0)$ is an ω-solution of (2.38) now shows that the definition

$$H^-(\kappa,\omega,\xi) := \xi + \mu(\kappa,\omega,\kappa,\xi)$$

furnishes the uniquely determined mapping H^- for which the difference $\lambda^{(2)}(\cdot;\kappa,\omega,H^-(\kappa,\omega,\xi)) - \lambda^{(1)}(\cdot;\kappa,\omega,\xi)$ is 1-quasibounded with respect to ω, for every $\kappa \in \mathbb{Z}$, $\omega \in \Omega$ and $\xi \in \mathbb{R}^{d^-}$. Moreover, the mapping H^- is measurable, continuous with respect to the last variable, and satisfies the required estimate.

In order to conclude the proof of the lemma let ν denote an arbitrary ω-solution of equation (2.37) and set $\xi_0 := \nu(\kappa_0)$, $\nu^*(k) := \lambda^{(2)}(k;\kappa_0,\omega,H^-(\kappa_0,\omega,\xi_0))$. Then the mapping ν^* is an ω-solution of (2.38) and $\nu^* - \nu$ is 1-quasibounded with respect to ω. Now let $\kappa \in \mathbb{Z}$ be arbitrary and $\xi^* := H^-(\kappa,\omega,\nu(\kappa))$. We already proved that ξ^* is the unique point in $\mathbb{R}^{d^-}$ for which the difference $\lambda^{(2)}(\cdot;\kappa,\omega,\xi^*) - \lambda^{(1)}(\cdot;\kappa,\omega,\nu(\kappa))$ is 1-quasibounded with respect to ω, and the identity $\lambda^{(1)}(k;\kappa,\omega,\nu(\kappa)) = \nu(k)$ immediately implies the 1-quasiboundedness (with respect to ω) of $\nu^* - \lambda^{(1)}(\cdot;\kappa,\omega,\nu(\kappa))$, i.e. we have $\nu^* = \lambda^{(2)}(\cdot;\kappa,\omega,\xi^*)$ and $H^-(\kappa,\omega,\nu(\kappa)) = \xi^* = \nu^*(\kappa)$. So, since $\kappa \in \mathbb{Z}$ has been chosen arbitrarily, the mapping $H^-(\cdot,\omega,\nu(\cdot))$ indeed is an ω-solution of (2.38), namely ν^*.   ◊

With the above lemma we now are able to present conditions furnishing the topological equivalence of certain nonlinear and linear random difference equations.

Proposition 2.3. *Consider the random difference equation*

$$x_{k+1} = A^-(k,\omega)x_k + f^-(k,\omega,x_k) \tag{2.40}$$

as well as the corresponding homogeneous linear random difference equation

$$x_{k+1} = A^-(k,\omega)x_k \tag{2.41}$$

with measurable mappings $A^- : \mathbb{Z} \times \Omega \to GL(\mathbb{R}^{d^-})$ *and* $f^- : \mathbb{Z} \times \Omega \times \mathbb{R}^{d^-} \to \mathbb{R}^{d^-}$. *Let* $\Phi^-(m,n,\omega)$ *denote the evolution operator of (2.41) and suppose that all ω-solutions of (2.40) exist uniquely on the whole of $\mathbb{Z}$, and that the corresponding general solution is measurable, as well as continuous with respect to the last variable. Finally assume that for arbitrary $m,n \in \mathbb{Z}$, $\omega \in \Omega$ and $x,\bar{x} \in \mathbb{R}^{d^-}$ we have:*

$$\|\Phi^-(m,n,\omega)\|_{n,m,\omega} \leq K\alpha_-^{m-n} \quad \text{for} \quad m \geq n,$$
$$\|f^-(k,\omega,x)\|_{k+1,\omega} \leq M \quad \text{and}$$
$$\|f^-(k,\omega,x) - f^-(k,\omega,\bar{x})\|_{k+1,\omega} \leq L\|x - \bar{x}\|_{k,\omega},$$

with some constants $0 < \alpha_- < 1$, $K \geq 1$ and $M \geq 0$, where $0 \leq L < \frac{1-\alpha_-}{K}$. Then there are measurable mappings $H^-,\tilde{H}^- : \mathbb{Z} \times \Omega \times \mathbb{R}^{d^-} \to \mathbb{R}^{d^-}$ with the following properties:

(a) *For arbitrary $\kappa \in \mathbb{Z}$ and $\omega \in \Omega$ the mappings $H^-(\kappa,\omega,\cdot)$ and $\tilde{H}^-(\kappa,\omega,\cdot)$ are continuous and inverse to each other, i.e. they are homeomorphisms on $\mathbb{R}^{d^-}$.*

(b) *For every ω-solution μ of (2.40) the mapping $H^-(\cdot,\omega,\mu(\cdot))$ is an ω-solution of (2.41). Similarly, if ν denotes an arbitrary ω-solution of (2.41), then $\tilde{H}^-(\cdot,\omega,\nu(\cdot))$ is an ω-solution of (2.40).*

In other words, equations (2.40) and (2.41) are topologically equivalent via H^- and $\tilde{H}^-$.

Proof. With $f_1(k,\omega,x) := f^-(k,\omega,x)$ and $f_2(k,\omega,x) := 0$ Lemma 2.4 furnishes a measurable mapping H^- having the properties contained in *(a)* and *(b)* above.

Conversely, with $f_1(k,\omega,x) := 0$ and $f_2(k,\omega,x) := f^-(k,\omega,x)$ Lemma 2.4 yields a measurable mapping $\tilde{H}^-$ satisfying everything claimed in *(a)* and *(b)* above.

So in order to complete the proof of the proposition we only have to verify that the mappings $H^-(\kappa,\omega,\cdot)$ and $\tilde{H}^-(\kappa,\omega,\cdot)$ are inverse to each other, for arbitrary $\kappa \in \mathbb{Z}$ and $\omega \in \Omega$. To that end, we apply Lemma 2.4 once again, yet this time with $f_1(k,\omega,x) := f^-(k,\omega,x)$ and $f_2(k,\omega,x) := f^-(k,\omega,x)$. If λ denotes the general solution of (2.40), then there is a unique mapping H^* such that for every $\kappa \in \mathbb{Z}$, $\omega \in \Omega$ and $\xi \in \mathbb{R}^{d^-}$ the difference $\lambda(\cdot;\kappa,\omega,H^*(\kappa,\omega,\xi)) - \lambda(\cdot;\kappa,\omega,\xi)$ is 1-quasibounded with respect to ω, namely $H^*(\kappa,\omega,\xi) \equiv \xi$ on $\mathbb{Z} \times \Omega \times \mathbb{R}^{d^-}$. On the other hand, the above definitions of H^- and $\tilde{H}^-$ show that for every $\kappa \in \mathbb{Z}$, $\omega \in \Omega$ and $\xi \in \mathbb{R}^{d^-}$ both $\Phi^-(\cdot,\kappa,\omega)H^-(\kappa,\omega,\xi) - \lambda(\cdot;\kappa,\omega,\xi)$ and $\lambda(\cdot;\kappa,\omega,\tilde{H}^-(\kappa,\omega,H^-(\kappa,\omega,\xi))) - \Phi^-(\cdot,\kappa,\omega)H^-(\kappa,\omega,\xi)$ are 1-quasibounded with respect to ω. But then the difference $\lambda(\cdot;\kappa,\omega,\tilde{H}^-(\kappa,\omega,H^-(\kappa,\omega,\xi))) - \lambda(\cdot;\kappa,\omega,\xi)$ is 1-quasibounded, too, for all $\kappa \in \mathbb{Z}$, $\omega \in \Omega$ and $\xi \in \mathbb{R}^{d^-}$ – and the uniqueness assertion of Lemma 2.4 implies $\tilde{H}^-(\kappa,\omega,H^-(\kappa,\omega,\xi)) \equiv H^*(\kappa,\omega,\xi) \equiv \xi$ on $\mathbb{Z} \times \Omega \times \mathbb{R}^{d^-}$. Similarly, the identity $H^-(\kappa,\omega,\tilde{H}^-(\kappa,\omega,\xi)) \equiv \xi$ on $\mathbb{Z} \times \Omega \times \mathbb{R}^{d^-}$ may be verified. This completes the proof of the proposition.   ◊

Remark 2.3. If the mapping f^- in Proposition 2.3 satisfies $f^-(k,\omega,0) = 0$ for arbitrary $k \in \mathbb{Z}$ and $\omega \in \Omega$, then the above proof immediately furnishes $H^-(\kappa,\omega,0) = 0$ and $\tilde{H}^-(\kappa,\omega,0) = 0$ for all $\kappa \in \mathbb{Z}$ and $\omega \in \Omega$. This is due to the fact that in this situation the trivial solution of equation (2.40) is the uniquely determined ω-solution which is 1-quasibounded with respect to ω.   □

Of course there is a dual version of Proposition 2.3 for the case that the linear equation satisfies conditions as in Lemma 2.2 with the additional assumption $\alpha_+ > 1$. However, due to lack of space we leave both the formulation and the proof to the reader.

Now we have gathered everything in order to present the Hartman-Grobman theorems for random difference equations. To begin with, we state and prove a random version of the *classical Hartman-Grobman theorem*.

Theorem 2.3. *Consider the random difference equation*

$$x_{k+1} = A(k,\omega)x_k + F(k,\omega,x_k) \qquad (2.42)$$

satisfying the assumptions (A1), (A2), (A3) and (A4) from the beginning of 2.3. Furthermore, assume $1 \notin (\alpha_{i,+} - \delta, \alpha_{i,-} + \delta)$, for all $i = 1, \ldots, p$.

Then there is some constant $L^* > 0$ such that for all $0 \le L < L^*$ the random difference equation (2.42) satisfies all the assumptions of Theorem 2.1 and is topologically equivalent to the homogeneous linear random difference equation

$$x_{k+1} = A(k,\omega)x_k \qquad (2.43)$$

via two mappings H and $\tilde{H}$, with $H(\kappa,\omega,0) = 0$ and $\tilde{H}(\kappa,\omega,0) = 0$ for $\kappa \in \mathbb{Z}$ and $\omega \in \Omega$. Even more, H maps any ω-solution of (2.42) which is contained in the random invariant fiber bundle $S^{i,j}(\omega)$ onto an ω-solution of (2.43) which lies in the corresponding invariant subspace bundle, and vice versa.

Proof. According to Theorem 2.2 there is a positive constant $L_1^* > 0$ such that for all $0 \le L < L_1^*$ equation (2.42) is topologically equivalent to the decoupled equation (2.32), via two mappings D and $\tilde{D}$. Furthermore, the proof of Proposition 2.2 furnishes some constant $0 < L^* \le L_1^*$ such that for all $0 \le L < L^*$ the estimate

$$\|F_i(k,\omega,\pi^{i,i}(k,\omega,x^i)) - F_i(k,\omega,\pi^{i,i}(k,\omega,\bar{x}^i))\|_{k+1,\omega} \le 2L\|x^i - \bar{x}^i\|_{k,\omega}$$

holds, for arbitrary $i = 1, \ldots, p$, $k \in \mathbb{Z}$, $\omega \in \Omega$ and $x^i, \bar{x}^i \in \mathbb{R}^{d_i}$, as well as $2L < \frac{\delta}{K}$. But then, the equation

$$x_{k+1}^i = A_i(k,\omega)x_k^i + F_i(k,\omega,\pi^{i,i}(k,\omega,x_k^i)) \qquad (2.44)$$

satisfies all the assumptions of Proposition 2.3 or its dual version, provided $1 \ge \alpha_{i,-} + \delta$ or $1 \le \alpha_{i,+} - \delta$, respectively – and therefore (2.44) is topologically equivalent to

$$x_{k+1}^i = A_i(k,\omega)x_k^i$$

via two mappings $H^{(i)}$ and $\tilde{H}^{(i)}$, where according to Remark 2.3 we have $H^{(i)}(\kappa,\omega,0) \equiv 0$ and $\tilde{H}^{(i)}(\kappa,\omega,0) \equiv 0$ on $\mathbb{Z} \times \Omega$. Finally, by combining the mappings D and $H^{(i)}$ appropriately we may construct the required mapping H – and since ω-solutions which are contained in some invariant subspace bundle of (2.32) are mapped onto this bundle by H, this completes the proof of Theorem 2.3.   ◊

It is obvious that in the so-called *hyperbolic* case, i.e. if $1 \notin [\alpha_{i,+}, \alpha_{i,-}]$ for all $i = 1, \ldots, p$, we may choose δ in (A1) in such a way that Theorem 2.3 can be applied. But what can be said in the non-hyperbolic case? The answer to this question is the subject of the following *generalized Hartman-Grobman theorem* for random difference equations.

Theorem 2.4. *Again, consider the random difference equation*

$$x_{k+1} = A(k, \omega)x_k + F(k, \omega, x_k) \qquad (2.45)$$

satisfying the assumptions (A1), (A2), (A3) and (A4) from the beginning of 2.3. Furthermore, assume $1 \in [\alpha_{i,+}, \alpha_{i,-}]$ for an $i \in \{1, \ldots, p\}$.

Then there is some constant $L^* > 0$ such that for all $0 \le L < L^*$ the random difference equation (2.45) satisfies all the assumptions of Theorem 2.1 and is topologically equivalent to the decoupled, partially linear random difference equation

$$x_{k+1} = A(k, \omega)x_k + \tilde{F}(k, \omega, x_k) \qquad (2.46)$$

where $\tilde{F}_i(k, \omega, x) := F_i(k, \omega, \pi^{i,i}(k, \omega, x^i))$ and $\tilde{F}_j(k, \omega, x) := 0$ for all $j \ne i$, via two mappings H and $\tilde{H}$, with $H(\kappa, \omega, 0) = 0$ and $\tilde{H}(\kappa, \omega, 0) = 0$ for $\kappa \in \mathbb{Z}$ and $\omega \in \Omega$. Even more, H maps any ω-solution of (2.45) which is contained in the random invariant fiber bundle $\mathcal{S}^{i,j}(\omega)$ onto an ω-solution of (2.46) which lies in the corresponding invariant subspace bundle, and vice versa.

Proof. The proof is completely analogous to the proof of Theorem 2.3 – yet now the i-th equation cannot be linearized any more.   ◊

We have reached the end of our investigations concerning nonautonomous random difference equations, and beginning with the next section the results obtained so far will be applied to random dynamical systems.

3. Random Dynamical Systems

3.1 Preliminaries and Hypotheses

Comparing the results of the last section to what we said about random dynamical systems in the introduction shows that there is still quite a gap to be closed. Whereas on the one hand, the connection between nonautonomous random difference equations and discrete-time random dynamical systems has not been established completely yet (apart from the remark at the end of Section 1), it is not at all clear either how to choose the norms depending on both time and chance. Even more, the alert reader might have spotted another problem. If we apply the results of the last section to the deterministic autonomous special case the constructed objects do depend explicitly on time – although one would expect them to be independent of it. Hence, it seems reasonable that if we only apply the results to those (very special) random difference equations generating discrete-time random dynamical systems we might miss some interesting additional properties of the constructed objects.

Fortunately, all of the above problems can be solved satisfactorily, and this subsection will be devoted to the first two questions. To begin with, let us try to gain more insight into the connection between random difference equations and discrete-time random dynamical systems.

Lemma 3.1. *Assume we are given an arbitrary metric dynamical system* $(\Omega, \mathcal{F}, \mathbb{P}, (\theta_n)_{n\in\mathbb{Z}})$, *as well as a random difference equation*

$$x_{k+1} = f(k, \omega, x_k) \qquad (3.1)$$

where $f : \mathbb{Z} \times \Omega \times \mathbb{R}^d \to \mathbb{R}^d$ *denotes some measurable mapping. Furthermore, suppose that all ω-solutions of (3.1) exist uniquely on $\mathbb{Z}$ and that the general solution λ in the sense of (2.3) is measurable. Then the following three assertions are equivalent:*

(a) *The mapping φ defined as $\varphi(k, \omega, x) := \lambda(k; 0, \omega, x)$ is a measurable discrete-time random dynamical system over* $(\Omega, \mathcal{F}, \mathbb{P}, (\theta_n)_{n\in\mathbb{Z}})$.
(b) *The general solution of (3.1) satisfies $\lambda(k; \kappa, \omega, x) = \lambda(k - \kappa; 0, \theta_\kappa \omega, x)$ for arbitrary $k, \kappa \in \mathbb{Z}$, $\omega \in \Omega$ and $x \in \mathbb{R}^d$.*
(c) *The mapping f satisfies $f(k, \omega, x) = f(0, \theta_k \omega, x)$ for all $k \in \mathbb{Z}$, $\omega \in \Omega$ and $x \in \mathbb{R}^d$.*

Proof. Since the proof boils down to the use of (2.4) and some easy inductive arguments we leave the verification of the lemma to the reader. ◊

Now let us turn our attention to the second question posed in the beginning of this section, i.e. how the norms should be chosen in our situation. The respective results dating from Boxler [9] and Dahlke [11] are summarized in the following proposition, which is a consequence of the multiplicative ergodic theorem (Theorem 1.1).

Proposition 3.1. *In the situation of Theorem 1.1 let $a > 0$ denote an arbitrary, but fixed, real constant such that the intervals $[\lambda_i - a, \lambda_i + a]$, $i = 1, \ldots, p$, are disjoint, and let $||\cdot||$ denote the Euclidean norm on $\mathbb{R}^d$. Furthermore, define for arbitrary $\omega \in \Omega$ and $x = x^1 + \ldots + x^p \in E_1(\omega) \oplus \ldots \oplus E_p(\omega)$ a new norm $||x||_\omega := \sqrt{||x^1||_\omega^2 + \ldots + ||x^p||_\omega^2}$, where for $u \in E_i(\omega)$ we set*

$$
||u||_\omega := \begin{cases} \left(\displaystyle\int_{-\infty}^{\infty} ||\Phi(t, \omega)u||^2 e^{-2(\lambda_i t + a|t|)} dt \right)^{\frac{1}{2}} & \text{for} \quad \mathbb{T} = \mathbb{R}, \\[3ex] \left(\displaystyle\sum_{n=-\infty}^{\infty} ||\Phi(n, \omega)u||^2 e^{-2(\lambda_i n + a|n|)} \right)^{\frac{1}{2}} & \text{for} \quad \mathbb{T} = \mathbb{Z}, \end{cases}
$$

and let $||x||_\omega := ||x||$ for all $\omega \notin \tilde{\Omega}$. Then the following holds:

(a) $||\cdot||_\omega$ *is a random norm on* $\mathbb{R}^d$, *i.e. the mapping* $(x,\omega) \mapsto ||x||_\omega$ *is measurable.*

(b) *For every* $\varepsilon > 0$ *there is a measurable mapping* $B_\varepsilon : \Omega \to [1,\infty)$ *such that both* $\frac{1}{B_\varepsilon(\omega)}||x|| \leq ||x||_\omega \leq B_\varepsilon(\omega)||x||$ *and* $B_\varepsilon(\theta_t\omega) \leq B_\varepsilon(\omega)e^{\varepsilon|t|}$ *hold for every choice of* $x \in \mathbb{R}^d$, $\omega \in \tilde{\Omega}$ *and* $t \in \mathbb{T}$.

(c) *For every* $\omega \in \tilde{\Omega}$, $i = 1, \ldots, p$, $x \in E_i(\omega)$ *and* $t \in \mathbb{T}$ *we have*

$$e^{\lambda_i t - a|t|} \leq ||\Phi(t,\omega)|_{E_i(\omega)}||_{\omega,\theta_t\omega} \leq e^{\lambda_i t + a|t|},$$

with $||\Phi(t,\omega)|_{E_i(\omega)}||_{\omega,\theta_t\omega} := \sup\{||\Phi(t,\omega)x||_{\theta_t\omega} : x \in E_i(\omega), ||x||_\omega \leq 1\}.$

Proof. The proof can be found in Arnold [2, Theorem 3.74]. ◊

It will turn out in a moment that the above random norms are exactly what is needed to apply the results of the last section to random dynamical systems. But first of all, we like to present the random dynamical systems that actually will be studied in this third section. Let φ denote a measurable random dynamical system on $\mathbb{R}^d$ over some ergodic metric dynamical system $(\Omega, \mathcal{F}, \mathbb{P}, (\theta_t)_{t\in\mathbb{T}})$, where $\mathbb{T} = \mathbb{Z}$ or $\mathbb{T} = \mathbb{R}$, and assume that the origin is a fixed point for φ, i.e. suppose $\varphi(t,\omega,0) = 0$ for all $t \in \mathbb{T}$ and $\omega \in \Omega$[7]. Furthermore, let Φ denote a measurable linear random dynamical system on $\mathbb{R}^d$ over $(\Omega, \mathcal{F}, \mathbb{P}, (\theta_t)_{t\in\mathbb{T}})$ and define the "nonlinear" part Ψ of φ by $\Psi(t,\omega,x) := \varphi(t,\omega,x) - \Phi(t,\omega)x$. As for the linear part Φ we assume that the following hypothesis holds.

(H1) *The measurable linear random dynamical system* Φ *is in block-diagonal form, i.e. we have* $\Phi(t,\omega)x = (\Phi_1(t,\omega)x^1, \Phi_2(t,\omega)x^2, \ldots, \Phi_p(t,\omega)x^p)$ *with respect to some splitting* $\mathbb{R}^d = \mathbb{R}^{d_1} \times \ldots \times \mathbb{R}^{d_p}$. *Moreover, suppose that all the assertions which are contained in Theorem 1.1 and Proposition 3.1 are valid for* Φ *and that the* i-*th Oseledets space is given by* $E_i(\omega) = \{0\} \times \ldots \times \{0\} \times \mathbb{R}^{d_i} \times \{0\} \times \ldots \times \{0\} \subset \mathbb{R}^d$, *for* $i = 1, \ldots, p$.

Basically, the above hypothesis (H1) is without loss of generality, since in the situation of Theorem 1.1 and Proposition 3.1 it is possible to block-diagonalize the given linear random dynamical system by measurably choosing bases in the Oseledets spaces. For more details we refer the reader to Wanner [30, pp. 114ff].

In view of our further proceeding we have to replace the random norm guaranteed by hypothesis (H1) by a new one. To that end, define a random norm on $\mathbb{R}^{d_i}$ via

$$||x^i||_{i,\omega} := ||(0, \ldots, 0, x^i, 0, \ldots, 0)||_\omega,$$

for arbitrary $x^i \in \mathbb{R}^{d_i}$ and $i = 1, \ldots, p$, and then let

$$||x||_\omega^* = ||(x^1, \ldots, x^p)||_\omega^* := ||x^1||_{1,\omega} + \ldots + ||x^p||_{p,\omega}$$

---

[7] One can show that this last assumption is without loss of generality – in contrast to the deterministic case (cf. Arnold [2, Proposition 1.23]).

for $x = (x^1, \ldots, x^p) \in \mathbb{R}^{d_1} \times \ldots \times \mathbb{R}^{d_p} = \mathbb{R}^d$. Obviously, the random norm $\|\cdot\|_\omega^*$ is equivalent to the original one, and the assertions of Proposition 3.1 remain valid with respect to this new norm. From now on, we will always use the above-defined random norm $\|\cdot\|_\omega^*$ on $\mathbb{R}^d$ – and therefore we omit the "$*$".

Finally, let us state the hypotheses for the nonlinear part Ψ of the random dynamical system φ.

(H2) *Assume that for arbitrary* $i = 1, \ldots, p$, $t \in [0, 1]$, $\omega \in \tilde{\Omega}$ *and* $x, \bar{x} \in \mathbb{R}^d$ *the estimates*

$$\|\Psi_i(t, \omega, x) - \Psi_i(t, \omega, \bar{x})\|_{\theta_t \omega} \leq L \|x - \bar{x}\|_\omega \quad and \quad \|\Psi_i(t, \omega, x)\|_{\theta_t \omega} \leq M$$

hold, with $M \geq 0$ *and* $0 \leq L < L^*$, *for some sufficiently small* L^*.

The assumptions contained in (H1) and (H2) are already all we need to directly apply the results of Section 2 to the random dynamical system φ. In order to conclude this subsection let us make this more explicit. For arbitrary $k \in \mathbb{Z}$, $\omega \in \Omega$ and $x \in \mathbb{R}^d$ let $A(k, \omega) := \Phi(1, \theta_k \omega)$ and $F(k, \omega, x) := \Psi(1, \theta_k \omega, x)$. Then it may easily be verified that the general solution λ of the random difference equation

$$x_{k+1} = A(k, \omega) x_k + F(k, \omega, x_k) \tag{3.2}$$

exists in the sense of (2.3), is measurable, and continuous with respect to the last variable. Moreover, Lemma 3.1 furnishes the identity

$$\lambda(k; \kappa, \omega, \xi) = \varphi(k - \kappa, \theta_\kappa \omega, \xi) ,$$

for arbitrary $k, \kappa \in \mathbb{Z}$, $\omega \in \Omega$ and $\xi \in \mathbb{R}^d$. In other words, in the case $\mathbb{T} = \mathbb{Z}$ the random dynamical system φ corresponds exactly to the general solution λ of (3.2), whereas in the case $\mathbb{T} = \mathbb{R}$ we may reconstruct only the values of φ at integral times.

Next, choose a constant $a > 0$ such that the intervals $[\lambda_i - a, \lambda_i + a]$, $i = 1, \ldots, p$, are disjoint, and in the hyperbolic case (introduced in Definition 1.3) further assume that $0 \notin [\lambda_i - a, \lambda_i + a]$ for $i = 1, \ldots, p$. If we define $\alpha_{i,+} := e^{\lambda_i - a}$ and $\alpha_{i,-} := e^{\lambda_i + a}$, then Lemma 3.1 and Proposition 3.1(c) imply for arbitrary $m, n \in \mathbb{Z}$ and $\omega \in \tilde{\Omega}$ the estimates

$$\|\hat{\Phi}_i(m, n, \omega)\|_{\theta_n \omega, \theta_m \omega} = \|\Phi_i(m - n, \theta_n \omega)\|_{\theta_n \omega, \theta_m \omega} \leq \alpha_{i,+}^{m-n} , \quad m \leq n ,$$

$$\|\hat{\Phi}_i(m, n, \omega)\|_{\theta_n \omega, \theta_m \omega} = \|\Phi_i(m - n, \theta_n \omega)\|_{\theta_n \omega, \theta_m \omega} \leq \alpha_{i,-}^{m-n} , \quad m \geq n ,$$

provided $\hat{\Phi}_i$ denotes the evolution operator of the homogeneous linear random difference equation $x_{k+1}^i = A_i(k, \omega) x_k^i$, where $A_i(k, \omega) := \Phi_i(1, \theta_k \omega)$. Finally choose (and then fix) an arbitrary positive constant δ with

$$0 < \delta < \min \left\{ \frac{\alpha_{1,+} - \alpha_{2,-}}{2} , \ldots , \frac{\alpha_{p-1,+} - \alpha_{p,-}}{2} , \alpha_{p,+} \right\} ,$$

where in the hyperbolic case we further require $1 \notin (\alpha_{i,+} - \delta, \alpha_{i,-} + \delta)$ for $i = 1, \ldots, p$. At this point, the reader may verify with the aid of (H1) and (H2) – but without any effort – that the random difference equation (3.2) actually satisfies everything contained in (A1), (A2), (A3) and (A4) from the beginning of Subsection 2.3, if we set $|| \cdot ||_{k,\omega} := || \cdot ||_{\theta_k \omega}$.

3.2 Random Invariant Manifolds

The time has come for harvesting! By now we are prepared well enough to apply all the results of Section 2 to the random dynamical systems introduced in the last subsection, and we begin by proving the existence of *random invariant manifolds*. But first of all, we like to adapt the notion of quasiboundedness to our new situation.

Definition 3.1. *Assume we are in the situation of Theorem 1.1 and Proposition 3.1, let $\mu : \mathbb{T} \to \mathbb{R}^d$ denote an arbitrary mapping, and let $\gamma \in \mathbb{R}^+$ be a positive constant. Then μ is called γ^+-quasibounded with respect to ω if for all $t \in \mathbb{T}$ with $t \geq 0$ the estimate*

$$||\mu(t)||_{\theta_t \omega} \leq C\gamma^t$$

holds, with some constant $C \geq 0$. Similar to Definition 2.1 the notion of γ^-- or γ-quasiboundedness may be defined if the above estimate holds for every $t \in \mathbb{T}$ with $t \leq 0$ or for all $t \in \mathbb{T}$, respectively.

Of course, there is nothing really exciting in this definition. Yet Proposition 3.1 shows that now quasiboundedness actually is connected to the "real" asymptotic behavior of the mapping μ, i.e. the asymptotic behavior with respect to the Euclidean norm. More precisely, we have the following lemma.

Lemma 3.2. *In the situation of Definition 3.1 the following is true:*

(a) *If μ is a γ^+-quasibounded mapping with respect to $\omega \in \tilde{\Omega}$, then for every $\beta > \gamma$ there is a $C_\beta \geq 0$ such that $||\mu(t)|| \leq C_\beta \cdot \beta^t$ for arbitrary $t \in \mathbb{T}$ with $t \geq 0$.*

(b) *Conversely, suppose that μ satisfies $||\mu(t)|| \leq C_\beta \cdot \beta^t$ for arbitrary $t \in \mathbb{T}$ with $t \geq 0$, where $\beta > 0$ and $C_\beta \geq 0$. Then μ is γ^+-quasibounded with respect to ω, for every choice of $\gamma > \beta$ and every $\omega \in \tilde{\Omega}$.*

Analogous assertions are valid for the notion of γ^--quasiboundedness if in (a) or (b) we choose $0 < \beta < \gamma$ or $0 < \gamma < \beta$, respectively.

Proof. All we have to do is to use the estimates contained in Proposition 3.1(b).   ◊

The already announced main result of this subsection is contained in the following theorem. On the one hand, it shows that Theorem 2.1 is tailor-made for constructing invariant sets for φ, whereas on the other hand the very special form of the random difference equation (3.2) enables us to derive information concerning these sets which is far beyond the scope of Theorem 2.1.

Theorem 3.1. *Assume we are given a measurable random dynamical system φ on $\mathbb{R}^d$ over some ergodic metric dynamical system $(\Omega, \mathcal{F}, \mathbb{P}, (\theta_t)_{t \in T})$, with fixed point 0. Assume further that there is a linear random dynamical system Φ such that the hypotheses (H1) and (H2) are satisfied with $\varphi(t, \omega, x) = \Phi(t, \omega)x + \Psi(t, \omega, x)$. Then for every choice of $1 \leq i \leq j \leq p$ with $(i, j) \neq (1, p)$ there is a random manifold $S^{i,j}(\omega) \subset \mathbb{R}^d$ with the following properties:*

(a) *We have $S^{i,j}(\omega) = \{x \in \mathbb{R}^d : (x^{<i}, x^{>j}) = s^{i,j}(\omega, x^{i \leq j})\}$ for some measurable mapping $s^{i,j} : \Omega \times \mathbb{R}^{d_i} \times \ldots \times \mathbb{R}^{d_j} \to \mathbb{R}^{d_1} \times \ldots \times \mathbb{R}^{d_{i-1}} \times \mathbb{R}^{d_{j+1}} \times \ldots \times \mathbb{R}^{d_p}$, the mappings $s^{i,j}(\omega, \cdot)$ are globally Lipschitz continuous for arbitrary $\omega \in \tilde{\Omega}$, and the corresponding Lipschitz constants converge to 0 as $L \to 0$.*

(b) *The random manifold $S^{i,j}(\omega)$ is invariant under the random dynamical system φ, i.e. for arbitrary $t \in T$ and $\omega \in \tilde{\Omega}$ we have*

$$\varphi(t, \omega) S^{i,j}(\omega) = S^{i,j}(\theta_t \omega) .$$

(c) *The random invariant manifold $S^{i,j}(\omega)$ may be characterized dynamically as*

$$S^{i,j}(\omega) = \{\xi \in \mathbb{R}^d : \varphi(\cdot, \omega, \xi) \text{ is } \gamma_1^+\text{- and } \gamma_2^-\text{-quasibounded}$$
$$\text{with respect to } \omega\}$$

for every $\omega \in \tilde{\Omega}$ and $\gamma_1 \in [\alpha_{1,-}+\delta, \infty)$ for $i = 1$, $\gamma_1 \in [\alpha_{i,-}+\delta, \alpha_{i-1,+}-\delta]$ for $i > 1$, $\gamma_2 \in [\alpha_{j+1,-}+\delta, \alpha_{j,+}-\delta]$ for $j < p$, $\gamma_2 \in (0, \alpha_{p,+}-\delta]$ for $j = p$.

If we finally define $S^{1,p}(\omega) := \mathbb{R}^d$ for $\omega \in \Omega$, then the assertions of (b) and (c) remain valid for $S^{1,p}(\omega)$.

Proof. According to (H1) and (H2) the random difference equation (3.2) satisfies all the assumptions of Theorem 2.1 if we replace there Ω by $\tilde{\Omega}$. Let $1 \leq i \leq j \leq p$ with $(i, j) \neq (1, p)$ be arbitrary, but fixed, and let $\hat{s}^{i,j}$ denote the mapping guaranteed by Theorem 2.1 which generates the random invariant fiber bundle $\hat{S}^{i,j}(\omega)$. If we now define the mapping $s^{i,j}$ as

$$s^{i,j}(\omega, x^i, \ldots, x^j) := \hat{s}^{i,j}(0, \omega, x^i, \ldots, x^j)$$

for arbitrary $\omega \in \tilde{\Omega}$ and $(x^i, \ldots, x^j) \in \mathbb{R}^{d_i} \times \ldots \times \mathbb{R}^{d_j}$, and extend this definition measurably onto the whole of Ω (e.g. by setting the mapping to 0 for all $\omega \in \Omega \setminus \tilde{\Omega}$), then the assertions contained in (a) and (c) are satisfied, if in (c) we replace $\varphi(\cdot, \omega, \xi)$ by the restriction $\varphi(\cdot, \omega, \xi)|_{\mathbb{Z}}$. As for the invariance claimed in (b) we begin by proving

$$\hat{s}^{i,j}(\kappa, \omega, \cdot) = \hat{s}^{i,j}(0, \theta_\kappa \omega, \cdot) \tag{3.3}$$

for arbitrary $\kappa \in \mathbb{Z}$ and $\omega \in \tilde{\Omega}$. To that end, let $\kappa \in \mathbb{Z}$, $\omega \in \tilde{\Omega}$ and $\xi \in \mathbb{R}^d$ be arbitrary with $(\kappa, \xi) \in \hat{\mathcal{S}}^{i,j}(\omega)$. Due to Theorem 2.1 the mapping $\lambda(\cdot; \kappa, \omega, \xi)$ is both γ_1^+- and γ_2^--quasibounded with respect to ω, where γ_1 and γ_2 may be chosen as in (c). Observing that Lemma 3.1(b) furnishes the identity

$$\gamma^{-k}||\lambda(k; 0, \theta_\kappa \omega, \xi)||_{\theta_k \theta_\kappa \omega} = \gamma^\kappa \gamma^{-(\kappa+k)}||\lambda(k+\kappa; \kappa, \omega, \xi)||_{\theta_{k+\kappa}\omega} \quad \text{for } k \in \mathbb{Z}$$

we immediately deduce that the mapping $\lambda(\cdot; 0, \theta_\kappa \omega, \xi)$ is both γ_1^+- and γ_2^--quasibounded with respect to $\theta_\kappa \omega$, i.e. we have $(0, \xi) \in \hat{\mathcal{S}}^{i,j}(\theta_\kappa \omega)$. But this inclusion finally implies $\hat{s}^{i,j}(\kappa, \omega, \xi^{i \leq j}) = (\xi^{<i}, \xi^{>j}) = \hat{s}^{i,j}(0, \theta_\kappa \omega, \xi^{i \leq j})$, and the verification of (3.3) is complete.

Now let $t \in \mathbb{Z}$, $\omega \in \tilde{\Omega}$ and $\xi \in \mathcal{S}^{i,j}(\omega)$ be arbitrary. Then the invariance of the fiber bundle $\hat{\mathcal{S}}^{i,j}(\omega)$ proved in Theorem 2.1, together with (3.3), furnishes $\varphi(t, \omega, \xi) \in \mathcal{S}^{i,j}(\theta_t \omega)$, thus we have $\varphi(t, \omega)\mathcal{S}^{i,j}(\omega) \subset \mathcal{S}^{i,j}(\theta_t \omega)$. Applying the cocycle property of φ eventually yields

$$\mathcal{S}^{i,j}(\theta_t \omega) = \varphi(t, \omega)\varphi(-t, \theta_t \omega)\mathcal{S}^{i,j}(\theta_t \omega) \subset$$
$$\subset \varphi(t, \omega)\mathcal{S}^{i,j}(\theta_{-t}\theta_t \omega) = \varphi(t, \omega)\mathcal{S}^{i,j}(\omega) ,$$

so the identity in (b) is true for integral values of t.

Whereas for $\mathbb{T} = \mathbb{Z}$ the proof of Theorem 3.1 is complete, in the case $\mathbb{T} = \mathbb{R}$ there is still some work to be done. More precisely, we have to show that the identity in (b) holds for arbitrary $t \in \mathbb{R}$ rather than only for $t \in \mathbb{Z}$, and that in (c) we may write $\varphi(\cdot, \omega, \xi)$ instead of $\varphi(\cdot, \omega, \xi)|_\mathbb{Z}$.

As for the completion of (b) let $\omega \in \tilde{\Omega}$ and $\xi \in \mathcal{S}^{i,j}(\omega)$ be arbitrary, but fixed. We already know that both $c_{\omega, \xi}^{(1)} := \sup\{\gamma_1^{-k}||\varphi(k, \omega, \xi)||_{\theta_k \omega} : k \in \mathbb{Z}_0^+\}$ and $c_{\omega, \xi}^{(2)} := \sup\{\gamma_2^{-k}||\varphi(k, \omega, \xi)||_{\theta_k \omega} : k \in \mathbb{Z}_0^-\}$ are finite, where γ_1 and γ_2 are chosen as above. Next, let $t \in [0, 1]$ be arbitrary, but fixed. Then Proposition 3.1(c) implies for every $\hat{\omega} \in \tilde{\Omega}$ the estimate $||\Phi(t, \hat{\omega})||_{\hat{\omega}, \theta_t \hat{\omega}} \leq c^{(3)} := \max\{\alpha_{1,-}, 1\}$, and with the help of (H2) and the cocycle property of φ we further deduce for arbitrary $k \in \mathbb{Z}_0^+$

$$||\varphi(k, \theta_t \omega, \varphi(t, \omega, \xi))||_{\theta_k \theta_t \omega} =$$
$$= ||\Phi(t, \theta_k \omega)\varphi(k, \omega, \xi) + \Psi(t, \theta_k \omega, \varphi(k, \omega, \xi))||_{\theta_t \theta_k \omega} \leq \quad (3.4)$$
$$\leq (c^{(3)} + pL)c_{\omega, \xi}^{(1)} \cdot \gamma_1^k ,$$

and similarly for $k \in \mathbb{Z}_0^-$

$$||\varphi(k, \theta_t \omega, \varphi(t, \omega, \xi))||_{\theta_k \theta_t \omega} \leq (c^{(3)} + pL)c_{\omega, \xi}^{(2)} \cdot \gamma_2^k . \quad (3.5)$$

Now the already proved dynamical characterization furnishes $\varphi(t, \omega, \xi) \in \mathcal{S}^{i,j}(\theta_t \omega)$, and together with the above-proved invariance for integral times we get

$$\varphi(\tau + t, \omega, \xi) = \varphi(\tau, \theta_t \omega, \varphi(t, \omega, \xi)) \in \mathcal{S}^{i,j}(\theta_\tau \theta_t \omega) = \mathcal{S}^{i,j}(\theta_{\tau+t}\omega)$$

for arbitrary $\tau \in \mathbb{Z}$ and $t \in [0, 1]$, i.e. we have $\varphi(t, \omega, \xi) \in S^{i,j}(\theta_t \omega)$ for all $t \in \mathbb{R}$ – which immediately yields the identity in *(b)*.

Finally, let us complete the proof of *(c)*. Obviously it suffices to show that for all $\omega \in \tilde{\Omega}$ and $\xi \in S^{i,j}(\omega)$ both $\{\gamma_1^{-t} \| \varphi(t, \omega, \xi) \|_{\theta_t \omega} : t \in \mathbb{R}_0^+\}$ and $\{\gamma_2^{-t} \| \varphi(t, \omega, \xi) \|_{\theta_t \omega} : t \in \mathbb{R}_0^-\}$ are bounded. But this is an easy consequence of the inequalities

$$\gamma_1^{-(k_1+t)} \| \varphi(k_1 + t, \omega, \xi) \|_{\theta_{k_1+t}\omega} \leq \max\{\gamma_1^{-1}, 1\} \cdot (c^{(3)} + pL) c_{\omega,\xi}^{(1)},$$

$$\gamma_2^{-(k_2+t)} \| \varphi(k_2 + t, \omega, \xi) \|_{\theta_{k_2+t}\omega} \leq \max\{\gamma_2^{-1}, 1\} \cdot (c^{(3)} + pL) c_{\omega,\xi}^{(2)},$$

for arbitrary $k_1 \in \mathbb{Z}_0^+$, $k_2 \in \mathbb{Z}^-$ and $t \in [0, 1]$, which follow from (3.4) and (3.5), respectively. $\diamond$

Remark 3.1. If $\hat{\pi}^{i,j}$ denotes the mapping from Remark 2.1 corresponding to $\hat{S}^{i,j}(\omega)$, then the definition $\pi^{i,j}(\omega, \xi) := \hat{\pi}^{i,j}(0, \omega, \xi)$, for arbitrary $\omega \in \tilde{\Omega}$ and $\xi \in \mathbb{R}^{d_i} \times \ldots \times \mathbb{R}^{d_j}$ (as well as $\pi^{i,j}(\omega, \xi) := 0$ for $\omega \notin \tilde{\Omega}$), allows an alternative representation of the random invariant manifold $S^{i,j}(\omega)$, namely $S^{i,j}(\omega) = \{\pi^{i,j}(\omega, \xi) : \xi \in \mathbb{R}^{d_i} \times \ldots \times \mathbb{R}^{d_j}\}$, for every $\omega \in \tilde{\Omega}$. $\square$

Theorem 3.1 shows that the random invariant subspaces $E_i(\omega)$ of the linear random dynamical system Φ survive a sufficiently small nonlinear perturbation, i.e. for the nonlinear random dynamical system φ we get random invariant manifolds $S^{i,i}(\omega)$ which in view of the terminology of Theorem 1.1 are called *Oseledets manifolds*. Likewise, the invariant direct sums $E_i(\omega) \oplus \ldots \oplus E_j(\omega)$ are turned into the random invariant manifolds $S^{i,j}(\omega)$. For discrete-time random dynamical systems of class C^1 these results were already obtained by Dahlke [11].

As we mentioned in the introduction, random dynamical systems generalize ordinary dynamical systems – we only have to consider a mapping φ being independent of ω. Applying Theorem 3.1 to such a mapping, where Ω consists of only one element and $\theta_t := \mathrm{id}_\Omega$ for $t \in \mathbb{T}$, yields invariant manifolds (independent of ω) for the dynamical system φ. Hence, we are able to reproduce well-known results on dynamical systems, which can be found for instance in Irwin [20]. Having this deterministic situation in mind, we like to introduce the following notions dating from Kelley [21]. Suppose that $p \geq 3$ and $\lambda_i = 0$ for some $1 < i < p$. Then $S^{<i}(\omega)$, $S^{\leq i}(\omega)$, $S^{i,i}(\omega)$, $S^{\geq i}(\omega)$ and $S^{>i}(\omega)$ are called *unstable, center-unstable, center, center-stable* and *stable random invariant manifold*, respectively. Whereas Carverhill [10] already proved the existence of random stable manifolds, the case of random center manifolds has been treated in Boxler [9].

To close this subsection, the reader may convince himself using Lemma 3.2 that the notions "stable" and "unstable" actually are justified in the following sense: Every ω-orbit of φ which is contained in the stable (or unstable) random invariant manifold approaches the origin exponentially as $t \to \infty$ (or $t \to -\infty$).

3.3 Asymptotic Phases

The next result of our linearization theory for random dynamical systems is concerned with the asymptotic behavior of those ω-orbits which are not contained in any of the random invariant manifolds of Theorem 3.1. For this we assign to every such ω-orbit *asymptotic phases* on certain random invariant manifolds, i.e. unique orbits reflecting the behavior of our given orbit as $t \to \infty$ or $t \to -\infty$.

Theorem 3.2. *Assume we are given a measurable random dynamical system φ on $\mathbb{R}^d$ over some ergodic metric dynamical system $(\Omega, \mathcal{F}, \mathbb{P}, (\theta_t)_{t \in T})$, with fixed point 0. Assume further that there is a linear random dynamical system Φ such that the hypotheses (H1) and (H2) are satisfied with $\varphi(t, \omega, x) = \Phi(t, \omega)x + \Psi(t, \omega, x)$. Then for every choice of $i \in \{1, \dots, p-1\}$ there are measurable mappings $P^{\leq i}, P^{>i} : \Omega \times \mathbb{R}^d \to \mathbb{R}^d$ such that for arbitrary $\omega \in \tilde{\Omega}$ the following holds:*

(a) *The mappings $P^{\leq i}(\omega) := P^{\leq i}(\omega, \cdot) : \mathbb{R}^d \to \mathbb{R}^d$ and $P^{>i}(\omega) := P^{>i}(\omega, \cdot) : \mathbb{R}^d \to \mathbb{R}^d$ are continuous with $P^{\leq i}(\omega)0 = P^{>i}(\omega)0 = 0$, as well as $P^{\leq i}(\omega)\mathbb{R}^d = \mathcal{S}^{\leq i}(\omega)$ and $P^{>i}(\omega)\mathbb{R}^d = \mathcal{S}^{>i}(\omega)$.*

(b) *The random mappings $P^{\leq i}(\omega)$ and $P^{>i}(\omega)$ map ω-orbits of φ onto ω-orbits of φ, i.e. for arbitrary $t \in T$ and $\xi \in \mathbb{R}^d$ we have both*

$$P^{\leq i}(\theta_t\omega)\varphi(t, \omega, \xi) = \varphi(t, \omega, P^{\leq i}(\omega)\xi) \quad and$$
$$P^{>i}(\theta_t\omega)\varphi(t, \omega, \xi) = \varphi(t, \omega, P^{>i}(\omega)\xi).$$

(c) *For every $\xi_0 \in \mathcal{S}^{\leq i}(\omega)$ the preimage $P^{\leq i}(\omega)^{-1}(\xi_0) \subset \mathbb{R}^d$ is the graph of a globally Lipschitz continuous mapping whose Lipschitz constant converges to 0 as $L \to 0$. Moreover, the dynamical characterization*

$$P^{\leq i}(\omega)^{-1}(\xi_0) = \{\xi \in \mathbb{R}^d : \varphi(\cdot, \omega, \xi) - \varphi(\cdot, \omega, \xi_0) \text{ is } \gamma^+\text{-quasibounded}$$
$$\text{with respect to } \omega\}$$

is true for all $\gamma \in [\alpha_{i+1,-} + \delta, \alpha_{i,+} - \delta]$. In particular we have $P^{\leq i}(\omega)\xi_0 = \xi_0$ and $P^{\leq i}(\omega)^{-1}(0) = \mathcal{S}^{>i}(\omega)$. For arbitrary $\eta_0 \in \mathcal{S}^{>i}(\omega)$ the preimage $P^{>i}(\omega)^{-1}(\eta_0)$ is the graph of a globally Lipschitz continuous mapping, too, where again the Lipschitz constant converges to 0 as $L \to 0$ and the dynamical characterization

$$P^{>i}(\omega)^{-1}(\eta_0) = \{\eta \in \mathbb{R}^d : \varphi(\cdot, \omega, \eta) - \varphi(\cdot, \omega, \eta_0) \text{ is } \gamma^-\text{-quasibounded}$$
$$\text{with respect to } \omega\}$$

holds with γ chosen as above. Especially this furnishes $P^{>i}(\omega)\eta_0 = \eta_0$ and $P^{>i}(\omega)^{-1}(0) = \mathcal{S}^{\leq i}(\omega)$.

Proof. Let $i \in \{1, \ldots, p-1\}$ be arbitrary, but fixed. If in Lemma 2.3 and Theorem 2.2 we replace Ω by $\tilde{\Omega}$, then (H1) and (H2) show that the difference equation (3.2) satisfies all the assumptions of the above lemma and theorem. Now let $\hat{P}^{\leq i}$ and $\hat{P}^{>i}$ denote the mappings guaranteed by Theorem 2.2, define the mappings $P^{\leq i}$ and $P^{>i}$ as

$$P^{\leq i}(\omega) := \hat{P}^{\leq i}(0, \omega, \cdot) \quad \text{and} \quad P^{>i}(\omega) := \hat{P}^{>i}(0, \omega, \cdot) ,$$

for arbitrary $\omega \in \tilde{\Omega}$, and extend this definition measurably onto the whole of Ω. Obviously, the assertions of both *(a)* and *(c)* are immediate consequences of Lemma 2.3 and the proof of Theorem 2.2, provided in *(c)* we replace the mappings contained in the dynamical characterizations by their restrictions to $\mathbb{Z}$. As for the proof of *(b)* we begin by verifying the identities

$$\hat{P}^{\leq i}(\kappa, \omega, \xi) = \hat{P}^{\leq i}(0, \theta_\kappa \omega, \xi) \quad \text{and} \quad \hat{P}^{>i}(\kappa, \omega, \xi) = \hat{P}^{>i}(0, \theta_\kappa \omega, \xi) \quad (3.6)$$

for every $\kappa \in \mathbb{Z}$, $\omega \in \tilde{\Omega}$ and $\xi \in \mathbb{R}^d$. So, let $\kappa \in \mathbb{Z}$, $\omega \in \tilde{\Omega}$ and $\xi \in \mathbb{R}^d$ be arbitrary, but fixed, and let λ denote the general solution of (3.2). According to Theorem 2.2 the point $\xi^* := \hat{P}^{\leq i}(\kappa, \omega, \xi)$ is the unique point of $\mathbb{R}^d$ with $(\kappa, \xi^*) \in \hat{S}^{\leq i}(\omega)$ and such that the difference $\lambda(\cdot; \kappa, \omega, \xi) - \lambda(\cdot; \kappa, \omega, \xi^*)$ is γ^+-quasibounded with respect to ω, for every choice of $\gamma \in [\alpha_{i+1,-} + \delta, \alpha_{i,+} - \delta]$[8]. Similar to the proof of Theorem 3.1 we may deduce $(0, \xi^*) \in \hat{S}^{\leq i}(\theta_\kappa \omega)$, as well as the γ^+-quasiboundedness with respect to $\theta_\kappa \omega$ of the difference $\lambda(\cdot; 0, \theta_\kappa \omega, \xi) - \lambda(\cdot; 0, \theta_\kappa \omega, \xi^*)$. Applying Theorem 2.2 once again eventually furnishes $\hat{P}^{\leq i}(\kappa, \omega, \xi) = \xi^* = \hat{P}^{\leq i}(0, \theta_\kappa \omega, \xi)$, as required. The assertion concerning $\hat{P}^{>i}$ can be verified analogously.

In the case $\mathbb{T} = \mathbb{Z}$ the proof of Theorem 3.2 already is complete, since the properties of $P^{\leq i}(\omega)$ and $P^{>i}(\omega)$ claimed in *(b)* follow immediately from (3.6) and Theorem 2.2*(c)*. But we still have to close some gaps for $\mathbb{T} = \mathbb{R}$. More precisely, we have to show that the identities in *(b)* are satisfied for arbitrary $t \in \mathbb{R}$ rather than only for $t \in \mathbb{Z}$ and that in the dynamical characterizations of *(c)* we may use the given mappings rather than the restrictions to $\mathbb{Z}$. Whereas the second problem can be solved similar to the proof of Theorem 3.1 using the inequality

$$\|\varphi(t, \omega, \eta) - \varphi(t, \omega, \bar{\eta})\|_{\theta_t \omega} \leq (\max\{\alpha_{1,-}, 1\} + pL)\|\eta - \bar{\eta}\|_\omega =: C_\varphi \|\eta - \bar{\eta}\|_\omega$$
$$(3.7)$$

for $t \in [0, 1]$, $\omega \in \tilde{\Omega}$ and $\eta, \bar{\eta} \in \mathbb{R}^d$ (which is an easy consequence of (H1) and (H2)), the verification of *(b)* is a little bit more involved. Let $t \in [-1, 0]$ and $\omega \in \tilde{\Omega}$ be arbitrary, but fixed, and define

$$P_t^{\leq i}(\omega) := \varphi(-t, \theta_t \omega) \circ P^{\leq i}(\theta_t \omega) \circ \varphi(t, \omega) .$$

Then *(a)* and Theorem 3.1 furnish

---

[8] Again, $\hat{S}^{\leq i}(\omega)$ denotes the random invariant fiber bundle of (3.2) guaranteed by Theorem 2.1.

$$P_t^{\leq i}(\omega)\mathbb{R}^d = \varphi(-t, \theta_t\omega)P^{\leq i}(\theta_t\omega)\mathbb{R}^d = \varphi(-t, \theta_t\omega)S^{\leq i}(\theta_t\omega) = S^{\leq i}(\omega) \ . \quad (3.8)$$

For $\xi \in \mathbb{R}^d$, $\hat{\omega} := \theta_t\omega$ and $\hat{\xi} := \varphi(t, \omega, \xi)$ the above dynamical characterization (which does not use the given mappings but rather their restrictions to $\mathbb{Z}$) implies the boundedness of $\{\gamma^{-k}\|\varphi(k, \hat{\omega}, \hat{\xi}) - \varphi(k, \hat{\omega}, P^{\leq i}(\hat{\omega})\hat{\xi})\|_{\theta_k\hat{\omega}} : k \in \mathbb{Z}_0^+\}$, and observing that with (3.7) we get the estimate

$$\begin{aligned}
&\|\varphi(k, \omega, \xi) - \varphi(k, \omega, P_t^{\leq i}(\omega)\xi)\|_{\theta_k\omega} = \\
= \ &\|\varphi(-t, \theta_k\hat{\omega}, \varphi(k, \hat{\omega}, \hat{\xi})) - \varphi(-t, \theta_k\hat{\omega}, \varphi(k, \hat{\omega}, P^{\leq i}(\hat{\omega})\hat{\xi}))\|_{\theta_{-t}\theta_k\hat{\omega}} \leq \\
\leq \ &C_\varphi \|\varphi(k, \hat{\omega}, \hat{\xi}) - \varphi(k, \hat{\omega}, P^{\leq i}(\hat{\omega})\hat{\xi})\|_{\theta_k\hat{\omega}}
\end{aligned}$$

we conclude that the set $\{\gamma^{-k}\|\varphi(k, \omega, \xi) - \varphi(k, \omega, P_t^{\leq i}(\omega)\xi)\|_{\theta_k\omega} : k \in \mathbb{Z}_0^+\}$ is bounded, too. But now, a second application of the above-proved dynamical characterization furnishes with (3.8) the identity $P_t^{\leq i}(\omega)\xi = P^{\leq i}(\omega)\xi$, i.e. we have $P^{\leq i}(\theta_t\omega)\varphi(t, \omega, \xi) = \varphi(t, \omega, P^{\leq i}(\omega)\xi)$ for arbitrary $t \in [-1, 0]$, $\omega \in \tilde{\Omega}$ and $\xi \in \mathbb{R}^d$, and together with the already proved part of *(b)* we finally calculate

$$P^{\leq i}(\theta_{t+\tau}\omega) \circ \varphi(t+\tau, \omega) = \varphi(t, \theta_\tau\omega) \circ P^{\leq i}(\theta_\tau\omega) \circ \varphi(\tau, \omega) = \varphi(t+\tau, \omega) \circ P^{\leq i}(\omega)$$

for $\tau \in \mathbb{Z}$, $t \in [-1, 0]$ and $\omega \in \tilde{\Omega}$. Since the corresponding identity for $P^{>i}$ can be proved analogously, this completes the proof of Theorem 3.2. $\quad\diamond$

Before we come to the last topic of Section 3, let us at least mention two further results which could have been included in the above theorem[9]. First of all, it is possible to extend the identities in *(a)* with the help of the dynamical characterizations contained in *(c)* in the following way: For arbitrary $\omega \in \tilde{\Omega}$ and $1 \leq i \leq \ell \leq j \leq p$ with $\ell < p$ we have $P^{\leq \ell}(\omega)S^{i,j}(\omega) = S^{i,\ell}(\omega)$, whereas for $\omega \in \tilde{\Omega}$ and $1 \leq i \leq \ell \leq j \leq p$ with $\ell > 1$ we get $P^{>\ell-1}(\omega)S^{i,j}(\omega) = S^{\ell,j}(\omega)$.

As for the second result, we begin by reminding the reader of the fact that all ω-orbits which are contained in the stable random invariant manifold converge to the origin exponentially as $t \to \infty$. Yet, having the well-known phase portrait of the autonomous ordinary differential equation $\dot{x} = -x$, $\dot{y} = -2y$ in mind it seems reasonable to expect that most orbits inside this manifold will approach the origin "tangent" to some random invariant sub-manifold. In fact, with the help of Theorem 3.2 one can prove the following: Let $\varphi(\cdot, \omega, \xi)$ denote an ω-orbit which is contained in $S^{i,j}(\omega) \setminus S^{\ell+1,j}(\omega)$, for some $i \leq \ell < j$. Then the identity

$$\lim_{t\to\infty} \frac{\mathrm{dist}(\varphi(t, \omega, \xi), S^{i,\ell}(\theta_t\omega))}{\|\varphi(t, \omega, \xi)\|} = 0$$

holds, i.e. for every $\varepsilon > 0$ there is a $\tau \in \mathbb{T}$ such that for all $t \geq \tau$ the estimate

---

[9] The proofs of these results can be found in Wanner [30, pp. 125ff] or – in a deterministic version – in Aulbach, Wanner [8].

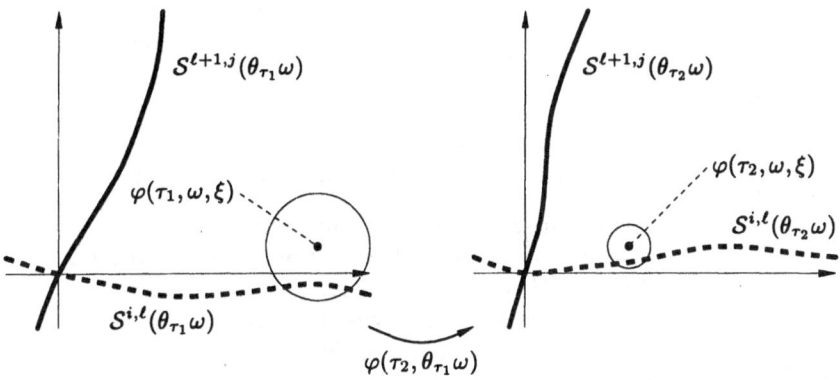

Fig. 3.1. Approaching the origin inside the stable random invariant manifold

$$\operatorname{dist}(\varphi(t,\omega,\xi), \mathcal{S}^{i,\ell}(\theta_t\omega)) \le \varepsilon \cdot \|\varphi(t,\omega,\xi)\|$$

is satisfied. In particular, if $\lim_{t\to\infty} \varphi(t,\omega,\xi) = 0$, which is the case for instance if $\lambda_i < 0$, the ω-orbit $\varphi(\cdot,\omega,\xi)$ approaches the origin "tangent" to the random invariant manifold $\mathcal{S}^{i,\ell}(\omega)$ as $t \to \infty$. (Cf. Figure 3.1, where we have chosen $\varepsilon = \frac{1}{5}$ and $\tau_2 > \tau_1 \ge \tau$, and where the two circles have radii $\varepsilon\|\varphi(\tau_1,\omega,\xi)\|$ and $\varepsilon\|\varphi(\tau_2,\omega,\xi)\|$, respectively.) Analogous statements can be made as $t \to -\infty$.

3.4 The Hartman-Grobman Theorems

In this final subsection of Section 3 we present the last two results of our linearization theory for random dynamical systems – the Hartman-Grobman theorems. First of all, let us state and prove a random version of the classical Hartman-Grobman theorem dealing with the hyperbolic case.

Theorem 3.3. *Assume we are given a measurable random dynamical system φ on $\mathbb{R}^d$ over some ergodic metric dynamical system $(\Omega, \mathcal{F}, \mathbb{P}, (\theta_t)_{t\in\mathbb{T}})$, with fixed point 0. Assume further that there is a linear random dynamical system Φ such that the hypotheses (H1) and (H2) are satisfied with $\varphi(t,\omega,x) = \Phi(t,\omega)x + \Psi(t,\omega,x)$ and that Φ is hyperbolic, i.e. none of its Lyapunov exponents vanishes. Then there is a measurable mapping $h : \Omega \times \mathbb{R}^d \to \mathbb{R}^d$ with the following properties:*

(a) For arbitrary $\omega \in \Omega$ the mapping $h(\omega) := h(\omega, \cdot)$ is a homeomorphism on $\mathbb{R}^d$ with $h(\omega)0 = 0$, and the mapping $h(\cdot)^{-1}(\cdot) : \Omega \times \mathbb{R}^d \to \mathbb{R}^d$ is measurable.

(b) For every $\omega \in \tilde{\Omega}$ the mapping $h(\omega)$ maps ω-orbits of φ onto ω-orbits of the linear random dynamical system Φ, i.e. for arbitrary $t \in \mathbb{T}$ and $\xi \in \mathbb{R}^d$ we have

$$\varphi(t,\omega,\xi) = h(\theta_t\omega)^{-1}\Phi(t,\omega)h(\omega)\xi.$$

(c) *For arbitrary $\omega \in \tilde{\Omega}$ and $1 \leq i \leq j \leq p$ the mapping $h(\omega)$ maps the random invariant manifold $S^{i,j}(\omega)$ onto the direct sum $E_i(\omega) \oplus \ldots \oplus E_j(\omega)$.*

In other words, in the hyperbolic case the nonlinear random dynamical system φ is topologically equivalent to the linear random dynamical system Φ.

Proof. If we apply Theorem 2.3 to the random difference equation (3.2), where again Ω is replaced by $\tilde{\Omega}$, and if $\hat{H}$ denotes the so-constructed mapping, then the definition $h(\omega) := \hat{H}(0, \omega, \cdot)$ for $\omega \in \tilde{\Omega}$, and $h(\omega) := \mathrm{id}_{\mathbb{R}^d}$ for $\omega \notin \tilde{\Omega}$ furnishes a random homeomorphism satisfying both (a) and (c).

As for the proof of (b) we have to go into the details of the proof of Theorem 2.3, mainly due to the fact that we consider the continuous-time case as well. To that end, define a mapping $\varphi^* : \mathbb{T} \times \Omega \times \mathbb{R}^d \to \mathbb{R}^d$ by $\varphi^*(t, \omega, \xi) := (\varphi_1^*(t, \omega, \xi^1), \ldots, \varphi_p^*(t, \omega, \xi^p))$ for arbitrary $t \in \mathbb{T}$, $\omega \in \Omega$ and $\xi \in \mathbb{R}^d$, where the i-th component φ_i^* is defined as $\varphi_i^*(t, \omega, \xi^i) := \Phi_i(t, \omega)\xi^i + \Psi_i(t, \omega, \pi^{i,i}(\omega, \xi^i))$ with the mapping $\pi^{i,i}$ from Remark 3.1. If we now combine the first part of the proof of Theorem 2.3 (which was nothing but an application of Proposition 2.2) and Theorem 3.2, we may construct a random homeomorphism $D(\omega)$ satisfying

$$D(\theta_t\omega) \circ \varphi(t, \omega) = \varphi^*(t, \omega) \circ D(\omega)$$

for arbitrary $t \in \mathbb{Z}$ and $\omega \in \tilde{\Omega}$, by setting $D(\omega) := \hat{D}(0, \omega, \cdot)$, where $\hat{D}$ denotes the mapping guaranteed by Proposition 2.2. Even more, according to the construction of $\hat{D}$ and Theorem 3.2 (b) we deduce that the above identity actually holds for all $t \in \mathbb{T}$ and $\omega \in \tilde{\Omega}$ – and this immediately implies that φ^* is a random dynamical system on $\mathbb{R}^d$, and the i-th component φ_i^* is a random dynamical system on $\mathbb{R}^{d_i}$, for $i = 1, \ldots, p$.

Next, let $i \in \{1, \ldots, p\}$ be arbitrary, but fixed. Applying Proposition 2.3 to the random difference equation generating φ_i^* furnishes a mapping $\hat{K}_i$ which maps the ω-solutions of this equation bijectively onto the ω-solutions of the corresponding linear equation. In addition, using the dynamical description of $\hat{K}_i$ one can prove the identity $\hat{K}_i(\kappa, \omega, \cdot) = \hat{K}_i(0, \theta_\kappa\omega, \cdot)$ for arbitrary $\kappa \in \mathbb{Z}$ and $\omega \in \tilde{\Omega}$ as in the proof of Theorem 3.2. If we now define $K_i(\omega) := \hat{K}_i(0, \omega, \cdot)$ it may easily be verified that for arbitrary $t \in \mathbb{Z}$, $\omega \in \tilde{\Omega}$ and $\xi^i \in \mathbb{R}^{d_i}$ the identity

$$K_i(\theta_t\omega)\varphi_i^*(t, \omega, \xi^i) = \Phi_i(t, \omega)K_i(\omega)\xi^i \tag{3.9}$$

holds, and that for every choice of $\omega \in \tilde{\Omega}$ and $\xi^i \in \mathbb{R}^{d_i}$ the point $K_i(\omega)\xi^i$ is the uniquely determined point in $\mathbb{R}^{d_i}$ for which the set

$$\{\|\Phi_i(k, \omega)K_i(\omega)\xi^i - \varphi_i^*(k, \omega, \xi^i)\|_{\theta_k\omega} : k \in \mathbb{Z}\} \tag{3.10}$$

is bounded. Recalling that the mapping $\hat{H}$ mentioned in the beginning of this proof was constructed by suitably combining $\hat{D}$ and $\hat{K}_i$ shows that in the case $\mathbb{T} = \mathbb{Z}$ the proof of Theorem 3.3 already is complete.

In order to complete the proof of the theorem for the case $\mathbf{T} = \mathbb{R}$ as well, we only have to prove the validity of (3.9) for arbitrary $t \in \mathbb{R}$. To that end, let $t \in [-1, 0]$ and $\omega \in \tilde{\Omega}$ be arbitrary, but fixed, and define

$$K_{i,t}(\omega) := \Phi_i(-t, \theta_t \omega) \circ K_i(\theta_t \omega) \circ \varphi_i^*(t, \omega) .$$

Then for arbitrary $k \in \mathbb{Z}$ and $\xi^i \in \mathbb{R}^{d_i}$ we deduce, using the abbreviations $\hat{\omega} := \theta_t \omega$ and $\hat{\xi}^i := \varphi_i^*(t, \omega, \xi^i)$, as well as (H1) and (H2), the estimate

$$\|\Phi_i(k, \omega)K_{i,t}(\omega)\xi^i - \varphi_i^*(k, \omega, \xi^i)\|_{\theta_k \omega} =$$
$$= \|\Phi_i(-t, \theta_k \hat{\omega})\Phi_i(k, \hat{\omega})K_i(\hat{\omega})\hat{\xi}^i - \varphi_i^*(-t, \theta_k \hat{\omega}, \varphi_i^*(k, \hat{\omega}, \hat{\xi}^i))\|_{\theta_k \omega} \le$$
$$\le \|\Phi_i(-t, \theta_k \hat{\omega})(\Phi_i(k, \hat{\omega})K_i(\hat{\omega})\hat{\xi}^i - \varphi_i^*(k, \hat{\omega}, \hat{\xi}^i))\|_{\theta_{-t}\theta_k \hat{\omega}} +$$
$$+ \|\underbrace{(\Phi_i(-t, \theta_k \hat{\omega}) - \varphi_i^*(-t, \theta_k \hat{\omega}, \cdot))}_{=-\Psi_i(-t, \theta_k \hat{\omega}, \pi^{i,i}(\theta_k \hat{\omega}, \cdot))} \varphi_i^*(k, \hat{\omega}, \hat{\xi}^i)\|_{\theta_{-t}\theta_k \hat{\omega}} \le$$
$$\le \max\{\alpha_{i,-}, 1\} \cdot \|\Phi_i(k, \hat{\omega})K_i(\hat{\omega})\hat{\xi}^i - \varphi_i^*(k, \hat{\omega}, \hat{\xi}^i)\|_{\theta_k \hat{\omega}} + M ,$$

which, together with the discussion concerning (3.10), finally establishes the boundedness of the set $\{\|\Phi_i(k, \omega)K_{i,t}(\omega)\xi^i - \varphi_i^*(k, \omega, \xi^i)\|_{\theta_k \omega} : k \in \mathbb{Z}\}$. According to the above-mentioned uniqueness of $K_i(\omega)\xi^i$ we further derive the identity $K_{i,t}(\omega)\xi^i = K_i(\omega)\xi^i$, and therefore $K_i(\theta_t \omega) \circ \varphi_i^*(t, \omega) = \Phi_i(t, \omega) \circ K_i(\omega)$ for all $t \in [-1, 0]$ and $\omega \in \tilde{\Omega}$. Together with (3.9) this completes the proof of Theorem 3.3. $\diamond$

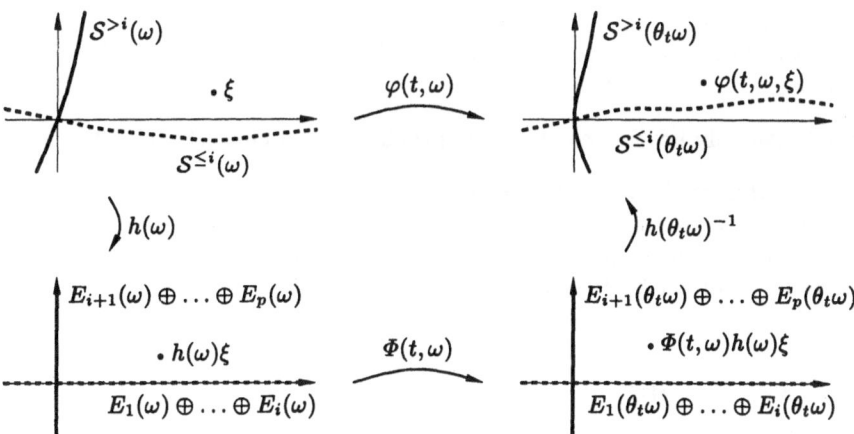

Fig. 3.2. The Hartman-Grobman theorem for random dynamical systems

Assertions *(b)* and *(c)* of the above theorem are depicted in Figure 3.2, where for the sake of clarity we only included the two random invariant manifolds $\mathcal{S}^{\le i}(\omega)$ and $\mathcal{S}^{>i}(\omega)$ – together with the corresponding random linear subspaces $E_1(\omega) \oplus \ldots \oplus E_i(\omega)$ and $E_{i+1}(\omega) \oplus \ldots \oplus E_p(\omega)$.

In the final result of our linearization theory for random dynamical systems we show that in the non-hyperbolic case only a partial linearization of φ can be achieved by means of a random homeomorphism.

Theorem 3.4. *Assume we are given a measurable random dynamical system φ on $\mathbb{R}^d$ over some ergodic metric dynamical system $(\Omega, \mathcal{F}, \mathbb{P}, (\theta_t)_{t \in \mathbb{T}})$, with fixed point 0. Assume further that there is a linear random dynamical system Φ such that the hypotheses (H1) and (H2) are satisfied with $\varphi(t, \omega, x) = \Phi(t, \omega)x + \Psi(t, \omega, x)$ and that $\lambda_i = 0$ for some index $i \in \{1, \ldots, p\}$. Then there is a measurable mapping $h : \Omega \times \mathbb{R}^d \to \mathbb{R}^d$ with the following properties:*

(a) *For arbitrary $\omega \in \Omega$ the mapping $h(\omega) := h(\omega, \cdot)$ is a homeomorphism on $\mathbb{R}^d$ with $h(\omega)0 = 0$, and the mapping $h(\cdot)^{-1}(\cdot) : \Omega \times \mathbb{R}^d \to \mathbb{R}^d$ is measurable.*

(b) *For every $\omega \in \tilde{\Omega}$ the mapping $h(\omega)$ maps ω-orbits of φ onto ω-orbits of the random dynamical system $\varphi^*(t, \omega, \xi) := \Phi(t, \omega)\xi + \Psi^*(t, \omega, \xi)$, where the new nonlinearity Ψ^* is defined by $\Psi_i^*(t, \omega, \xi) := \Psi_i(t, \omega, \pi^{i,i}(\omega, \xi^i))$ and $\Psi_j^*(t, \omega, \xi) := 0$ for $j \neq i$, and $\pi^{i,i}$ denotes the mapping defined in Remark 3.1. In other words, for arbitrary $t \in \mathbb{T}$ and $\xi \in \mathbb{R}^d$ we have*

$$\varphi(t, \omega, \xi) = h(\theta_t \omega)^{-1} \varphi^*(t, \omega, h(\omega)\xi) .$$

(c) *For arbitrary $\omega \in \tilde{\Omega}$ and $1 \leq i \leq j \leq p$ the mapping $h(\omega)$ maps the random invariant manifold $S^{i,j}(\omega)$ onto the direct sum $E_i(\omega) \oplus \ldots \oplus E_j(\omega)$.*

So, in the non-hyperbolic case the nonlinear random dynamical system φ is topologically equivalent to the decoupled, partially linear random dynamical system φ^*.

Proof. We only have to combine the proofs of Theorems 2.4 and 3.3.   ◊

To close this section, let us note that whereas the components φ_j^* with $j \neq i$ of the decoupled random dynamical system φ^* are linear, the remaining nonlinear component φ_i^* exactly describes the behavior of φ on the random invariant manifold $S^{i,i}(\omega)$, i.e. on the random center manifold.

4. Local Results

4.1 The Discrete-Time Case

There is one huge drawback of the linearization theory developed in the last section: Hypothesis (H2) is so strong that probably all interesting systems will not satisfy it. Even worse, (H2) is formulated by means of random norms, so in practice it is almost impossible to decide whether or not a given random dynamical system satisfies these assumptions. Of course, this is not at all surprising. The hypotheses of the last section were tailor-made for implying global results, and in all interesting nonlinear situations the best we can hope for are local ones.

This section is devoted to demonstrating how local results may be obtained for large classes of random dynamical systems using the global results of Section 3, and we begin in this subsection by considering the case of discrete-time random dynamical systems.

The basic idea for deriving local results from global ones is well-known. Starting from a random dynamical system φ with fixed point 0 we shall construct a new random dynamical system $\tilde{\varphi}$ which coincides with φ on some random neighborhood of the origin, and in addition satisfies all the assumptions of the last section. In order to construct $\tilde{\varphi}$ in the discrete-time case we need the following lemma.

Lemma 4.1. Let $(\tilde{\Omega}, \tilde{\mathcal{F}})$ denote an arbitrary measurable space and $f : \tilde{\Omega} \times \mathbb{R}^d \to \mathbb{R}^d$ a measurable mapping such that $f(\omega, \cdot)$ is continuous for all $\omega \in \tilde{\Omega}$ with $f(\omega, 0) = 0$ and

$$\lim_{(x,y) \to (0,0)} \frac{\|f(\omega, x) - f(\omega, y)\|}{\|x - y\|} = 0 . \tag{4.1}$$

Furthermore, let $\|\cdot\|_\omega$ denote a random norm on $\mathbb{R}^d$ satisfying the assertions of Proposition 3.1(b), and let $L > 0$ be an arbitrary positive constant. Then there are measurable mappings $\varrho : \tilde{\Omega} \to \mathbb{R}^+$ and $\tilde{f} : \tilde{\Omega} \times \mathbb{R}^d \to \mathbb{R}^d$ with the following properties:

(a) If we define a random neighborhood of 0 by $U(\omega) := \{x \in \mathbb{R}^d : \|x\| < \varrho(\omega)\}$, then the identity $f(\omega, x) = \tilde{f}(\omega, x)$ holds for all $\omega \in \tilde{\Omega}$ and $x \in U(\omega)$.

(b) For every $\omega \in \tilde{\Omega}$ and $x, y \in \mathbb{R}^d$ we have both $\|\tilde{f}(\omega, x) - \tilde{f}(\omega, y)\|_{\theta_1 \omega} \leq L\|x - y\|_\omega$ and $\|\tilde{f}(\omega, x)\|_{\theta_1 \omega} \leq 1$.

Proof. Let $B := B_1 : \tilde{\Omega} \to [1, \infty)$ denote the mapping of Proposition 3.1(b) for $\varepsilon = 1$, and define a measurable mapping $L^* : \tilde{\Omega} \to \mathbb{R}^+$ by $L^*(\omega) := \frac{L}{e B(\omega)^2} \leq \frac{L}{B(\omega) B(\theta_1 \omega)}$. Then because of (4.1) and the required continuity of $f(\omega, \cdot)$ the definition

$$\varrho^*(\omega) := \max\{\varepsilon \in (0, 1] : \|f(\omega, x) - f(\omega, y)\| \leq L^*(\omega)\|x - y\|$$
$$\text{for all } \|x\| \leq \varepsilon, \|y\| \leq \varepsilon\}$$

is admissible, i.e. the set on the right-hand side is not empty and actually has a maximum. Moreover, it may easily be verified that the measurability of f and L^* implies the measurability of $\varrho^* : \tilde{\Omega} \to (0, 1]$. Finally, let $\varrho(\omega) := \min\left\{\varrho^*(\omega), \frac{B(\omega)}{L}\right\} > 0$ for $\omega \in \tilde{\Omega}$, and for $\varepsilon > 0$ let $r_\varepsilon : \mathbb{R}^d \to \mathbb{R}^d$ denote the radial retraction defined by $r_\varepsilon(x) := x$ for $\|x\| \leq \varepsilon$, and $r_\varepsilon(x) := \frac{\varepsilon}{\|x\|} x$ for $\|x\| > \varepsilon$, which is globally Lipschitz continuous with Lipschitz constant 1. If we now define the required mapping $\tilde{f} : \tilde{\Omega} \times \mathbb{R}^d \to \mathbb{R}^d$ as

Thomas Wanner

$$\tilde{f}(\omega, x) := f(\omega, r_{\varrho(\omega)}(x)) \quad \text{for arbitrary} \quad \omega \in \tilde{\Omega}, x \in \mathbb{R}^d \,,$$

then $\tilde{f}$ obviously is measurable and satisfies *(a)*. Furthermore, for arbitrary $\omega \in \tilde{\Omega}$ and $x, y \in \mathbb{R}^d$ Proposition 3.1*(b)* furnishes

$$\|\tilde{f}(\omega, x) - \tilde{f}(\omega, y)\|_{\theta_1 \omega} \le B(\theta_1 \omega) L^*(\omega) \|r_{\varrho(\omega)}(x) - r_{\varrho(\omega)}(y)\| \le L \|x - y\|_\omega$$

as well as

$$\|\tilde{f}(\omega, x)\|_{\theta_1 \omega} \le B(\theta_1 \omega) L^*(\omega) \|r_{\varrho(\omega)}(x)\| \le \frac{L\varrho(\omega)}{B(\omega)} \le 1 \,,$$

and the proof of Lemma 4.1 is complete. ◊

Now we are ready to present the above-mentioned construction of the random dynamical system $\tilde{\varphi}$ which satisfies all the assumptions of the last section, for certain discrete-time random dynamical systems φ.

Proposition 4.1. *Assume we are given a discrete-time continuous random dynamical system φ on $\mathbb{R}^d$ over some ergodic metric dynamical system $(\Omega, \mathcal{F}, \mathbb{P}, (\theta_k)_{k \in \mathbb{Z}})$, with fixed point 0. Assume further that there is a linear random dynamical system Φ satisfying hypothesis (H1) from the beginning of Section 3, and that the nonlinear part Ψ defined via $\varphi(k, \omega, x) = \Phi(k, \omega)x + \Psi(k, \omega, x)$ satisfies*

$$\lim_{(x,y) \to (0,0)} \frac{\|\Psi(1, \omega, x) - \Psi(1, \omega, y)\|}{\|x - y\|} = 0$$

for arbitrary $\omega \in \tilde{\Omega}$. Then there is a measurable mapping $\varrho : \Omega \to \mathbb{R}^+$ and a discrete-time random dynamical system $\tilde{\varphi}(k, \omega, x) = \Phi(k, \omega)x + \tilde{\Psi}(k, \omega, x)$ over $(\Omega, \mathcal{F}, \mathbb{P}, (\theta_k)_{k \in \mathbb{Z}})$ satisfying both (H1) and (H2), as well as

$$\varphi(k, \omega, x) = \tilde{\varphi}(k, \omega, x) \tag{4.2}$$

for $\omega \in \tilde{\Omega}$, $x \in U(\omega)$ and $k \in I_{\max}(\omega, x) := \{k_{\min}(\omega, x), \dots, k_{\max}(\omega, x)\}$, where $U(\omega)$ is defined as in Lemma 4.1(a) and

$$
\begin{aligned}
k_{\min}(\omega, x) &:= \inf\{k \in \mathbb{Z}_0^- : \varphi(\kappa, \omega, x) \in U(\theta_\kappa \omega) \text{ for } \kappa = k, \dots, 0\} \le 0 \,, \\
k_{\max}(\omega, x) &:= \sup\{k \in \mathbb{Z}_0^+ : \varphi(\kappa, \omega, x) \in U(\theta_\kappa \omega) \text{ for } \kappa = 0, \dots, k\} \ge 0 \,.
\end{aligned}
$$

Moreover, for every $\omega \in \tilde{\Omega}$ we have both

$$\lim_{x \to 0} k_{\min}(\omega, x) = -\infty \quad \text{and} \quad \lim_{x \to 0} k_{\max}(\omega, x) = \infty \,. \tag{4.3}$$

In particular, the last two identities imply for every choice of $\omega \in \tilde{\Omega}$ and $N \in \mathbb{N}$ the existence of a neighborhood $U_N(\omega) \subset \mathbb{R}^d$ of the origin such that the equality in (4.2) holds for all $k \in \{-N, \dots, N\}$ and $x \in U_N(\omega)$.

Proof. According to Lemma 3.1 the mapping $\varphi(\cdot, \omega, \xi)$ is the unique ω-solution of the initial value problem

$$x_{k+1} = A(\theta_k\omega)x_k + f(\theta_k\omega, x_k) \quad , \quad x_0 = \xi, \tag{4.4}$$

where $A(\omega) := \Phi(1, \omega)$ and $f(\omega, x) := \Psi(1, \omega, x)$. Now choose $L > 0$ as in (H2), and construct the mapping $\tilde{f}$ by first applying Lemma 4.1 to $f|_{\tilde{\Omega} \times \mathbb{R}^d}$, and then setting the obtained mapping to 0 on $(\Omega \setminus \tilde{\Omega}) \times \mathbb{R}^d$. Furthermore, let $\varrho : \Omega \to \mathbb{R}^+$ denote the mapping guaranteed by Lemma 4.1, measurably extended to the whole of Ω. Similar to the proof of Lemma 2.2 one can show that the initial value problem

$$x_{k+1} = A(\theta_k\omega)x_k + \tilde{f}(\theta_k\omega, x_k) \quad , \quad x_0 = \xi \tag{4.5}$$

has a uniquely determined solution $\tilde{\varphi}(\cdot, \omega, \xi)$ on the whole of $\mathbb{Z}$, for every choice of $\omega \in \Omega$ and $\xi \in \mathbb{R}^d$. Lemma 5.1 and the uniform contraction principle imply the measurability of $\tilde{\varphi}$, as well as the continuity with respect to the last variable, and Lemma 3.1 shows that $\tilde{\varphi}$ actually is a discrete-time random dynamical system. Furthermore, due to Lemma 4.1 *(a)* the ω-solutions of (4.4) and (4.5) coincide on $I_{\max}(\omega, \xi)$ for arbitrary $\omega \in \tilde{\Omega}$. If we finally define the nonlinear part $\tilde{\Psi}$ as in the formulation of Proposition 4.1, then – together with $\tilde{\Psi}(1, \omega, x) = \tilde{f}(\omega, x)$ – Lemma 4.1 *(b)* furnishes the validity of (H2).

In order to conclude the proof we only have to verify the limits in (4.3). Because of symmetry, we only consider the second one. For this, let $n \in \mathbb{N}$ and $\omega \in \tilde{\Omega}$ be arbitrary, but fixed. According to the assumed continuity of the mappings $\varphi(k, \omega, \cdot)$ and the identity $\varphi(k, \omega, 0) = 0$ there is an $\varepsilon > 0$ such that for all $x \in \mathbb{R}^d$ with $\|x\| < \varepsilon$ and all $\kappa = 0, \ldots, n$ the estimate $\|\varphi(\kappa, \omega, x)\| < \varrho(\theta_\kappa\omega)$ holds, i.e. for arbitrary $x \in \mathbb{R}^d$ with $\|x\| < \varepsilon$ we have $k_{\max}(\omega, x) \geq n$. This already proves the claim. ◊

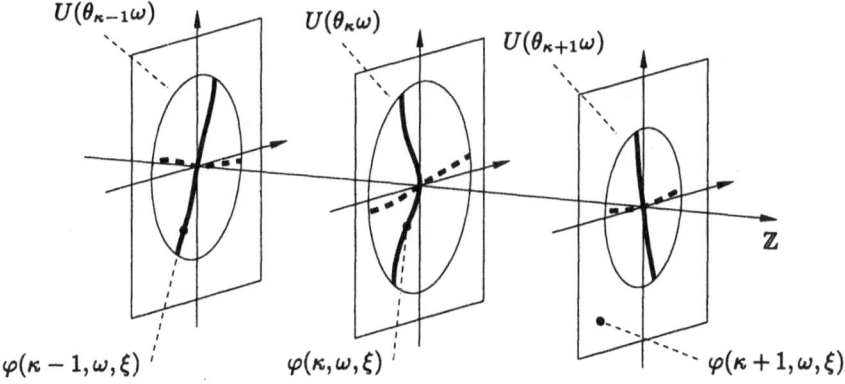

Fig. 4.1. Local random invariant manifolds

To conclude this subsection, let us demonstrate how Proposition 4.1 can be used to obtain local results for discrete-time random dynamical systems φ satisfying the above assumptions[10]. All we have to do is to apply the results of Section 3 to the new random dynamical system $\tilde{\varphi}$, "restrict" everything to $U(\omega)$, and then formulate these local results for φ. More precisely, this means for instance that an ω-orbit $\varphi(\cdot, \omega, \xi)$ is contained in the (local) random invariant manifold $\tilde{S}^{i,j}(\omega) \cap U(\omega)$ only as long as it remains in the random neighborhood $U(\omega)$[11]. In other words, we have $\varphi(k, \omega, \xi) \in \tilde{S}^{i,j}(\theta_k \omega) \cap U(\theta_k \omega)$ for all $k \in I_{\max}(\omega, \xi)$. (Cf. Figure 4.1. There we assume $\kappa = k_{\max}(\omega, \xi) < \infty$.) Analogously, in the Hartman-Grobman theorem the identity $\varphi(k, \omega, \xi) = h(\theta_k \omega)^{-1} \Phi(k, \omega) h(\omega) \xi$ in general is only valid for $k \in I_{\max}(\omega, \xi)$, with $\omega \in \tilde{\Omega}$ and $\xi \in U(\omega)$ – and the limits in (4.3) show that the nearer ξ is to the origin, the larger the interval $I_{\max}(\omega, \xi)$ will be.

4.2 The Continuous-Time Case

Establishing local results from global ones has been relatively straightforward in the discrete-time case. Unfortunately, things are not that simple if we consider $\mathbb{T} = \mathbb{R}$, mainly due to the following observation. In the last subsection the new random dynamical system $\tilde{\varphi}$ has been constructed by first cutting off the nonlinear part of the generator of φ, i.e. the random difference equation (4.4), and then by letting the so-constructed new equation generate $\tilde{\varphi}$ through its general solution. Thus, it should not be too surprising that in the continuous-time case as well the new random dynamical system $\tilde{\varphi}$ has to be constructed by means of modifying the corresponding infinitesimal generator, and here the problems actually begin[12]. Whereas in the discrete-time case there was only one kind of generator, in the continuous-time case we have two of them – both *random differential equations* and *stochastic differential equations* – and they have to be dealt with separately.

In this subsection we shall only consider the case of random dynamical systems generated by random differential equations, i.e. by certain ordinary differential equations depending measurably on some parameter ω. But, why do we exclude stochastic differential equations, which cannot be interpreted "pathwise" (for fixed ω) as ordinary differential equations? Well, for cutting off a stochastic differential equation generating φ we have to use random norms leading to so-called "non-adapted" objects which cannot be handled

---

[10] Note that the assumptions of Proposition 4.1 are automatically satisfied for smooth random dynamical systems of class C^1 with fixed point 0 if we choose $\Phi(k, \omega) := D\varphi(k, \omega, 0)$. Here, $D\varphi(k, \omega, 0)$ denotes the Jacobian of $\varphi(k, \omega)$ in the origin.

[11] Here, $\tilde{S}^{i,j}(\omega)$ denotes a global random invariant manifold for $\tilde{\varphi}$ according to Theorem 3.1.

[12] Of course the best way would be to construct $\tilde{\varphi}$ on the level of the random dynamical system, i.e. without reverting to an infinitesimal generator. Yet, even in the deterministic case it is not at all clear how this could be done.

within the existing theory of stochastic integration[13]. In other words, the problem of suitably cutting off a stochastic differential equation is still open.

As for random differential equations things are much more easier, and they will be the subject of the rest of this subsection. To fix the notation, let $(\Omega, \mathcal{F})$ denote an arbitrary measurable space and let $f : \mathbb{R} \times \Omega \times \mathbb{R}^d \to \mathbb{R}^d$ be measurable. Then we call

$$\dot{x} = f(t, \omega, x) \tag{4.6}$$

a *random differential equation*, and an absolutely continuous mapping $\lambda : I \to \mathbb{R}^d$ defined on some interval $I \subset \mathbb{R}$ is called ω-*solution of (4.6)* if the mapping $f(\cdot, \omega, \lambda(\cdot))$ is locally integrable, and for all $t, s \in I$ the identity $\lambda(t) - \lambda(s) = \int_s^t f(\tau, \omega, \lambda(\tau))d\tau$ holds, or equivalently, if $\dot{\lambda}(t) = f(t, \omega, \lambda(t))$ for almost all $t \in I$.

Similar to Lemma 3.1 the following proposition gives conditions under which a random differential equation actually generates a random dynamical system. This result has been taken from Arnold [2, 3], where even more can be found on the connection between random dynamical systems and random differential equations.

Proposition 4.2. *Let $(\Omega, \mathcal{F}, \mathbb{P}, (\theta_t)_{t \in \mathbb{R}})$ be an arbitrary ergodic metric dynamical system, and let $f : \Omega \times \mathbb{R}^d \to \mathbb{R}^d$ be measurable. Suppose that $f(\omega, \cdot)$ is continuous for all $\omega \in \Omega$ and that for every compact subset $K \subset \mathbb{R}^d$ the inclusion*

$$\sup_{x \in K} \|f(\omega, x)\| + \sup_{x, y \in K, x \neq y} \frac{\|f(\omega, x) - f(\omega, y)\|}{\|x - y\|} \in L^1(\Omega, \mathcal{F}, \mathbb{P})$$

holds. Finally, let $\|f(\omega, x)\| \leq a(\omega)\|x\| + b(\omega)$ for arbitrary $\omega \in \Omega$ and $x \in \mathbb{R}^d$, where $a, b \in L^1(\Omega, \mathcal{F}, \mathbb{P})$. Then there is a θ_t-invariant set $\hat{\Omega} \in \mathcal{F}$ with $\mathbb{P}(\hat{\Omega}) = 1$ such that – after having redefined f outside $\hat{\Omega}$ by $f(\omega, \cdot) := 0$ – the maximal solution $\varphi(\cdot, \omega, \xi)$ of the initial value problem

$$\dot{x} = f(\theta_t \omega, x) \quad , \quad x(0) = \xi$$

generates a continuous-time continuous random dynamical system on $\mathbb{R}^d$ over $(\Omega, \mathcal{F}, \mathbb{P}, (\theta_t)_{t \in \mathbb{R}})$.

Proof. The proof can be found in Arnold [2, Theorem 2.15].   ◊

¿From now on we shall follow the lines of the last subsection, and our aim is to derive a complete analogue to Proposition 4.1. However, this time we have to be very careful in constructing the mapping ϱ which defines the random neighborhood $U(\omega)$. If we would simply use Lemma 4.1 to cut off the nonlinear part of the generator we could not rule out an identity of the form

---

[13] To remedy this problem a non-anticipative calculus is on its way. For more details we refer the reader to Arnold, Imkeller [5].

$$\inf_{-\varepsilon \le t \le \varepsilon} \varrho(\theta_t \omega) = 0 \quad \text{for arbitrary} \quad \varepsilon > 0$$

for some $\omega \in \tilde{\Omega}$. But such an identity would render an analogue to Proposition 4.1 completely useless, since then every ω-orbit $\varphi(\cdot, \omega, \xi)$ would immediately leave $U(\omega)$. So, rather than using Lemma 4.1 we have to apply a more sophisticated cut off technique, which essentially boils down to the following lemma.

Lemma 4.2. *Let $(\tilde{\Omega}, \tilde{\mathcal{F}}, \tilde{\mathbb{P}}, (\theta_t)_{t \in \mathbb{R}})$ denote a metric dynamical system, and consider a measurable mapping $f : \tilde{\Omega} \times \mathbb{R}^d \to \mathbb{R}^d$ such that $f(\omega, \cdot)$ is continuous with $f(\omega, 0) = 0$ for all $\omega \in \tilde{\Omega}$, and that both*

$$\lim_{(x,y) \to (0,0)} \frac{\|f(\omega, x) - f(\omega, y)\|}{\|x - y\|} = 0 \tag{4.7}$$

and

$$\sup_{\|x\| \le c, \|y\| \le c, x \ne y} \frac{\|f(\omega, x) - f(\omega, y)\|}{\|x - y\|} \in L^1(\tilde{\Omega}, \tilde{\mathcal{F}}, \tilde{\mathbb{P}}) \tag{4.8}$$

hold, for some $c > 0$. Furthermore, let $\| \cdot \|_\omega$ denote a random norm on $\mathbb{R}^d$ satisfying the assertions of Proposition 3.1(b), and let $L_0 > 0$ be an arbitrary constant. Then there are measurable mappings $\varrho : \tilde{\Omega} \to (0, c]$, $L : \tilde{\Omega} \to \overline{\mathbb{R}}^+$ and $\tilde{f} : \tilde{\Omega} \times \mathbb{R}^d \to \mathbb{R}^d$, as well as a θ_t-invariant set $\hat{\Omega} \in \tilde{\mathcal{F}}$ with $\tilde{\mathbb{P}}(\hat{\Omega}) = 1$ such that the following holds:

(a) If we define a random neighborhood of 0 by

$$U(\omega) := \{x \in \mathbb{R}^d : \|x\| < \varrho(\omega)\},$$

then the identity $f(\omega, x) = \tilde{f}(\omega, x)$ holds for all $\omega \in \hat{\Omega}$ and $x \in U(\omega)$. Additionally, we have $\tilde{f}(\omega, x) = 0$ for $\omega \in \tilde{\Omega} \setminus \hat{\Omega}$ and $x \in \mathbb{R}^d$.

(b) For arbitrary $\omega \in \hat{\Omega}$ and $x, y \in \mathbb{R}^d$ we have

$$\|\tilde{f}(\omega, x) - \tilde{f}(\omega, y)\|_\omega \le L(\omega)\|x - y\|_\omega \text{ and } \|\tilde{f}(\omega, x)\|_\omega \le L(\omega)c,$$

as well as $\int_0^1 L(\theta_s \omega) ds \le L_0$.

(c) For all $\omega \in \hat{\Omega}$ and $a, b \in \mathbb{R}$ with $a \le b$ the estimate $\inf_{a \le t \le b} \varrho(\theta_t \omega) > 0$ holds, i.e. the mapping $\varrho(\theta.\omega)$ is locally bounded away from 0.

Proof. The proof of this lemma is very technical, so we refrain from presenting all the details. (The reader who is nevertheless interested in these details may consult [30, pp. 151ff].) If we define a mapping $L^* : \tilde{\Omega} \times (0, c] \to \overline{\mathbb{R}}_0^+$ by

$$L^*(\omega, \varepsilon) := \sup_{\|x\| \le \varepsilon, \|y\| \le \varepsilon, x \ne y} \frac{\|f(\omega, x) - f(\omega, y)\|}{\|x - y\|},$$

then L^* is measurable and $L^*(\omega, \cdot)$ is increasing, for arbitrary $\omega \in \tilde{\Omega}$. Furthermore, (4.7) and (4.8) imply $\lim_{\varepsilon \to 0} L^*(\omega, \varepsilon) = 0$ for every $\omega \in \tilde{\Omega}$, as

well as $L^*(\cdot, \varepsilon) \in L^1(\tilde{\Omega}, \tilde{\mathcal{F}}, \tilde{\mathbb{P}})$ for $\varepsilon \in (0, c]$. This last inclusion furnishes –
with the help of Lemma 2.14 in Arnold [2] – a θ_t-invariant set $\hat{\Omega} \in \tilde{\mathcal{F}}$ with
$\tilde{\mathbb{P}}(\hat{\Omega}) = 1$ such that $L^*(\theta.\omega, \varepsilon) : \mathbb{R} \to \overline{\mathbb{R}}_0^+$ is locally integrable for all $\omega \in \hat{\Omega}$
and $\varepsilon \in (0, c]$.

Now let $B := B_1 : \tilde{\Omega} \to [1, \infty)$ denote the measurable mapping of Proposition 3.1(b). Then $B(\theta.\omega)$ is locally bounded, and therefore we may define
another measurable mapping $L^{**} : \hat{\Omega} \times (0, c] \to \mathbb{R}_0^+$ by

$$L^{**}(\omega, \varepsilon) := \int_0^1 B(\theta_s\omega)^2 L^*(\theta_s\omega, \varepsilon) ds < \infty.$$

It is not too difficult to check that the mapping $L^{**}(\omega, \cdot)$ is increasing with
$\lim_{\varepsilon \to 0} L^{**}(\omega, \varepsilon) = 0$ and that $L^{**}(\theta.\omega, \varepsilon)$ is continuous, for $\omega \in \hat{\Omega}$ and $\varepsilon \in (0, c]$.

The next step towards constructing the mapping ϱ is the definition of a
measurable mapping $\varrho^* : \hat{\Omega} \to (0, c]$ by $\varrho^*(\omega) := \sup\{\varepsilon \in (0, c] : L^{**}(\omega, \varepsilon) \leq L_0\}$. One can prove that for arbitrary $\omega \in \hat{\Omega}$ and $a, b \in \mathbb{R}$ with $a \leq b$ the
estimate $\inf_{a \leq t \leq b} \varrho^*(\theta_t\omega) > 0$ holds, and that for every $\omega \in \hat{\Omega}$ the mapping
$\varrho^*(\theta.\omega)$ is upper semicontinuous. Finally, define

$$\varrho(\omega) := \frac{1}{2} \cdot \inf_{-1 \leq t \leq 0} \varrho^*(\theta_t\omega) > 0$$

for $\omega \in \hat{\Omega}$. Then the upper semicontinuity of ϱ^* shows that in the above
infimum we may replace $t \in [-1, 0]$ by $t \in [-1, 0] \cap \mathbb{Q}$, furnishing the measurability of ϱ.

As for the proof of (c) let $\varrho_0 > 0$ be such that $\varrho^*(\theta_t\omega) \geq \varrho_0$ for all
$t \in [a - 1, b]$. Then for every $s \in [-1, 0]$ and $t \in [a, b]$ we first get $\varrho^*(\theta_s\theta_t\omega) \geq \varrho_0$, and further we deduce $\varrho(\theta_t\omega) = \frac{1}{2} \inf_{-1 \leq s \leq 0} \varrho^*(\theta_s\theta_t\omega) \geq \frac{1}{2}\varrho_0 > 0$ for
$t \in [a, b]$.

In order to conclude the proof of Lemma 4.2 we only have to verify the
assertions contained in (a) and (b). To that end, define

$$\tilde{f}(\omega, x) := f(\omega, r_{\varrho(\omega)}(x)) \quad \text{for arbitrary} \quad \omega \in \hat{\Omega}, x \in \mathbb{R}^d,$$

where r_ε again denotes the radial retraction on $\mathbb{R}^d$, as well as $L(\omega) := B(\omega)^2 L^*(\omega, \varrho(\omega))$ for $\omega \in \hat{\Omega}$. Then the mappings $\tilde{f} : \hat{\Omega} \times \mathbb{R}^d \to \mathbb{R}^d$ and
$L : \hat{\Omega} \to \overline{\mathbb{R}}_0^+$ are measurable, and the assertions contained in (a) are satisfied
if we set $\tilde{f}$ to 0 on $(\tilde{\Omega} \setminus \hat{\Omega}) \times \mathbb{R}^d$, and extend L and ϱ measurably to the whole
of $\tilde{\Omega}$.

As for (b) we note that the first two estimates may be derived as in the
proof of Lemma 4.1. For the proof of the remaining inequality the definition
of $\varrho(\theta_s\omega)$ furnishes $\varrho(\theta_s\omega) \leq \frac{1}{2}\varrho^*(\theta_\tau\theta_s\omega)$ for arbitrary $\omega \in \hat{\Omega}$, $s \in [0, 1]$ and
$\tau \in [-1, 0]$, and with $\tau = -s$ we further deduce $\varrho(\theta_s\omega) \leq \frac{1}{2}\varrho^*(\omega)$ for all
$\omega \in \hat{\Omega}$, $s \in [0, 1]$. Finally, since $L^*(\omega, \cdot)$ is increasing the definition of ϱ^*
implies

$$\int_0^1 L(\theta_s\omega)ds \le \int_0^1 B(\theta_s\omega)^2 L^*(\theta_s\omega, \tfrac{1}{2}\varrho^*(\omega))ds \le L_0$$

for $\omega \in \hat{\Omega}$, and the proof is complete. ◇

With this lemma at hand we now directly proceed to the continuous-time analogue of Proposition 4.1.

Proposition 4.3. *Assume we are given a continuous-time continuous random dynamical system φ on $\mathbb{R}^d$ over some ergodic metric dynamical system $(\Omega, \mathcal{F}, \mathbb{P}, (\theta_t)_{t\in\mathbb{R}})$ with fixed point 0, which is generated by the random differential equation $\dot{x} = A(\theta_t\omega)x + f(\theta_t\omega, x)$, with $A : \Omega \to \mathrm{L}(\mathbb{R}^d)$, $f : \Omega \times \mathbb{R}^d \to \mathbb{R}^d$, where $\|A\| \in L^1(\Omega, \mathcal{F}, \mathbb{P})$ and f satisfies the assumptions of Proposition 4.2. Suppose further that*

$$\lim_{(x,y)\to(0,0)} \frac{\|f(\omega,x) - f(\omega,y)\|}{\|x - y\|} = 0$$

and that $f(\omega,0) = 0$ for all $\omega \in \Omega$. Finally, the linear part Φ of φ generated by the linear equation $\dot{x} = A(\theta_t\omega)x$ should satisfy hypothesis (H1). Then there is a θ_t-invariant set $\hat{\Omega} \in \mathcal{F}$ with $\mathbb{P}(\hat{\Omega}) = 1$, as well as a measurable mapping $\varrho : \Omega \to \mathbb{R}^+$ and a continuous-time continuous random dynamical system $\tilde{\varphi}(t,\omega,x) = \Phi(t,\omega)x + \tilde{\Psi}(t,\omega,x)$ over $(\Omega, \mathcal{F}, \mathbb{P}, (\theta_t)_{t\in\mathbb{R}})$ satisfying both (H1) and (H2), as well as

$$\varphi(t,\omega,x) = \tilde{\varphi}(t,\omega,x) \tag{4.9}$$

for $\omega \in \hat{\Omega}$, $x \in U(\omega)$ and $t \in I_{\max}(\omega,x) := (t_{\min}(\omega,x), t_{\max}(\omega,x))$, where $U(\omega)$ is defined as in Lemma 4.2(a) and

$$t_{\min}(\omega,x) := \inf\{t \in \mathbb{R}_0^- : \varphi(\tau,\omega,x) \in U(\theta_\tau\omega) \text{ for all } t \le \tau \le 0\} \le 0,$$
$$t_{\max}(\omega,x) := \sup\{t \in \mathbb{R}_0^+ : \varphi(\tau,\omega,x) \in U(\theta_\tau\omega) \text{ for all } 0 \le \tau \le t\} \ge 0.$$

Moreover, for every $\omega \in \hat{\Omega}$ we have both

$$\lim_{x\to 0} t_{\min}(\omega,x) = -\infty \quad and \quad \lim_{x\to 0} t_{\max}(\omega,x) = \infty. \tag{4.10}$$

In particular, the last two identities imply for every choice of $\omega \in \hat{\Omega}$ and $T \in \mathbb{R}^+$ the existence of a neighborhood $U_T(\omega) \subset \mathbb{R}^d$ of the origin such that the equality in (4.9) holds for all $t \in (-T,T)$ and $x \in U_T(\omega)$.

Proof. According to (H1) there is a positive constant $\gamma \in \mathbb{R}^+$ such that for all $\omega \in \tilde{\Omega}$ and $t \in \mathbb{R}_0^+$ the estimate

$$\|\Phi(t,\omega)\|_{\omega,\theta_t\omega} \le e^{\gamma t} \tag{4.11}$$

holds. Now choose $L_0 > 0$ small enough to ensure that the constant $e^\gamma \left(L_0 + L_0^2 e^{L_0}\right) > 0$ satisfies the needs of (H2), let $\tilde{f}$ denote the mapping guaranteed by Lemma 4.1 for $f|_{\tilde{\Omega}\times\mathbb{R}^d}$, which has been set to 0 on $(\Omega\setminus\tilde{\Omega})\times\mathbb{R}^d$, and let ϱ and L denote the corresponding mappings, which have been extended measurably onto the whole of Ω.

It may easily be verified that the mapping $A(\cdot) \cdot + \tilde{f}(\cdot, \cdot)$ satisfies all the assumptions of Proposition 4.2, i.e. this mapping generates a continuous random dynamical system $\tilde{\varphi}^{14}$. Furthermore, Lemma 4.2(a) furnishes

$$\varphi(t, \omega, x) = \tilde{\varphi}(t, \omega, x) \quad \text{for arbitrary} \quad \omega \in \hat{\Omega}, x \in U(\omega), t \in I_{\max}(\omega, x) \ .$$

Next we like to prove that the nonlinear part $\tilde{\Psi}$ of $\tilde{\varphi}$ satisfies the assumptions contained in (H2), with $\tilde{\Omega}$ replaced by $\hat{\Omega}$. To that end, let $\tilde{\Phi}(t, s, \omega)$ denote the evolution operator of the linear random differential equation $\dot{x} = A(\theta_t \omega)x$ (cf. Amann [1]). Then an application of Theorem 2.13 in Arnold [2] yields the identity $\tilde{\Phi}(t, s, \omega) = \Phi(t - s, \theta_s \omega)$, and together with (4.11) we deduce

$$\|\tilde{\Phi}(t, s, \omega)\|_{\theta_s \omega, \theta_t \omega} \le e^{\gamma(t-s)} \quad \text{for arbitrary} \quad t \ge s \ , \ \omega \in \hat{\Omega} \ . \qquad (4.12)$$

Since $\tilde{\varphi}(\cdot, \omega, x)$ solves the initial value problem $\dot{x} = A(\theta_t \omega)x + \tilde{f}(\theta_t \omega, x)$, $x(0) = x$, the well-known variation of constants formula implies

$$\tilde{\varphi}(t, \omega, x) = \tilde{\Phi}(t, 0, \omega)x + \int_0^t \tilde{\Phi}(t, s, \omega)\tilde{f}(\theta_s \omega, \tilde{\varphi}(s, \omega, x))ds \ ,$$

and finally

$$\tilde{\Psi}(t, \omega, x) = \int_0^t \tilde{\Phi}(t, s, \omega)\tilde{f}(\theta_s \omega, \tilde{\Phi}(s, 0, \omega)x + \tilde{\Psi}(s, \omega, x))ds \ , \qquad (4.13)$$

for arbitrary $t \in \mathbb{R}$, $\omega \in \hat{\Omega}$ and $x \in \mathbb{R}^d$. Using the last two estimates in Lemma 4.2(b) (e.g. for $c = 1$) and (4.12) we now deduce the estimate

$$\|\tilde{\Psi}(t, \omega, x)\|_{\theta_t \omega} \le \int_0^t e^{\gamma(t-s)} L(\theta_s \omega)ds \le e^\gamma \int_0^1 L(\theta_s \omega)ds \le L_0 e^\gamma \ ,$$

for $t \in [0, 1]$, $\omega \in \hat{\Omega}$ and $x \in \mathbb{R}^d$, i.e. $\tilde{\Psi}$ satisfies the boundedness condition of (H2). Furthermore, a standard application of Gronwall's lemma yields – with the aid of the first inequality in Lemma 4.2(b), as well as (4.12) and (4.13) – after some straightforward calculations

$$\|\tilde{\Psi}(t, \omega, x) - \tilde{\Psi}(t, \omega, \bar{x})\|_{\theta_t \omega} \le e^\gamma \left(L_0 + L_0^2 e^{L_0}\right) \|x - \bar{x}\|_\omega \ ,$$

for every $t \in [0, 1]$, $\omega \in \hat{\Omega}$ and $x, \bar{x} \in \mathbb{R}^d$, and the choice of L_0 from the beginning of the proof shows that $\tilde{\Psi}$ indeed satisfies all the assumptions contained in (H2).

In order to conclude the proof of the proposition we only have to verify the identities in (4.10). For this, let $\omega \in \hat{\Omega}$ and $T \in \mathbb{R}^+$ be arbitrary, but fixed.

---

[14] Without loss of generality there is no need for a further redefinition – otherwise we pass to a smaller set $\hat{\Omega}$.

Due to Lemma 4.2*(c)* we have $\varrho_0 := \inf_{t \in [-T,T]} \varrho(\theta_t \omega) > 0$. Furthermore, the continuity of $\varphi(\cdot, \omega, \cdot)$ shows – having the identity $\varphi(\cdot, \omega, 0) \equiv 0$ on $\mathbb{R}$ in mind – that there is an $\varepsilon > 0$ such that for every $\xi \in \mathbb{R}^d$ with $\|\xi\| < \varepsilon$ the estimate $\|\varphi(t, \omega, \xi)\| < \varrho_0 \leq \varrho(\theta_t \omega)$ holds for all $t \in [-T, T]$, i.e. for arbitrary $\xi \in \mathbb{R}^d$ with $\|\xi\| < \varepsilon$ we have $t_{\min}(\omega, \xi) \leq -T$ and $t_{\max}(\omega, \xi) \geq T$. This completes the proof of the proposition.   ◊

With the above Proposition 4.3 it is finally possible to obtain local results for random dynamical systems which are generated by random differential equations as well, similar to the proceeding presented at the end of the last subsection.

5. Appendix

In this appendix we like to present three simple fixed point results that were used in Section 2. The first one is an easy consequence of Banach's fixed point theorem and shows that under natural conditions a uniquely determined fixed point of a parameter dependent contraction depends measurably on the parameter if the contraction does.

Lemma 5.1. *Let $(A, \mathcal{A})$ be a measurable space and let $T : \mathbb{R}^d \times A \to \mathbb{R}^d$ be a measurable mapping such that $T(\cdot, a)$ is a contraction on $\mathbb{R}^d$ with respect to some norm $\| \cdot \|_a$ for every $a \in A$. Let $t(a) \in \mathbb{R}^d$ denote the uniquely determined fixed point of $T(\cdot, a)$, where $a \in A$. Then the mapping $t : A \to \mathbb{R}^d$ is measurable.*

Proof. Defining recursively $x_0(a) := 0$ and $x_{n+1}(a) := T(x_n(a), a)$ for $a \in A$ and $n \in \mathbb{N}_0$ yields measurable mappings $x_n : A \to \mathbb{R}^d$ for $n \in \mathbb{N}_0$. Furthermore, since on $\mathbb{R}^d$ all norms are equivalent, Banach's fixed point theorem furnishes $\lim_{n \to \infty} x_n(a) = t(a)$ for every $a \in A$. But then t is the pointwise limit of measurable mappings, thus it is measurable, too.   ◊

The next lemma deals with parameter dependent contractions on certain function spaces. Again we will derive conditions furnishing the measurable dependence of the corresponding fixed points.

Lemma 5.2. *Let $(A, \mathcal{A})$ and $(P, \mathcal{P})$ be measurable spaces, and assume we are given a family of Banach spaces $B_p \subset \{\nu : A \to \mathbb{R}^d \mid \nu \text{ is measurable}\}$, $p \in P$, with norms $\| \cdot \|_p$ such that for every $p \in P$ and $a \in A$ the evaluation mapping $B_p \ni \nu \mapsto \nu(a) \in \mathbb{R}^d$ is continuous. Moreover, let $T_p : B_p \to B_p$, $p \in P$, be a family of operators satisfying the following two conditions:*

(a) For every $p \in P$ the mapping T_p is a contraction on B_p with respect to $\| \cdot \|_p$.

(b) For every measurable mapping $\mu : A \times P \to \mathbb{R}^d$ satisfying $\mu(\cdot, p) \in B_p$ for all $p \in P$ the mapping $\hat{\mu} : A \times P \to \mathbb{R}^d$ defined as $\hat{\mu}(a, p) := (T_p \mu(\cdot, p))(a)$ is measurable, too.

Let $\nu^*(\cdot,p) \in B_p$ denote the unique fixed point of T_p for $p \in P$. Then the so-defined mapping $\nu^* : A \times P \to \mathbb{R}^d$ is measurable.

Proof. Define a sequence $(\nu_n)_{n \in \mathbb{N}_0}$ of mappings recursively via $\nu_0(a,p) := 0$ and $\nu_{n+1}(a,p) := (T_p\nu_n(\cdot,p))(a)$ for all $a \in A$, $p \in P$ and $n \in \mathbb{N}_0$. Then *(b)* implies the measurability of $\nu_n : A \times P \to \mathbb{R}^d$ for $n \in \mathbb{N}_0$, as well as $\nu_n(\cdot,p) \in B_p$ for every $p \in P$. Moreover, according to *(a)* and Banach's fixed point theorem we have $\nu^*(\cdot,p) = \lim_{n\to\infty} \nu_n(\cdot,p)$ in B_p, for arbitrary $p \in P$, and the continuity of the evaluation mappings furnishes $\nu^*(a,p) = \lim_{n\to\infty} \nu_n(a,p)$ in $\mathbb{R}^d$, for arbitrary $a \in A$ and $p \in P$. Hence, ν^* is the pointwise limit of a sequence of measurable mappings, thus measurable.   ◇

Our final fixed point result considers a situation similar to the one in the preceding lemma. This time, however, we wish to give conditions implying the continuous dependence on the pair $(a,p) \in A \times P$.

Lemma 5.3. *Let A and P be arbitrary metric spaces, and assume we are given a Banach space $B \subset \{\nu : A \to \mathbb{R}^d \mid \nu \text{ is continuous}\}$ such that for every $\nu \in B$ and $a \in A$ the estimate $\|\nu(a)\| \le c(a)\|\nu\|$ holds with a locally bounded mapping $c : A \to \mathbb{R}_0^+$. Moreover, let $T_p : B \to B$, $p \in P$, be a family of operators satisfying the following two conditions:*

(a) For every $p \in P$ and $\nu_1, \nu_2 \in B$ we have $\|T_p\nu_1 - T_p\nu_2\| \le C\|\nu_1 - \nu_2\|$, where the real constant $0 \le C < 1$ is independent of $p \in P$.

(b) For every continuous mapping $\mu : A \times P \to \mathbb{R}^d$ such that $\mu(\cdot,p) \in B$ for all $p \in P$ and $\{\|\mu(\cdot,p)\| : p \in P\}$ is bounded, the mapping $\hat{\mu} : A \times P \to \mathbb{R}^d$ defined by $\hat{\mu}(a,p) := (T_p\mu(\cdot,p))(a)$ is continuous, and $\{\|\hat{\mu}(\cdot,p)\| : p \in P\}$ is bounded.

Let $\nu^*(\cdot,p)$ denote the unique fixed point of T_p for $p \in P$. Then the mapping $\nu^* : A \times P \to \mathbb{R}^d$ is continuous.

Proof. Let X denote the linear space of all continuous mappings $\mu : A \times P \to \mathbb{R}^d$ such that the set $\{\|\mu(\cdot,p)\| : p \in P\}$ is bounded and $\mu(\cdot,p) \in B$ for all $p \in P$. It is routine to check that if we define $\||\mu\|| := \sup\{\|\mu(\cdot,p)\| : p \in P\}$ for $\mu \in X$, then X is a Banach space. Now define a mapping T by $(T\mu)(a,p) := (T_p\mu(\cdot,p))(a)$ for $a \in A$, $p \in P$ and $\mu \in X$. According to *(b)* the mapping $T\mu$ is contained in X for every $\mu \in X$, i.e. we have $T : X \to X$. Let $\mu_1, \mu_2 \in X$ be arbitrary. Then *(a)* furnishes for all $p \in P$

$$\|(T\mu_1)(\cdot,p) - (T\mu_2)(\cdot,p)\| \le C\|\mu_1(\cdot,p) - \mu_2(\cdot,p)\| \le C\||\mu_1 - \mu_2\|| \,,$$

and therefore

$$\||T\mu_1 - T\mu_2\|| \le C\||\mu_1 - \mu_2\|| \,.$$

Because of $0 \le C < 1$ the mapping T is a contraction on X, hence it has a unique fixed point μ^*. But our above construction implies that for arbitrary $p \in P$ we have $\mu^*(\cdot,p) = T_p\mu^*(\cdot,p)$, and therefore $\mu^* = \nu^*$. This completes the proof of the lemma.   ◇

References

1. H. Amann, *Ordinary Differential Equations*. De Gruyter, Berlin – New York (1990).
2. L. Arnold, *Random Dynamical Systems*. In preparation.
3. L. Arnold, *Generation of random dynamical systems*. Report Nr. 280, Institut für Dynamische Systeme, Universität Bremen (1993).
4. L. Arnold, H. Crauel, *Random dynamical systems*, in L. Arnold, H. Crauel, J.-P. Eckmann (eds.), *Lyapunov Exponents*. Lecture Notes in Mathematics 1486, Springer, Berlin – Heidelberg (1991).
5. L. Arnold, P. Imkeller, *Anticipative problems in multiplicative ergodic theory*. Preprint, Universität Bremen (1993).
6. B. Aulbach, *A reduction principle for nonautonomous differential equations*. Archiv der Mathematik **39** (1982), 217–232.
7. B. Aulbach, *Hierarchies of invariant manifolds*. Journal of the Nigerian Mathematical Society **6** (1987), 71–89.
8. B. Aulbach, Th. Wanner, *Invariant Fiber Bundles and Topological Equivalence in Dynamical Processes*. In preparation.
9. P. Boxler, *A stochastic version of center manifold theory*. Probability Theory and Related Fields **83** (1989), 509–545.
10. A. Carverhill, *Flows of stochastic dynamical systems: ergodic theory*. Stochastics **14** (1985), 273–317.
11. S. Dahlke, *Invariante Mannigfaltigkeiten für Produkte zufälliger Diffeomorphismen*. Dissertation, Universität Bremen (1989).
12. D. M. Grobman, *Homeomorphisms of systems of differential equations*. Doklady Akademii Nauk SSSR **128** (1959), 880.
13. D. M. Grobman, *The topological classification of the vicinity of a singular point in n-dimensional space*. Mathematics of the USSR – Sbornik **56** (1962), 77–94.
14. P. Hartman, *A lemma in the theory of structural stability of differential equations*. Proceedings of the American Mathematical Society 11 (1960), 610–620.
15. P. Hartman, *On the local linearization of differential equations*. Proceedings of the American Mathematical Society 14 (1963), 568–573.
16. P. Hartman, *Ordinary Differential Equations*. Birkhäuser, Boston – Basel – Stuttgart (1982).
17. S. Hilger, *Ein Maßkettenkalkül mit Anwendung auf Zentrumsmannigfaltigkeiten*. Dissertation, Universität Würzburg (1988).
18. S. Hilger, *Generalized theorem of Hartman-Grobman on measure chains*. Preprint (1992).
19. M. W. Hirsch, C. C. Pugh, M. Shub, *Invariant Manifolds*. Lecture Notes in Mathematics 583, Springer, Berlin – Heidelberg – New York (1977).
20. M. C. Irwin, *Smooth Dynamical Systems*. Academic Press, London (1980).
21. A. Kelley, *The stable, center-stable, center, center-unstable, unstable manifolds*. Journal of Differential Equations 3 (1967), 546–570.
22. A. Kelley, *Stability of the center-stable manifold*. Journal of Mathematical Analysis and Applications **18** (1967), 336–344.
23. U. Kirchgraber, K. J. Palmer, *Geometry in the Neighborhood of Invariant Manifolds of Maps and Flows and Linearization*. Longman Scientific and Technical, London (1990).
24. V. I. Oseledets, *A multiplicative ergodic theorem. Lyapunov characteristic numbers for dynamical systems*. Transactions of the Moscow Mathematical Society **19** (1968), 197–231.

25. K. J. Palmer, *A generalization of Hartman's linearization theorem*. Journal of Mathematical Analysis and Applications **41** (1973), 753–758.
26. K. J. Palmer, *Linearization near an integral manifold*. Journal of Mathematical Analysis and Applications **51** (1975), 243–255.
27. V. A. Pliss, *Principal reduction in the theory of stability of motion*. Izvestiya Akademii Nauk SSSR, Seriya Matematika **28** (1964), 1297–1324 (Russian).
28. A. N. Shoshitaishvili, *Bifurcations of topological type at singular points of parametrized vector fields*. Functional Analysis and its Applications **6** (1972), 169–170.
29. A. N. Shoshitaishvili, *Bifurcations of topological type at singular points of parametrized vector fields*. Tr. Semin. I. G. Petrovskii 1 (1975), 279–309 (Russian).
30. Th. Wanner, *Zur Linearisierung zufälliger dynamischer Systeme*. Dissertation, Universität Augsburg (1993).